An Australian Phanerozoic Timescale

An Australian Phanerozoic Timescale

Edited by G.C. Young and J.R. Laurie

Melbourne
OXFORD UNIVERSITY PRESS

OXFORD UNIVERSITY PRESS AUSTRALIA

Oxford New York
Athens Auckland Bangkok Bombay
Calcutta Cape Town Dar es Salaam Delhi
Florence Hong Kong Istanbul Karachi
Kuala Lumpur Madras Madrid Melbourne
Mexico City Nairobi Paris Singapore
Taipei Tokyo Toronto

and associated companies in
Berlin Ibadan

OXFORD is a trade mark of Oxford University Press

First published 1996

National Library of Australia
Cataloguing-in-Publication data:

An Australian Phanerozoic timescale.
Bibliography.
Includes index.
ISBN 0 19 553951 6.

1. Geology, Stratigraphic. 2. Paleontology — Australia.
3. Geological time. I. Young, G. C. (Gavin C.). II. Laurie, J. (John)

559.4

Designed by R.T.J. Klinkhamer
Typeset by Desktop Concepts P/L, Melbourne
Printed by Australian Print Group
Published by Oxford University Press,
253 Normanby Road, South Melbourne, Australia

Contents

SECTION ONE OVERVIEW

SECTION ONE

Overview

1.1

Introduction

G.C. Young

The compilation of an AGSO Phanerozoic Timescale volume originated with preparation of a set of biostratigraphic charts for the *Palaeogeographic Atlas of Australia* project, and its successor, the *Phanerozoic History of Australia* project. Both projects were carried out by the Australian Geological Survey Organisation (AGSO) (at that time the Bureau of Mineral Resources, BMR) with funding support from the petroleum industry through the Australian Petroleum Industry Research Association (APIRA). The charts were subsequently revised and updated, and published together with explanatory text in 1989–91 as a series of ten *BMR Records* (nos 1989/31–40). The aim was to provide a set of calibrated biostratigraphic charts specifically for use in the Australasian region, in recognition of the fact that good age control is a prerequisite to interpreting the development and structure of the sedimentary basins that host our major petroleum, coal, and sedimentary mineral deposits.

A reliable time framework underpins our understanding of all historical aspects of the geosphere, from the analysis of patterns of distribution of the earth's resources through time, or the determination of the evolutionary history of today's animals and plants, to developing models of the interactive factors that determined past patterns of global change, and have shaped the modern Australian environment. For the Phanerozoic (approximately the last 545 million years, the period of 'visible life'), the most effective method of establishing such a time framework has been through the study of fossils (**palaeontology**), and their stratigraphic distribution in sedimentary rocks (**biostratigraphy**).

In Australia, as elsewhere, the presence of sedimentary rocks belonging to each of the ten geological periods was first established only when appropriate fossils were discovered. Preliminary palaeontological investigations of Australian sedimentary basins were used during the nineteenth century to establish the age of major suites of stratified rocks, and this provided a framework for more detailed biostratigraphic research. A good example of the significance and application of biostratigraphic studies is the use of graptolites to understand the stratigraphic distribution of gold in the Ordovician of the Victorian goldfields at the beginning of this century. The Victorian sequence of graptolite zones is now one of the most finely subdivided in the world, and is used as an international standard for the Ordovician Period. Today, the age control of Australian sedimentary sequences provided by biostratigraphic research is based on a wide range of organisms from microscopic remains like unicellular diatoms and foraminifera, or the pollen and spores of plants, to the macroscopic and megascopic, such as the larger invertebrates, and vertebrate remains from fishes to mammals.

However, many groups widely used in the northern hemisphere, which are well represented in Australia, have not yet had the detailed research needed for their full utilisation in biochronology.

Recent years have also seen a rapid growth in other methods of measuring geological time, using radioactive decay of mineral elements, reversals in the Earth's magnetic field, isotopic signatures of significant global events recorded in the sedimentary record, global changes in sea-level and climate, and other data. Ideally, all these methods need to be integrated to provide a reliable timescale based on a comprehensive database, and the current volume is a first attempt to do this. Numerical calibration of the timescale is of variable reliability, but current research on the Phanerozoic Timescale at AGSO is now focused on integrating detailed biological age control with high-precision isotopic dating provided by the Sensitive High Resolution Ion MicroProbe (SHRIMP), an isotopic dating facility developed at the Australian National University (ANU) in Canberra as a joint ANU–AGSO venture. This work is still in its early stages, but already, for some parts of the geological record (mainly Late Palaeozoic), large revisions have resulted, with international implications.

An Australian Phanerozoic Timescale is divided into two sections. **Section One** is an overview providing general information for each of the geological periods, and summarising methods of geochronology, magnetostratigraphy, biochronology, and numerical calibration of major boundaries for the whole of the Phanerozoic. The detailed time framework on which these overviews are based is presented in **Section Two**, which provides explanatory text for a set of ten biostratigraphic charts for each of the geological periods of the Phanerozoic. These summaries give more information on the development of rocks belonging to each geological period in the Australasian region. Each chart integrates isotopic and other data (magnetic reversal, eustasy curves) used for age control with fossil zonal schemes, and shows the relationship of Australian zones to standard international timescales and their numerical calibration, where this information is available. Emphasis has been given to local schemes derived from the biostratigraphic study of Australasian sedimentary sequences. These should be applicable to our region, including parts of south-eastern and eastern Asia, which share with Australia their geological origins during the Phanerozoic as fragments of the southern supercontinent of Gondwana.

All the biostratigraphic charts have been extensively updated and revised since the production of *BMR Records* 1989/31–40, but this is a continuing process. The enclosed charts incorporate recent geochronological and biostratigraphic data from the specialist scientific literature up to late 1994, and unpublished information from research in biochronology, isotopic dating, and magnetostratigraphy conducted by AGSO, and collaborating institutions and agencies. The detail of treatment and reliability of the charts varies for different parts of the geological timescale, and different groups of fossils require much further research before their potential for age control in Australian sequences is realised. It will be some time before we have developed a fully integrated chronological scale comparable to those available in the Northern Hemisphere.

A geological timescale incorporates two different types of scale: a relative scale based on rock sequences, the 'chronostratic scale', with subdivisions defined in relatively complete reference sections (boundary stratotypes), and a scale based on units of duration, the 'chronometric scale' (Harland et al. 1990). As outlined by Harland et al., subdivisions of the chronostratic scale depend on conventions to be agreed rather than discovered, whereas the development of the chronometric scale is a matter of discovery or estimation rather than agreement. However, both scales are subject to continuing reappraisal, and the charts presented here bring together the latest decisions regarding subdivisions of the chronostratic scale (principally under the International Union of Geological Sciences

Subcommission on Stratigraphy) with a numerical calibration (chronometric scale) based on latest research results integrating biochronology, isotopic dating, and other methods of age control and correlation.

Following Harland et al. (1990), we do not employ a rigid distinction between terminology for time and rock units (e.g. a rock 'system' deposited during a 'period' of the geological timescale). The traditional hierarchy of subdivision of the timescale is followed. Thus, the Phanerozoic **Eon** is subdivided into Palaeozoic, Mesozoic and Cainozoic **eras**, which are further subdivided into the geological **periods**, during which rocks of the corresponding geological **systems** were deposited. Major subdivisions (normally Early, Middle and Late) of each period and rock system are called **series**, which are further subdivided into **stages**. Subdivision of stages is provided by zonal schemes for various fossil groups, and the stages themselves are generally defined in terms of a standard zonation for particularly useful fossil groups. Following accepted practice, all boundary discussions relate to the base rather than the top of subdivisions of the timescale.

Abbreviations used in this work, other than those explained in the text, are:

CZ	conodont zone
DNAG	Decade of North American Geology
DSDP	Deep Sea Drilling Project
FAD	first appearance datum
FO	first occurrence
GSSP	Global stratotype section and point
GTS89	timescale of Harland et al. (1990, Fig. 1.7)
ICS	International Commission on Stratigraphy, IUGS
IGC	International Geological Congress
IUGS	International Union of Geological Sciences
Ka	thousands of years (age; also used for durations)
LAD	last appearance datum
LO	last occurrence
Ma	millions of years (age; also used for durations)
NSW	New South Wales
NT	Northern Territory
ODP	Ocean Drilling Program
PNG	Papua New Guinea
WA	Western Australia

1.2

The Phanerozoic Timescale

Compiled by G.C. Young and J.R. Laurie

This section provides a brief overview of the history and current status of major subdivisions, for each of the geological periods. Strata belonging to all major subdivisions of the Phanerozoic Eon are found in most of the major Australian sedimentary basins (Fig. 1). More detail on their development for each geological period will be found in Section Two.

1.2.1 PALAEOZOIC ERA

Cambrian Period

The Cambrian System was first named by Adam Sedgwick (in Sedgwick & Murchison 1835) for strata exposed in North Wales. The name is derived from the Roman *Cumbria* (for North Wales), and in modern usage applies essentially to Sedgwick's (1852) revised concept of his Lower Cambrian Series. The Cambrian Period is marked by the first abundant occurrence in the geological record of body fossils of multicellular animals (the metazoans). Some groups, such as trilobites and brachiopods, occur in great diversity and are widely distributed in shallow marine rocks of this age throughout the world.

The Cambrian Period is subdivided into **Early**, **Middle** and **Late** series, and in Australia twelve stages are recognised (see Fig. 2). The Global Stratotype Section and Point for the base of the Cambrian, recently ratified by the IUGS, is on the Burin Peninsula of eastern Newfoundland. This section contains three globally correlatable trace fossil zones occurring slightly earlier than the first trilobite faunas. In the stratotype section the Precambrian–Cambrian boundary is defined at the base of the ichnofossil zone of *Phycodes pedum*.

Ordovician Period

Lapworth (1879) introduced the term 'Ordovician' as a new system to resolve the controversy over strata in the North Wales succession that Sedgwick had included as Cambrian, but R.I. Murchison regarded as belonging to his Silurian System (see below). In Britain acceptance of the new name was slow, and one of the earliest usages of the term 'Ordovician' was by the Australian-born and trained palaeontologist T.S. Hall (1897), in the first of a series of classic papers on graptolites from the Victorian goldfields.

The Ordovician Period may be divided into **Early** and **Late** series, or into the five series recognised in the type area of North Wales (**Tremadoc, Arenig, Llanvirn, Caradoc, Ashgill**).

The approximate level for the base of Lapworth's Ordovician in the type area of North Wales (the base of the Tremadoc Series) is represented by a brief hiatus, and a more complete

Figure 1: Map of Australia showing location of geological basins, blocks and provinces (from Palfreyman 1984).

Cambrian–Ordovician boundary stratotype section is currently being sought in other parts of the world. Pelagic trilobites and graptolites have long usage in biostratigraphy of this part of the geological column, but in recent years a developing conodont zonation has assumed greater importance. Currently, an internationally correlatable conodont datum is being considered at sections in Jilin Province, north-eastern China, and in western Newfoundland, Canada. This datum lies close to the base of the *Cordylodus lindstromi* conodont zone, which is recognised in Australian sequences, and is used here as a provisional index fossil for defining the base of the Ordovician (see Nicoll 1990, 1991; Shergold & Nicoll 1992).

Silurian Period

The Silurian Period (named after a Welsh Borderland tribe) was erected by R.I. Murchison in 1839 to include a series of rocks claimed as Cambrian by Sedgwick. Murchison's 'Lower Silurian' was subsequently incorporated in Lapworth's Ordovician System, and his 'Upper Silurian' was not formally recognised as a geological period until officially adopted by the IGC in 1960. In the ensuing 30 years all the subdivisions (series and stages) of the Silurian Period have been formally defined and internationally agreed. The major subdivisions comprise four series (see Fig. 2) informally divided between **Early** and **Late Silurian.**

The base of the Silurian is defined at the base of the *Parakidograptus acuminatus* Zone in the Dobb's Linn section in Scotland (Cocks 1985). Recent reviews of the development of Silurian rocks in different regions of the world are given by Holland & Bassett (1989) and Bassett et al. (1991).

Devonian Period

The Devonian system of rocks was defined and named by Sedgwick & Murchison (1839) for a sequence of marine strata in Devon that lay above Cambrian and beneath Carboniferous strata. These strata were subsequently shown to be facies equivalents of the non-marine Old Red Sandstone of Scotland. Biochronological subdivision of the marine Devonian was based on work in Europe, the type area of all the Devonian stages.

The Devonian Period is divided into **Early**, **Middle** and **Late** series, and seven stages (see Fig. 2). All of these boundaries are now formally defined and ratified by the ICS, and all but one have GSSP adopted.

The base of the Devonian Period was defined in the Silurian–Devonian boundary stratotype section at Klonk, near Prague, Czechoslovakia, at the first appearance of the graptolite *Monograptus uniformis* (see Martinsson 1977).

Carboniferous Period

This was the first geological system to be established, with the name 'Carboniferous' being first used by Conybeare & Phillips (1822) for the 'Coal-measures, Carboniferous limestone, and Old red sandstone' of England and Wales. The Old Red Sandstone was later removed when the Devonian System was erected by Sedgwick & Murchison (1839). In western Europe the **Lower** and **Upper** Carboniferous are known as the **Dinantian** and **Silesian**, but in North America the terms **Mississippian** and **Pennsylvanian** are established for lower and upper 'subperiods', and the term 'Carboniferous' is rarely used.

The base of the Carboniferous was defined in 1979 on the first appearance of the conodont *Siphonodella sulcata* within the evolutionary lineage from *S. praesulcata* to *S. sulcata*, a level slightly below the lowermost record of the goniatite *Gattendorfia subinvoluta* previously used as an index macrofossil for this boundary (Paproth 1980). The Global Boundary Stratotype is the La Serre section in the Montagne Noire of southern France (Flajs & Feist 1988).

Permian Period

Murchison (1841) erected the Permian System to accommodate the extensive rock successions he encountered overlying Carboniferous strata during a visit to Russia. The name is derived from the city of Perm, on the flanks of the Ural Mountains.

The Permian is subdivided into **Early** and **Late** series, and nine stages (Fig. 2).

The position of the base of the Permian is not yet formally decided by the International Carboniferous–Permian Boundary Working Group. Traditionally, the boundary has been recognised at the base of the fusulinid *Pseudoschwagerina* Zone, which corresponds to the base of the Asselian Stage. This is now the agreed Russian boundary stratotype for the base of the Permian System in the Urals, and corresponds on the fusulinid scale to the base of the *Sphaeroschwagerina vulgaris–S. fusiformis* Zone

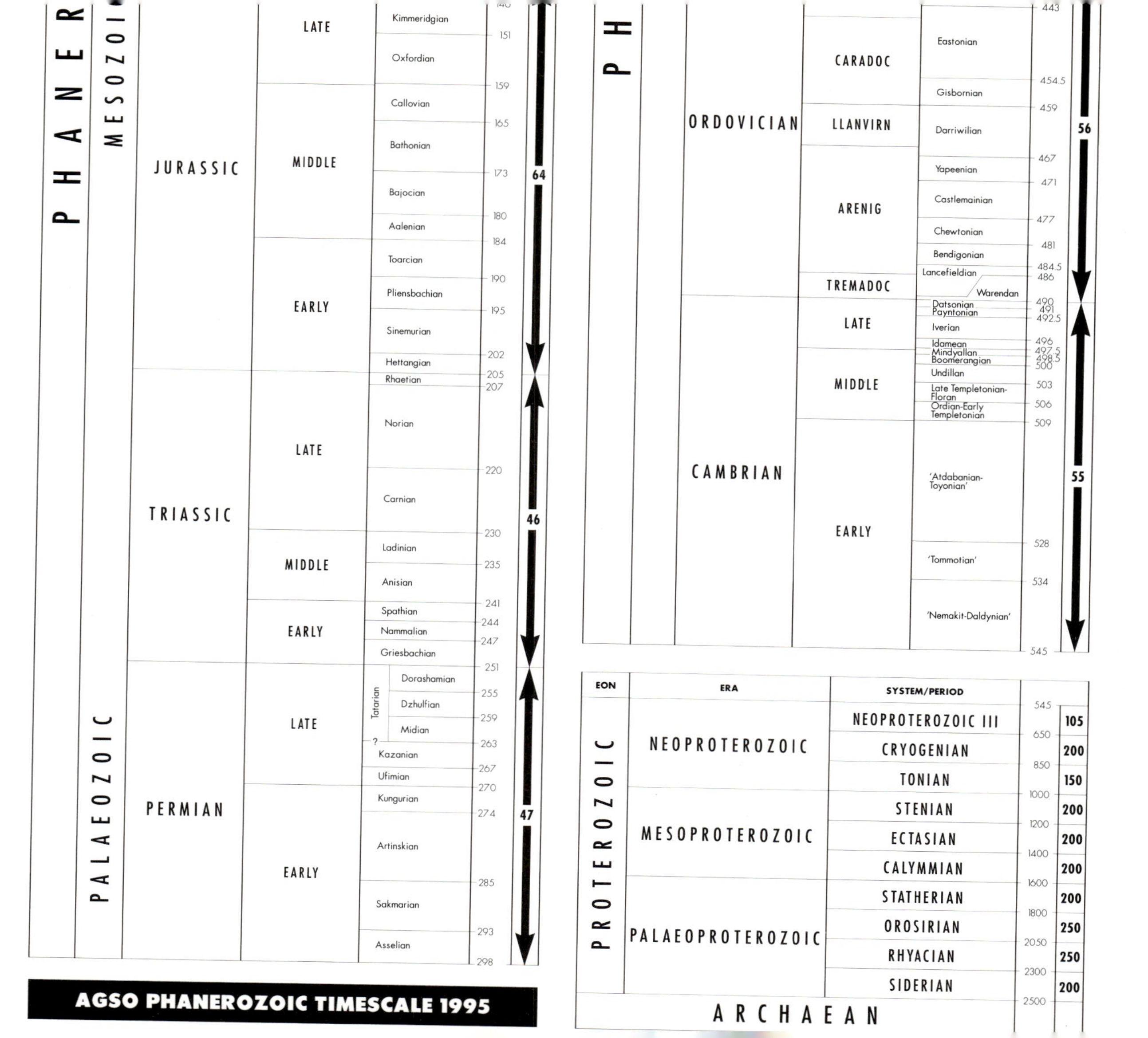

igure 2: This chart is a copy of the coloured version of the *AGSO Phanerozoic Timescale 1995: Wallchart and* *xplanatory Notes.*

EON	ERA	PERIOD SUBPERIOD		SERIES/EPOCH		STAGE	ESTIMATED AGE Ma	Duration
ZOIC	CAINOZOIC	QUATERNARY		PLEISTOCENE		Calabrian	1.78	65
		TERTIARY	NEOGENE	PLIOCENE		Piacenzian	3.56	
						Zanclean	5.32	
				MIOCENE	L	Messinian	7.12	
						Tortonian	11.2	
					M	Serravallian	14.8	
						Langhian	16.4	
					E	Burdigalian	20.52	
						Aquitanian	23.8	
			PALAEOGENE	OLIGOCENE	L	Chattian	28.5	
					E	Rupelian	33.7	
				EOCENE	L	Priabonian	37	
						Bartonian	41.3	
					M	Lutetian	49	
					E	Ypresian	54.8	
				PALEOCENE	L	Thanetian	57.9	
						Selandian	60.9	
					E	Danian	65	
		CRETACEOUS		LATE		Maastrichtian	73	76
						Campanian	83	
						Santonian	87	
						Coniacian	89	
						Turonian	91	
						Cenomanian	97.5	
				EARLY		Albian	108	
						Aptian	115	
						Barremian	123	
						Hauterivian	130	
						Valanginian	135	
						Berriasian		

EON	ERA	PERIOD SUBPERIOD			SERIES/EPOCH	STAGE	ESTIMATED AGE Ma	Duration
NEROZOIC	PALAEOZOIC						298	
		CARBONIFEROUS	PENNSYLVANIAN	SILESIAN	STEPHANIAN	Gzhelian	302	56
						Kasimovian	305	
					WESTPHALIAN	Moscovian	311.5	
						Bashkirian	314	
			MISSISSIPIAN		NAMURIAN	Serpukhovian	325	
				DINANTIAN	VISEAN	Brigantian	328.5	
						Asbian	330.5	
						Holkerian	335	
						Arundian	341.5	
						Chadian	344	
					TOURNAISIAN	Ivorian	348	
						Hastarian	354	
		DEVONIAN			LATE	Famennian	364.5	56
						Frasnian	369	
					MIDDLE	Givetian	378	
						Eifelian	384	
					EARLY	Emsian	399.5	
						Pragian	404.5	
						Lochkovian	410	
		SILURIAN			PRIDOLI		414	24
					LUDLOW	Ludfordian	417	
						Gorstian	420	
					WENLOCK	Homerian	422.5	
						Sheinwoodian	425	
					LLANDOVERY	Telychian	428	
						Aeronian	431	
						Rhuddanian	434	
						Bolindian		

Arduino in 1759 because he considered that they were separate from and in most cases were derived from the underlying Secondary rocks. In 1810, Brongniart applied the term to strata overlying the Cretaceous chalk in the Paris Basin, and he and Cuvier, at about the same time, described that sequence in great detail.

Lyell, in 1833, divided the Tertiary into epochs, which, with modifications introduced by workers such as Beyrich (1854) and Schimper (1874), remains the scheme in use today. Lyell's subdivision of the Tertiary was based on studies of the fossil molluscs by himself and Deshayes. The subdivisions were based on the percentage of species in common with the modern fauna.

The base of the Tertiary has not yet been defined by international agreement, but it marks one of the most profound levels of extinction in the fossil record.

Quaternary Period

In 1829, Desnoyers classified some marine, lacustrine and alluvial sediments from the Seine Basin under the name 'Quaternary' because he considered them to be younger than the Tertiary rocks. Reboul, in 1833, discussed the Quaternary Period and noted that it was characterised by fossils very similar to those now living. He considered that the Quaternary had two subdivisions: an historical part and a prehistoric, post-Tertiary part. At about the same time it was realised that the widespread erratics in the Alps and northern Europe were the result of deposition from extensive ice sheets. In 1839, Lyell proposed the **Pleistocene** Period for this 'ice age', which post-dated his established Pliocene of the Tertiary and preceded his **Recent**. In 1854, Morlot followed Reboul's usage of 'Quaternary', but recognised the glacial sequences as part of the division.

The base of the Pleistocene, and hence the Quaternary, was decided by the ICS in the Vrica section near Crotone, Calabria, in Italy. The boundary lies above the Last Appearance Datum (LAD) of *Discoaster broweri* and below the First Appearance Datum (FAD) of *Geophyrocapsa oceanica* and *Globgerinoides tenellus*.

1.3

Biochronology

G.C. Young

1.3.1 INTRODUCTION

Study of the fossil record over several centuries has provided the principal means of placing geological strata in a chronological sequence, and all the major subdivisions of the Phanerozoic Timescale are characterised by their own fossil groups. Fossils can be used in this way because, unlike most geological processes, which are cyclic or repetitive, biological evolution is unidirectional. This means that a species, once extinct, can never reappear, and the vast array of life forms that have evolved since the first appearance of organic remains in the geological record some 3400 Ma (million years) ago have provided an enormous database for use in biological age control of sedimentary strata (biochronology).

Some fossil groups stand out as having particular utility in geological age control, usually those that are readily preserved and widely distributed, with a complex morphology that underwent rapid evolutionary change. An early example is given by the ammonites used by d'Orbigny (1840–42) to subdivide the Jurassic Period into ten stages, each with three to eight ammonite zones. As noted above, in Australia graptolites formed the basis for subdividing the Ordovician into its ten stages. The more obvious 'macrofossils' were used by the early biostratigraphers, but, mainly during this century, techniques for extracting various types of microfossils from sedimentary rocks were developed, and many of these groups have proved particularly effective in high-resolution age control. Groups such as foraminifera, coccoliths, and radiolaria are widely used as biostratigraphic tools in Mesozoic and younger strata. Conodonts are an extremely useful fossil group for marine rocks during most of the Palaeozoic Era, and various types of acid-resistant plant microfossils known collectively as palynomorphs, have been applied in age control over most of the Phanerozoic.

The utility of specific fossil groups for age control depends on the sedimentary environment, and a continuing difficulty relates to integrating biozonations worked out for rocks deposited in the two major aquatic environments, that is, the problem of correlating between marine and non-marine strata. Some groups are more effective for this purpose because their remains or products are widely dispersed, a good example being the pollen grains and spores of land plants, which are often abundant in fluviatile or lacustrine sediments, but are also dispersed into shallow marine sequences. Other groups, such as fishes, lived in or moved between both marine and freshwater environments, and similar macro-remains or micro-remains (e.g. bones, spines or ichthyoliths, and scales) may occur in marine, marginal marine, or non-marine sequences. Some microfossil groups like ostracods or diatoms occupy a wide range of aquatic habitats, from ponds, rivers, and lakes, to salt marshes, seas and oceans. However, different groups of species are normally characteristic of each habitat, and assemblages of such groups preserved in sediments can also be useful indicators of palaeoenvironment. Although most aquatic animal groups represented in the fossil record could tolerate only a narrow range of salinity, it is well established that differences in faunal assemblages in marine depositional environments may be good indicators of water depth, turbidity, sediment type, and other palaeoenvironmental parameters (e.g. radiolaria in deeper water siliceous rocks, graptolites in black shales). These aspects must be taken into account when fossil groups are used for biozonation. For example, for Devonian conodonts two zonations have been developed to accommodate the numerous facies-related assemblages, one based on icriodid and polygnathid species typifying the shallow water facies, and another for palmatolepids and ancyrodellids, which tend to occur in deeper water sediments, and may have been pelagic.

1.3.2 METHODS OF CHRONOMETRIC CALIBRATION OF FOSSIL ZONES

Traditionally, biostratigraphic zonations have been seen to provide a 'relative' timescale, because they describe sequences of intervals whose durations in years are generally unknown, and may be different for different zones. To determine rates of geological processes one needs an 'absolute' timescale, that is, one measured in equal increments. The normal timescale calibrations are in years, thousands of years (Ka), or millions of years (Ma), and for this purpose isotopic dating techniques have provided the major data (see below). However, the view that the problem of quantifying stage and zonal duration times is simply a matter of obtaining more precise isotopically determined ages overlooks the fact that any equal-increment timescale can be used for relative rate analysis (subsidence and deposition rates, etc.). In recent years there have been many new developments in the quantitative aspects of biostratigraphic analysis, a topic of central importance to the proper integration of biostratigraphic and geochronologic datasets. Overviews and discussions of quantitative biostratigraphic applications in stratigraphic correlation were provided by Edwards (1982), Harper (1984), and Gradstein et al. (1985). General concepts and methods in biostratigraphy were previously summarised in Kauffman & Hazel (1977). Hughes (1989) has proposed new stratal correlation techniques to incorporate biochronological information, and the unitary association method as developed by Guex (1991) uses aspects of graph theory to express complex biostratigraphic information in data matrices, which can then be analysed by computer using algorithms to resolve conflicting stratigraphic relationships among species (see also Gradstein 1985*a*; Boulard 1993). The problem of placing confidence intervals on biostratigraphic ranges has been addressed by Strauss & Sadler (1989) and Marshall (1990). These new developments have been spawned by rapid advances in computer technology, on the one hand, and the increased precision required of biochronology resulting from new methods of event stratigraphy, on the other.

For much of the Palaeozoic and Mesozoic, the precision of biochronological age control far exceeds that currently provided by isotopic data (e.g. Gradstein et al. 1985; Murphy 1987). One aspect of timescale research that has achieved some attention in these parts of the geological column involves the development of non-annual equal-increment timescales, which can be used for solving rate problems in the absence of a reliable radiometrically based numerical scale. Most of these equal-increment analyses are adaptations of Shaw's (1964) method of graphic correlation for calibrating zonal duration time.

Examples include the subdivision of the Ordovician and Silurian based on an analysis of graptolitic black shales by Churkin et al. (1977), who assumed that uniform lithology in deep-water settings implied a relatively complete sedimentary record with uniform sedimentation rates, or at least rates that could be considered uniform within the limits of error of the analysis (see Shaw 1964). However, underlying assumptions concerning completeness of the sedimentary and fossil records, and assessments of biostratigraphic gaps remain topics of debate (e.g. Sadler 1981; Sadler & Dingus 1982; Harper & Crowley 1985; McKinney 1986*a,b*; Springer & Lilje 1988). Churkin et al. (1977) estimated graptolite zone durations ranging from 0.2 Ma to 7.7 Ma, a much greater resolution than provided by any other method. They tied their estimates to the available radiometric scale, but this is not necessary for the solution of rate problems (e.g. Murphy 1987). Assessment of and modifications to the graphic correlation method have been considered, among others, by Miller (1977), Brower & Bussey (1985) and Edwards (1989). Edwards (1984, 1985) used a hypothetical dataset to test the accuracy of the method. More recent applications include analysis of Early Devonian graptolite–conodont zones (Murphy 1977; Murphy & Berry 1983), and conodont-based chronostratigraphies for the Upper Ordovician (Sweet 1984) and Silurian (Kleffner 1989). Fordham (1992) has extended this to conodont zones from mid-Ordovician to Tournaisian by integrating 'tie-points' based on isotopic data from Harland et al. (1990). Graphic correlation and analysis of sedimentation rates and zonal durations are an important area of ongoing research for Palaeozoic strata, where there is currently a best general precision of about 1% on isotopically determined ages (J. Claoué-Long, pers. comm.). This translates to an error of approximately ± 2.5 Ma for the Permian, increasing to about ± 5 Ma as a best precision for the Cambrian. Such margins of error often encompass a number of biozones, a problem that is accentuated in older rocks. For example, the Devonian Period, with an estimated duration of about 56 Ma, is currently subdivided into over fifty zones and subzones based on conodonts and ammonoids, which gives an average zonal duration of about 1 Ma. It will be some time before the isotopic database can make a significant contribution to subdivisions of the timescale with this degree of detail.

1.3.3 MAJOR FOSSIL GROUPS USED IN AGE CONTROL OF SEDIMENTARY SEQUENCES

Brief comments are given here on the age, ecological preferences, and biostratigraphic applications of some of the fossil groups most widely used for biozonation over different parts of the geological record. Fossils are often subdivided into three size categories, each with its own methods of collection, preparation, and study. **Nannofossils** are the smallest, normally in the 5–60 µm size range, and require scanning electron microscopy for their detailed analysis. The term **microfossil** is generally applied to discrete fossil remains whose study requires the use of a microscope throughout. Microfossils (e.g. spores, pollen, algal cysts, acritarchs) with organic walls, as opposed to those with mineralised skeletons, are known collectively as palynomorphs, and again have their own special extraction and preparation techniques. The term **macrofossil** is applied to larger remains, the study of which may or may not include microscope work. Various chemical and mechanical extraction and preparation techniques have been developed for the large number of macrofossil groups that have application in biostratigraphy. The following groups are arranged alphabetically, and more detail on individual zonations for the different geological periods is given in Section Two.

Acritarchs (Precambrian–Recent) These microfossils are, by definition, of unknown botanical affinity. Usually hollow cysts, with a variety of external sculptural patterns, they range in size from 20 to 500 μm and are particularly well represented in the Early–Middle Palaeozoic (Ordovician–Devonian) and Mesozoic (Jurassic) marine strata, particularly those deposited in offshore environments. Their wide geographic distribution suggests a planktonic mode of life. They have been used extensively for age control particularly in Late Precambrian and Early Palaeozoic strata that lack other biostratigraphically useful fossils.

Ammonoids (Devonian–Cretaceous) These are shelled cephalopod molluscs, which were usually pelagic and were widely distributed in Palaeozoic and Mesozoic seas. They have been studied in detail because of their great utility in biostratigraphic zonation of marine Palaeozoic and Mesozoic strata. The ammonoid zonation in the Late Devonian (about 16 Ma duration) is subdivided into some thirty-six zones, and Jurassic ammonites are organised into more than sixty zones.

Brachiopods (Cambrian–Recent) The Brachiopoda were one of the major invertebrate macrofossil groups in the marine benthos of Palaeozoic seas, and they have been widely used in biostratigraphy for well over a century, with particular application in Palaeozoic rocks. Details of brachiopod zonations applicable to Australasian sedimentary sequences are given in the charts in Section Two. Unlike some other shelly invertebrates (e.g. bivalved molluscs and gastropods), the Brachiopoda never invaded fresh water, although one group (the lingulid brachiopods) were specialised for brackish water environments.

Chitinozoans (Ordovician–Permian) These vase-shaped microfossils are of uncertain affinity. The vesicles have a two-layered chitinous wall that is unusually resistant to oxidation and thermal alteration and, in some rocks such as slates, they may be the only preserved fossils. They are mostly 150–300 μm long, although some are smaller and others up to five times larger. They are known only from marine sediments, and are common in shallow-water shales and siltstones, but also occur in limestones, dolomites, graptolitic shales, slates and cherts, and were apparently planktonic. They have been most useful biostratigraphically in rocks of Ordovician to Devonian age, and are much less common in late Palaeozoic strata.

Coccoliths (Late Triassic–Recent) These tiny calcareous scales (5–15 μm in diameter) form a protective armour around unicellular protists, which today represent a major part of the oceanic phytoplankton. Coccoliths accumulate on the ocean floor in vast numbers to form deep-sea oozes and fossil chalks, but at depths greater than 3000–4000 m few survive carbonate dissolution. In modern subtropical and tropical regions they may form up to 25% by weight of the sediment, and similar proportions are encountered in Tertiary and Cretaceous rocks. They underwent their first major radiation in the Jurassic, and with the development of scanning electron microscopy they have become a major biostratigraphic tool for Mesozoic and Cainozoic oceanic sediments.

Conodonts (Cambrian–Triassic) These are microscopic (0.1–5.0 mm) phosphatic tooth-like structures that are often widespread and abundant in marine rocks of late Cambrian to late Triassic age, and probably belonged to some relative of the chordates. They underwent rapid evolutionary change, and are extremely useful biostratigraphic tools in Palaeozoic strata. Many facies-related conodont assemblages are recorded, and some may have been benthonic and others planktonic. They are often associated with graptolites, radiolarians and fish remains. In addition, conodont colour alteration has been usefully applied to maturation studies for the petroleum exploration industry.

Tentaculitids (Ordovician–Devonian) This is an extinct marine macrofossil group of uncertain affinity represented by calcareous cone-like fossils ranging in size from 15 to 80 mm. They have proved to be of great biostratigraphic utility during the Devonian.

Diatoms (Cretaceous–Recent) These are unicellular algae with a siliceous wall that occur in a wide range of habitats from ponds, rivers and lakes, to salt marshes, seas and oceans. Most are in the size range 20–200 μm. They may be benthonic or planktonic, and are restricted to the photic zone. Under optimal conditions their remains can accumulate in large numbers in the sediment, to form the silica deposits called diatomite. Diatoms have been widely used as biostratigraphic zonal indices in the Cretaceous and Tertiary, in both marine and non-marine successions.

Dinoflagellate cysts (Silurian–Recent) Dinoflagellates are single-celled organisms generally between 20 and 400 μm long, which have formed an important component of the marine phytoplankton since at least mid-Mesozoic times. Their life-cycle involves an encysting stage, and it is the organic-walled cysts, often called dinocysts, that are preserved as fossils. Palaeozoic acritarchs may represent an earlier phase of dinoflagellate evolutionary history, but their major radiation began in the late Triassic, and dinocysts have been widely used for age control and biozonation of Mesozoic and Cainozoic marine strata. A few genera occur in both fresh and salt water, but the majority of species are marine.

Foraminifera (Cambrian–Recent) These are unicellular animals that secrete a single or multi-chambered shell (test) of calcareous or siliceous minerals or agglutinated particles. They are both benthonic and planktonic, and today foraminiferal tests can make up the bulk of some marine sediments, with planktonic forams and coccoliths together comprising about 80% of modern carbonate deposition in seas and oceans. Foraminifera are important as biostratigraphic indicators in marine rocks of late Palaeozoic, Mesozoic and Cainozoic age. Foraminifera with hard tests were rare before the Devonian Period, and the fusulinids of the late Carboniferous and Permian were the first subgroup with wide utility in biostratigraphy. In addition to their extensive use in intercontinental correlation of late Mesozoic and Cainozoic sequences, they provide information on depth of deposition, palaeotemperature and palaeosalinity.

Graptolites (Cambrian–Devonian) These are extinct colonial organisms, with a chitinous skeleton, that were probably planktonic, and are commonly found in Early Palaeozoic black shales. Their wide distribution and rapid evolutionary change have made them important biostratigraphic zonal indicators, particularly for Ordovician and Silurian strata.

Ostracods (Cambrian–Recent) These are small crustaceans enclosed in a bivalved calcareous or chitinous shell; today they occupy almost all aquatic habitats, including many species adapted to brackish water, an environment inimical to many other organisms. They have been widely used for biozonation of marine strata on a local or regional scale, and also as palaeoenvironmental indicators (for palaeosalinity, water depth, etc.). Palaeozoic ostracods reached maximum diversity in the Devonian and early Carboniferous, when the first freshwater forms are also recorded. A second major radiation occurred in the Jurassic. Modern ostracods are abundant as both benthonic and pelagic organisms, with major subgroups characterising the different aquatic environments.

Radiolarians (Cambrian–Recent) These marine zooplankton include forms with an internal skeleton of opaline silica resistant to dissolution; today they form skeletal accumulations as radiolarian oozes at water depths of 3000–4000 m in the equatorial Pacific, and as fossils are frequently found in chert horizons. Radiolarians underwent major radiations in the Devonian–Carboniferous, Jurassic–Cretaceous, and Quaternary. They have been most widely used in biostratigraphic correlation of oceanic sediments of Mesozoic and Cainozoic age, but in recent years have also proved significant in age control of Palaeozoic strata, particularly cherts and other deeper water sediments in folded terranes.

Spores (Ordovician–Recent) and pollen (Carboniferous–Recent) These are reproductive micro-organs of vascular plants that have organic walls that are resistant to temperature, pressure, desiccation, and microbial attack, and are therefore often preserved in sediments. Spores and pollen grains may be produced by terrestrial plants in enormous quantities, and airborne 'miospores' (with a size range of about 5–200 µm) may occur up to 650 m above the land surface, and drift distances of up to 1700 km, although most settle within a few kilometres from their source. Very little of this material reaches the oceans by aerial dispersal, but they are washed off the land surface to be incorporated in river and lake sediments, or carried by currents into estuaries and coastal marine environments. The main biostratigraphic utility of spores and pollen grains is in terrestrial, lacustrine, fluviatile and deltaic sediments, but they are also important in correlating marine and non-marine strata. Their fossil record first becomes important in the Silurian with the establishment of a significant terrestrial vegetation. Pollen analysis gives indications of past vegetation cover, and palynomorphs are useful in petroleum exploration for maturation studies.

Trilobites (Cambrian–Permian) This is the best known group of extinct arthropods, and trilobites occupied a dominant role in the marine benthos during the Early Palaeozoic, but decreased in abundance in the Silurian and Devonian Periods and Late Palaeozoic. They are known only from marine rocks, and are one of the most important invertebrate macrofossil groups for biostratigraphy of Cambrian and Ordovician strata, with some thirty trilobite zones represented in the Middle–Late Cambrian. Some groups, like the minute agnostids, were pelagic and are widely distributed; consequently they are very useful for intercontinental correlation of Early Palaeozoic marine strata.

Vertebrates (Ordovician–Recent) The vertebrate skeleton of calcium phosphate is chemically resistant, and remains of aquatic vertebrates, mainly fishes, may be abundant and diverse in both marine and freshwater deposits. 'Ichthyoliths' (spines) and 'microvertebrates' (scales and teeth) are often associated with other phosphatic microfossils (e.g. conodonts) in acid residues. In Palaeozoic strata they may be particularly abundant in marginal or non-marine depositional environments where conodonts are rare or absent. Microvertebrates are currently being utilised for age control in both marine and non-marine strata, and are valuable tools in establishing marine–non-marine correlations. In some groups (e.g. sharks, Palaeozoic thelodonts) teeth and dermal denticles are shed throughout life, to be incorporated in large numbers in bottom sediments. Fishes were also the major inhabitants of freshwater environments during the Silurian and Devonian, and they have been widely used for biostratigraphy of Old Red Sandstone facies sediments throughout the world. Calcareous otoliths (ear stones) are first known in the fossil record from Devonian strata, but only become common in the mid-Mesozoic, and are often abundant in marine Tertiary rocks, where they have been widely used for biostratigraphic zonation.

1.4

Isotopic Geochronology

J.C. Claoué-Long

1.4.1 INTRODUCTION

A revolution is overtaking the numerical time calibration of Phanerozoic stratigraphy. A combination of new dating technologies (notably $^{40}Ar/^{39}Ar$), and new interpolation procedures linked to magnetostratigraphy as a measure of sea-floor spreading rates, is now providing ages precise to ±0.1% for parts of the Mesozoic and Cainozoic scales. In the Palaeozoic, and parts of the Mesozoic, refinements of zircon U/Pb dating (and of the $^{40}Ar/^{39}Ar$ method where preservation of minerals allows) are now resolving first-order age problems with precisions down to 1%. However, interpolations between available measured ages remain inherently less confident below the Oxfordian, where the magnetic record is fragmentary, and the database of reliably measured ages remains sparse.

This chapter briefly reviews research in progress that is driving the changes in analysis and thinking, and progresses to outline the basis of the time calibration in these biostratigraphic charts.

1.4.2 NEW METHODS OF RADIOMETRIC DATING

The database from which previously published timescales were constructed is dominated by age measurements made in the 1970s and before, mainly applications of the K/Ar, U/Pb and Rb/Sr isotopic systems to a variety of minerals and whole rock samples (cf. Odin 1982; Snelling 1985; Harland et al. 1990). Since the mid-1980s, the field of geochronology has changed dramatically: totally new methods of isotopic analysis have been developed and the field is now led by rapid means of studying isotopic compositions in very small samples. The new ages have vastly improved levels of precision and confidence, and permit new interpretations of pre-existing age measurements. Dating problems that have defied resolution for several decades are now proving amenable to satisfactory isotopic study.

Foremost among these new developments is the virtual replacement of conventional K/Ar dating by refinements of $^{40}Ar/^{39}Ar$ analysis as the method of choice for dating Phanerozoic rocks. Heating of potassium-bearing minerals by laser or in low-blank furnaces, micro-volume gas extraction, and new low-blank, high-resolution mass spectrometers, now combine to give $^{40}Ar/^{39}Ar$ dating an internal reproducibility better than 1%, sometimes better than 0.1%. Automation of the analytical process, which can take as little as 20 minutes per single determination, means that multiple replicate ages are readily obtained, and this has proved the key to identifying multiple age components in samples where inherited crystals and alteration would otherwise obscure the primary magmatic age.

Another development is the increasing use of U/Pb zircon dating in Phanerozoic stratigraphy. Zircon is a U-bearing trace mineral in most volcanic rocks that is very resistant to alteration and hence useful in dating older (e.g. Palaeozoic) systems. Miniaturised low-blank methods now make it possible to date individual crystals in those conventional zircon laboratories where crystals are chemically dissolved to release their U and Pb for analysis. More radically, the Sensitive High Resolution Ion MicroProbe (SHRIMP) at the Australian National University has been developed to probe, and date, individual growth zones within single crystals. This microbeam instrument eliminates the need for chemical dissolution of samples, and takes as little as 20 minutes per single determination. As with $^{40}Ar/^{39}Ar$ dating, therefore, replicate measurements are quickly obtained for the several age components present in any rock, and this permits geological histories to be interpreted with confidence.

There are other developments in dating technology, but $^{40}Ar/^{39}Ar$ and zircon U/Pb dating are singled out here because they have opened the way to dating samples that previous isotopic techniques could not approach successfully. Both methods build up a picture of all the age components that are present in a sample (inherited, igneous, metamorphic, hydrothermal, alteration) and so permit, in favourable cases, the unravelling of igneous crystallisation ages from the overprint of other processes. This greatly improves the geological relevance of the dates obtained. Complex dating prospects, such as volcanic ash altered to bentonitic clay, can now be dated if zircon (for U/Pb study) or sanidine ($^{40}Ar/^{39}Ar$) has survived; so samples can be chosen for dating according to their stratigraphic importance rather than for the more restricting criterion of pristine chemical preservation.

1.4.3 INTER-LABORATORY ACCURACY IN DATING

It is important to realise that these refinements in analytical method are internal to each technique; improvements in the 'absolute' accuracy of ages do not necessarily follow. This is because age measurements are never 'absolute', but are always relative to a value ascribed to a standard composition used as a laboratory reference. For example, dating by $^{40}Ar/^{39}Ar$ depends on conversion of ^{39}K to ^{39}Ar in a reactor, a process that depends in part on the neutron fluence. This is gauged from a monitor of 'known' age in the reactor, which allows an irradiation coefficient to be calculated and applied to the samples. Unfortunately, there is presently no available monitor mineral whose age is known with an accuracy approaching today's 0.1% internal reproducibility of $^{40}Ar/^{39}Ar$ measurement. Different laboratories reference to ages variously put between 513 Ma and 524 Ma for the widely used MMhb-1 hornblende standard, for example. U/Pb zircon ages measured by SHRIMP similarly depend on the age assigned to a standard zircon that is probed together with the sample during an analytical run.

Within any single dating method, some stability is possible if all data are referenced to an 'agreed' value for a widely distributed standard. Thus, the Tertiary timescale of Berggren et al. (in press *a*, *b*) is based mainly on $^{40}Ar/^{39}Ar$ ages recalculated to conform to stated values for the irradiation monitor minerals MMhb-1 and Fish Canyon sanidine. This achieves internal consistency, with all of the $^{40}Ar/^{39}Ar$ dates being very precise relative to each other, even though the 'absolute' accuracy of the standard and the calculated ages remain a problem.

Other parts of the Phanerozoic timescale are constructed from several isotopic chronometers, which opens the issue of intercalibration between different radiometric dating methods that have different half-life measurements and different reference materials. This is a poorly understood area beyond the scope of this book, and intercalibration of the available chronometers must surely be a priority for future timescale research. The reader is therefore cautioned that the timescale is based on a mix of ages measured by different chronometers whose comparison is not fully explored. It is a challenge now for geochronologists to match the internal reproducibility of the latest dating methods with confidence in their comparability and 'absolute' accuracy.

1.4.4 THE PHILOSOPHY OF INTERPOLATION

It is not usual for a biozone of interest to be dated directly in place. Usually, the age of a stratigraphic position is measured indirectly, interpolated from an isotopic age for a more-or-less conveniently placed volcanic horizon on the basis of correlation and some attribute of the intervening sedimentary record: a thickness of sediment, a number of fossil zones, a number of magnetic intervals. Implicit in these 'hourglass' measures of time is the assumption of a known rate of sedimentation, evolution, or sea-floor spreading. These rates in their turn rely on measurements of time, and the circularity of the reasoning appears complete. The arguments are iterative rather than truly circular, however, because each new age measurement improves estimates of what are reasonable rates of processes such as sea-floor spreading, and these in turn improve the interpolation procedure. The calibration of the numerical, biostratigraphic, and other scales is therefore improved by iteration, and the apparent circularity of reasoning is an essential part of the method.

The logic of uncertainty in timescale calibration is therefore not simple. In calibrating numerical time with biostratigraphy, magnetostratigraphy, event stratigraphy, and other measures of the progress of events, we are attempting to mesh interdependent observations. One measure necessarily influences the rest, and vice versa. The notion of uncertainty must try to take account of this, and of the correlation-dependence of the calibration process.

The most objective attempts to apply this thinking have been those of Harland et al. (1982, 1990), who applied standardised interpolation and error-minimisation procedures to a database of age information for the entire Phanerozoic. Some stability is achieved in the Cainozoic scale, where the database of age information is most comprehensive and confident, but the Mesozoic and Palaeozoic scales remain controversial in places owing to the relative scarcity of good data and the lack of a magnetostratigraphic comparison. Recently Agterberg (1990) proposed refinements to the mathematical interpolation procedures and these have been applied in the Mesozoic by Gradstein et al. (in press). Interpolation procedures remain inherently more subjective below the Oxfordian, and in the Cretaceous Magnetically Quiet Period, where the usefulness of magnetostratigraphy for independent correlation and interpolation is restricted.

The net uncertainties in timescales that result from these interpolation studies are almost certainly underestimated. This is because they take account of only one of the two sources of error in measuring stratigraphic ages: the well understood laboratory errors in numerical age measurement that are reported as an ordinary aspect of laboratory analysis and are readily incorporated into the calibration process. They do not include the equally important uncertainties in the biostratigraphic position, and correlation, of the horizons that have been dated, for the simple reason that these biostratigraphic uncertainties remain difficult to quantify. It is a challenge for future timescale calibrations to begin applying systematic error estimates to biostratigraphic correlation, to match the routinely available estimates of laboratory measurement uncertainty.

1.4.5 THE NUMERICAL TIMESCALE TODAY

The numerical timescale is therefore now in a state of rapid improvement. Existing databases of age information are being refined, and in places superseded, by the application of new thinking and new dating technologies. Processes of interpolating between measurements are becoming more sophisticated and less subjective, and are based on ever-more detailed stratigraphic correlations. In the Cainozoic, the numerical scale has been stable and uncontroversial — at least to first order — for some time and is now progressing to very useful levels of precision: in places the envelope of debate has narrowed to less than 1 Ma. In the Mesozoic, the scale is being stabilised now by a rapid increase in the database of $^{40}Ar/^{39}Ar$ ages, and by better understanding of interpolation procedures. In the Palaeozoic, the ages of major horizons and boundaries are now being measured with confidence, often for the first time, by applications of new dating methods that are better able to deal with complex alteration histories in older rocks. However, interpolation between the sparse reliable dates in the Palaeozoic remains a subjective process of estimation based, as much as anything, on perceptions of the relative durations of biozones. While these improvements progress in different parts of the Phanerozoic, the numerical scale will be based on very different qualities of information in different parts of the column. The scale adopted in this volume is therefore an interim measure.

In the **Cambrian** and **Ordovician**, new zircon dates are superseding the pre-existing age database. The quantity of new information is very small, but the recalibration of time to biozones is so radical in places that it is no longer realistic to use the earlier, less-certain age constraints. We therefore attempt to scale the Cambrian and Ordovician to the sparse new database of zircon ages. This part of our scale therefore departs significantly from existing published timescales.

In the **Silurian** and **Devonian**, no significant new data have emerged since the thorough timescale review of McKerrow et al. (1985). We therefore adopt that calibration, with minor amendments.

In the **Carboniferous**, a significant new database of zircon U/Pb and $^{40}Ar/^{39}Ar$ ages is published and in press (Hess & Lippolt 1986; Claoué-Long et al. in press; Roberts et al. in press *a*, *b*), and the scale is largely constrained by these. This part of our scale therefore differs significantly from previous compilations.

In the **Permian**, detailed zircon dating is now in progress, pending the outcome of which, subdivision of the period is conjectural because reliable age information is lacking.

In the **Triassic, Jurassic** and **Cretaceous**, a recent attempt at integration of available radiometric ages with systematic interpolation procedures by Gradstein et al. (1994) arrived too late to be incorporated in this discussion. The earlier compilations by Kennedy & Odin (1982), Hallam et al. (1985) and Odin (1985*b*) have therefore been applied.

In the **Tertiary** we follow the suggestions of Berggren et al. (in press *a*, *b*), who have thoroughly reviewed the most recent biostratigraphic and numerical age information.

1.5

Magnetostratigraphy

M. Idnurm, C. Klootwijk, H. Théveniaut and A. Trench

1.5.1 INTRODUCTION

Four techniques to develop magnetic polarity timescales were discussed in a seminal paper by Irving & Pullaiah (1976), namely the direct method, marine method, timescale method, and stratotype method. The direct method plots polarities of individual rock units directly against radiometric age of the rocks. The marine method compiles successions of marine magnetic anomalies with age tie-ins provided from radiometric dating of oceanic basalts or from biostratigraphic control on directly overlying sediments. The timescale method plots polarities of individual rock units through conversion of chronostratigraphic ages into absolute ages according to any preferred timescale. The stratotype method plots polarity successions against lithostratigraphic and/or biostratigraphic control.

The *direct method* has limited practical applicability; it is restricted to rocks that can be dated radiometrically and its resolution is limited by the precision and accuracy of radiometric methods. The *marine method* has provided some highly detailed polarity sequences for the Cainozoic and part of the Mesozoic. Its applicability is limited by the resolution of marine magnetic modelling techniques, the accuracy and precision of the biostratigraphic and radiometric tie-ins, and the absence of pre-Middle Jurassic ocean floor. The *timescale method* is a practical way of integrating results from individual studies, often carried out for other than magnetostratigraphic purposes. Its applicability is limited, however, by the absence of unequivocal stratigraphic superposition evidence and, for earlier results in particular, uncertainties about the primary origin of individual remanences. Nowadays, when magnetisations of virtually all lithologies can be measured with highly sensitive cryogenic magnetometers, the *stratotype method* has evolved as the preferred method for development of magnetostratigraphies. It provides unequivocal superposition control, better control on primary origin of the magnetisations, and its time resolution is limited only by lithostratigraphic and biostratigraphic detail of the studied sections.

Development of a reference magnetostratigraphy for the Phanerozoic, and for the Palaeozoic in particular, is hampered by the limited number of studies carried out to date that are dedicated entirely to magnetostratigraphy. Therefore, data of diverse sources, diverse quality, diverse reliability, and diverse resolution need to be integrated, resulting in various levels of detail and reliability for the reference magnetostratigraphy.

In several recent magnetostratigraphic compilations for the Early Palaeozoic, Allan Trench and co-workers (1991, 1993) classified original data and compiled magnetostratigraphies at three quality levels: fundamental, supplemental or unconstrained:

- *Fundamental* data, which come from dedicated magnetostratigraphic studies on continuous, well-controlled stratigraphic sections with a duration of several zones or stages (spanning at least several million years), and with reversal boundaries that are demonstrated to represent stratigraphic levels.
- *Supplemental* data, which come from well dated rocks that cover only a limited stratigraphic interval and generally display only a single magnetic polarity.
- *Unconstrained* data, which have inadequate biostratigraphic control and indications for a secondary origin rather than primary origin of the magnetisation.

Trench's fundamental and supplemental data equate most closely with Irving & Pullaiah's (1976) stratotype and timescale data. Unconstrained data are not useful for magnetostratigraphic application. This threefold approach to data compilation is equally applicable to the Late Palaeozoic, and the essence of Trench's technique has been followed throughout the Palaeozoic.

Nomenclature The International Subcommission on Stratigraphic Classification of the International Union of Geological Sciences proposed a standard magnetostratigraphic nomenclature (Anonymous 1979; Cox in Harland et al. 1982, 1990; see Hailwood 1989 for overview; Salvador 1994). The subcommission recommended against the use of the terms 'event' and 'epoch' for description of **time** derived from geomagnetic polarity, and proposed instead the terms 'polarity subchron' (10^4–10^5 yr) and 'polarity chron' (10^5–10^6 yr), with 'polarity superchron' (10^6–10^7 yr) describing a longer polarity interval in time. Corresponding chronostratigraphic terms for description of rocks formed during these time intervals are: 'polarity subchronozone', 'polarity chronozone' and 'polarity superchronozone'. Corresponding lithostratigraphic terms describing magnetic polarity units within rock units are: 'polarity subzone', 'polarity zone' and 'polarity superzone'. The term 'magnetozone' is nowadays used most commonly for polarity zone, and 'magnetic chron' likewise for polarity chron.

1.5.2 CAMBRIAN TO SILURIAN

Allan Trench

Introduction

The establishment of a magnetic polarity timescale for the Early Palaeozoic remains in its infancy although significant progress has been made in recent years. Early Palaeozoic rocks present a number of problems to magnetostratigraphic work beyond those of younger strata. These difficulties can be summarised as follows:

- The absence of preserved Palaeozoic oceanic crust precludes comparison of palaeomagnetic data with an independent magnetic anomaly pattern from which a high-resolution polarity sequence can be deduced. The Mesozoic–Tertiary oceanic magnetic anomalies have played a vital role in the development of polarity timescales for these periods (Harland et al. 1990).
- The likelihood of magnetic overprinting having completely destroyed the primary magnetic signature of a rock becomes greater with increasing rock age given the increasingly complex geological history affecting the rock. As such, geologically 'ideal' sections for magnetostratigraphic work may preserve no evidence of the palaeomagnetic field at the time of rock deposition (e.g. Ordovician–Silurian boundary section at Anticosti Island, Quebec; see Seguin & Petryk 1986, Ripperdan 1990).
- Plate configurations have changed dramatically since the Early Palaeozoic, making the identification of normal and reverse polarities equivocal for many plates and, in particular, for marginal continental terranes. For example, several competing palaeogeographies exist for Late Precambrian to Cambrian times (e.g. Piper 1987; Moores 1991; Kirschvink 1991; McKerrow et al. 1992).

The above constraints have resulted in very few dedicated magnetostratigraphic studies having been attempted on Early Palaeozoic rocks. Some notable exceptions include the works of Kirschvink & Rozanov (1984) on Cambrian limestones of the Siberian platform, Ripperdan & Kirschvink (1992) on the Cambrian–Ordovician boundary carbonates of Black Mountain, Australia, and Torsvik & Trench (1991) on Lower to Middle Ordovician carbonates from Sweden.

Despite the lack of magnetostratigraphic studies of Early Palaeozoic rocks, numerous palaeomagnetic studies have been undertaken with the aim of deducing continental palaeopoles and therefore clarifying continental reconstructions. These studies, although designed for other purposes, provide useful ancillary information from which at least the polarity bias of the Early Palaeozoic magnetic field can be deduced. Syntheses of the Early Palaeozoic magnetic field polarity have been undertaken by Irving & Pullaiah (1976), Khramov et al. (1965); Khramov (1987); Khramov & Rodionov 1981) and Trench et al. (1991, 1992, 1993).

Hailwood (1989) has detailed the magnetostratigraphic nomenclature as adopted in studies of Mesozoic, Tertiary and Recent sections. Given our present rudimentary knowledge of the Early Palaeozoic magnetic field, however, Trench et al. (1991, 1993) adopted an informal nomenclature for the subdivision of polarity intervals. In this scheme, reversals were named depending on the stratigraphic interval in which they occur (e.g. A[R] = Arenig, reversed; L[N2] = Llanvirn–Llandeilo, normal interval 2; W[N] = Wenlock, normal interval; LL[M] = Llandovery, mixed polarity interval). On a detailed scale when numerous reversals are present, Kirschvink & Rozanov (1984) opted to name reversals using a sequential lettering (A–T) and sign convention ('+' = normal,'–' = reverse).

Data quality for Early Palaeozoic magnetostratigraphic purposes has been assessed using the tripartite division of palaeomagnetic data (see above; after Trench et al. 1991).

Several conclusions may be drawn from the summaries presented below. When compared with equivalent studies of Cambrian and Silurian rocks, Ordovician strata reveal less numerous reversals of the Earth's magnetic field. This can be interpreted to indicate a lower reversal frequency during Ordovician time than either Cambrian or Silurian time. A preliminary estimate of reversal frequency for the Ordovician Period is of the order of one reversal per five million years (Trench et al. 1992).

Varying reversal frequency between Cambrian, Ordovician and Silurian time segments implies that Early Palaeozoic magnetostratigraphy will serve as a high-precision correlation tool only in times of rapid rate of reversals of the Earth's magnetic field (e.g. Precambrian– Cambrian boundary). Conversely, the comparative simplicity of segments of the Early Palaeozoic polarity record suggests that correlation of disparate sections should be possible with the appropriate magnetic zone with relative ease. Further establishment of the Cambrian, Ordovician and Silurian magnetostratigraphic timescale and subsequent correlation of sections should resolve palaeogeographic uncertainties affecting many Early Palaeozoic continental blocks.

Cambrian

A detailed compilation of all palaeomagnetic polarity data from Cambrian strata has yet to be carried out and is beyond the scope of the present summary. Uncertainties for the Cambrian are heightened by disputes regarding the absolute age of the Cambrian–Ordovician and particularly the Cambrian–Precambrian boundary (Harland et al. 1990; this study). Likewise, palaeogeographic uncertainties are greater in Cambrian time than either Ordovician or Silurian times and preclude unambiguous polarity definition. In this regard, no attempt has been made in this summary to reconcile the different authors' polarity interpretations for Cambrian time with the plethora of Cambrian reconstructions presently available. For example, Kirschvink (1991) has revised the polarity classification adopted during his previous studies of the Cambrian of the Siberian Platform (Kirschvink & Rozanov 1984).

Stratigraphy-parallel polarity reversals within Cambrian rocks have been described as follows:

- Kirschvink et al. (1991) report detailed magnetostratigraphic and carbon isotope results from probable Precambrian–Cambrian boundary sections in Morocco, Siberia and South China.
- Kirschvink (1978*a*,*b*) documents palaeomagnetic results from Late Precambrian and Early Cambrian sediments of the Amadeus Basin, Central Australia. Samples were collected at approximately 1 m intervals through an 800 m thick succession. An unconformity test (Kirschvink 1978*b*) is argued to support a magnetisation age equivalent to the stratigraphic age of these rocks. The Early Cambrian interval of the Arumbera Sandstone is represented by mixed polarities that extend upwards into the Todd River Dolomite and Eninta Sandstone, the former having a definite Early Cambrian biostratigraphic age (Shergold 1991*a*).
- Kirschvink & Rozanov (1984) report numerous magnetozones from Late Precambrian and Early Cambrian strata of the Siberian Platform. Stratigraphy-parallel reversals that could be correlated with archaeocyathid zones were observed. Reliable correlation of these data with coeval results from Australia remains equivocal.
- Klootwijk (1980) reports magnetic polarity data from Early Cambrian rocks of the Flinders Ranges of South Australia and Middle to early Late Cambrian rocks of the Amadeus Basin. Shergold (1989) summarised the polarities with respect to the biostratigraphic record. Mixed polarities are identified from the Lower Cambrian Ajax Limestone with the possible correlative section, the Wilkawillina Limestone and Oraparinna Shale, displaying only reverse polarity. Results from the Moodlatana Formation are reverse, with the Pantapinna Sandstone displaying both polarities. Results from the Amadeus Basin are reverse, within the Middle Cambrian upper Giles Creek Formation and lower Shannon Formation.
- Ripperdan & Kirschvink (1992) and Ripperdan et al. (1992) report detailed magnetostratigraphic and carbon isotope studies of the Black Mountain section in Queensland, proposed as representative of the Cambrian–Ordovician boundary on biostratigraphic grounds.
- Zhao et al. (1992) revealed multicomponent magnetisations in Cambrian (and Ordovician) sedimentary units from several localities in the North China Block. The higher unblocking temperature components pass field tests and are interpreted to approximate the depositional age of the sequences. Zhao et al. interpret the Cambrian strata to be predominantly reversely polarised with intermittent zones of mixed polarity.

Ordovician

Magnetic polarity data from Ordovician rocks are summarised in Figures 3 and 4. Fundamental data (Fig. 3) reveal a probable correlation of polarity results between detailed sections of carbonates on the Baltic (Torsvik & Trench 1991) and south Siberian (Khramov et al. 1965; Rodionov 1966) platforms. The latter section, along the banks of the Lena River, is currently under reinvestigation using contemporary analytical methods and superconducting magnetometer technology (Torsvik & Tait, pers. comm. 1993). Supplemental data (Fig. 4) are consistent with the results of the detailed platform successions and yield evidence of the polarity bias in their own right. An interesting outcome of this analysis suggests that polarity reversals occurred less frequently in Ordovician times than, for example, in Cainozoic times (Fig. 5, after Trench et al. 1992).

The following characteristics of the Ordovician magnetic field are evident (Figs 3, 4):

- All Arenig rocks have thus far yielded reversely polarised magnetisations.
- Several reversals of the geomagnetic field occurred during Llanvirn–Llandeilo times.
- The early part of the Caradoc Series corresponds to an interval of normal polarity. The latter part of the series is characterised by a reversely polarised field.
- Ashgill rocks have yielded normal polarities to date, although short reverse polarity zones that have yet to be unambiguously identified may occur.
- The above observations imply a geomagnetic field of predominantly reverse polarity during Early Ordovician times (Tremadoc–Llanvirn), succeeded by a predominantly normal polari-

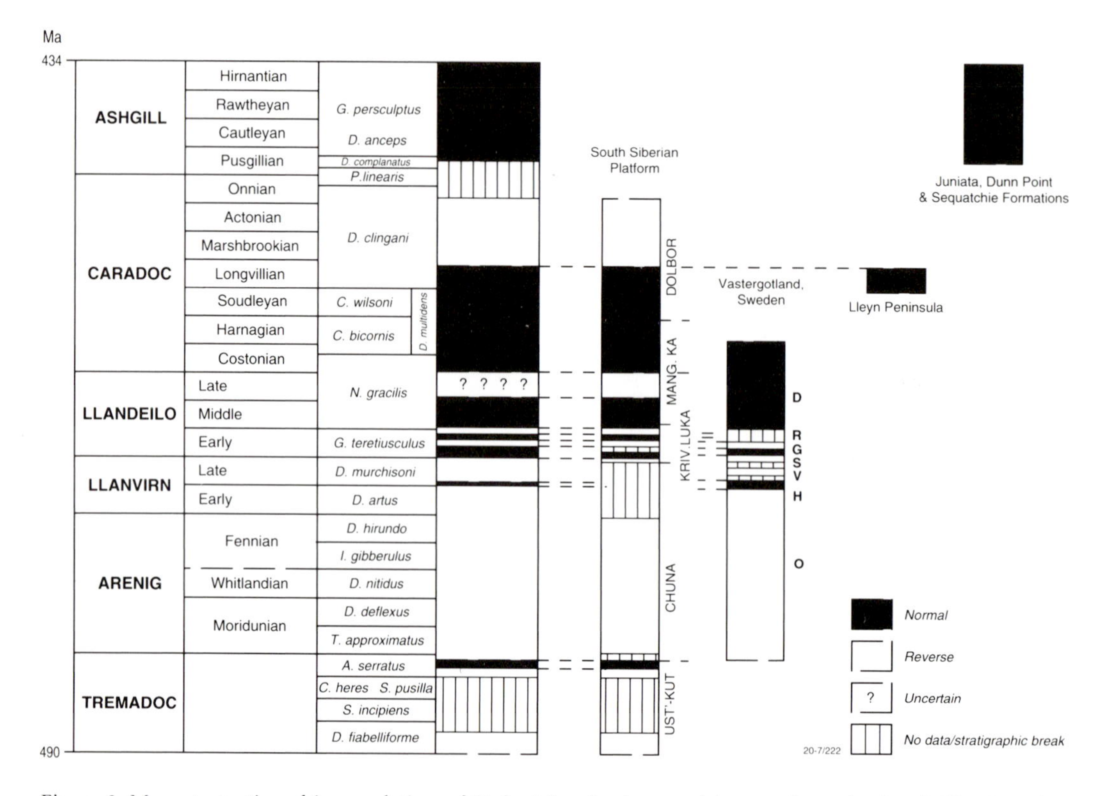

Figure 3: Magnetostratigraphic correlation of Ordovician *fundamental* datasets from the South Siberian Platform and southern Sweden (after Trench et al. 1991). Data from the Lleyn Peninsula, Wales (Thomas 1976), and the Juniata (Miller & Kent 1989), Dunn Point (Van der Voo & Johnson 1985) and Sequatchie Formations (Morrison 1983; Morrison & Ellwood 1986) of eastern North America are also included as they further constrain the composite stratigraphy. The inferred polarity for Ashgill time is based on normally magnetised palaeomagnetic data from the Dunn Point, Juniata and Sequatchie Formations. If postulated reversely magnetised sites within the Juniata and Sequatchie Formations are truly distinguishable from Alleghenian remagnetisations, then a short period (or periods) of reverse polarity may exist within the Ashgill. Left-hand columns refer to stratigraphic elements of the Ordovician and the adjacent right-hand column represents the composite magnetostratigraphic record. Faunal subdivisions are European zones. Subdivisions of the Arenig Series are taken from Fortey & Owens (1987). Siberian stages are indicated next to the appropriate column. Swedish formations (lettered) from Torsvik & Trench (1991) are as follows: D, Dalby Limestone; R, Ryd Limestone; G, Gullhogen Limestone; S, Skovde Limestone; V, Vämb Limestone; H, Holen Limestone; O, Orthoceras Limestone.

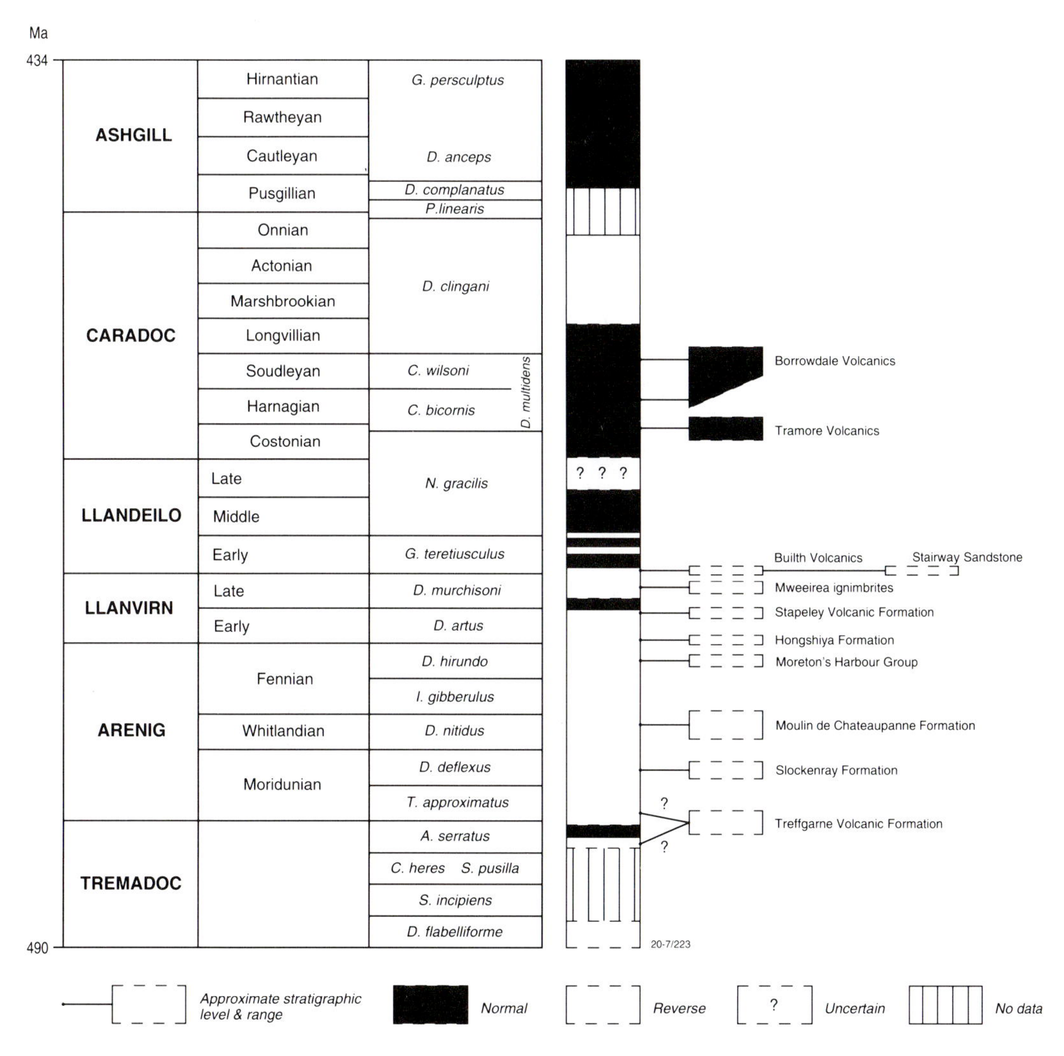

Figure 4: Composite Ordovician magnetostratigraphy (from Fig. 3) with the addition of supplementary datasets. Individual studies are as follows: Borrowdale Volcanics (Faller et al. 1977); Tramore Volcanics (Deutsch 1980); Builth Volcanics (Briden & Mullan 1984; Trench et al. 1991); Stairway Sandstone (Embleton 1972), Mweelrea Ignimbrites (Deutsch & Somayajulu 1973; Morris et al. 1973), Stapeley Volcanic Formation (McCabe & Channell 1990); Hongshiya Formation (Fang et al. 1990); Moreton's Harbour Group (Johnson et al. 1991); Moulin de Chateaupanne Formation (Perroud et al. 1985); Slockenray Formation (Trench et al. 1988); Treffgarne Volcanic Formation (Torsvik & Trench 1991; Trench et al. 1991).

ty field in Late Ordovician times (Llandeilo–Ashgill).

Magnetic polarity data that have recently become available from Ordovician strata include results from shales and limestones of the North China Block (Zhao et al. 1992) and the Bluffer Pond Formation of the northern Appalachians (Potts et al. 1993). The data from the North China Block pass reversal, baked contact, and fold tests (Mesozoic folding) and demonstrate evidence of an overall similarity in the magnetic polarity pattern in the stratigraphic succession

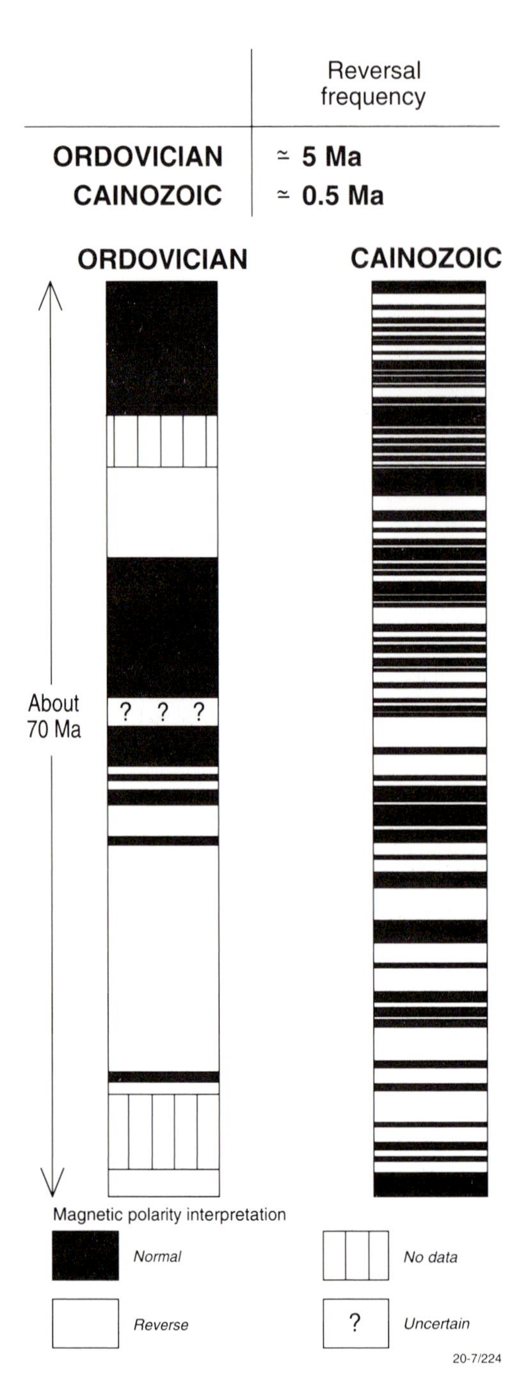

Figure 5: Comparison of Ordovician reversal pattern with Cainozoic reversal pattern indicating the relative simplicity of Ordovician polarities as presently determined.

at sampled localities (Fig. 6). Unfortunately, the apparent polar wander path for the North China Block is poorly constrained for the Early Palaeozoic and Zhao et al. (1992) were unable to unambiguously establish the magnetic polarity identity of their data. Comparison with the composite Ordovician magnetic stratigraphy (Fig. 3) might suggest that the North China data display the opposite polarity to that put forward by Zhao et al. (1992). Potts et al. (1993) present evidence for normal polarity of the Caradocian (Late Ordovician) palaeofield, consistent with the synthesis of previous studies presented by Trench et al. (1991).

Silurian

Magnetic polarity data for Silurian times are summarised in Figure 7 (after Trench et al. 1993). The following points of interest are noted:

- Palaeomagnetic data sets regarded as of primary origin are not yet available from a continuous Silurian stratigraphic succession of any great duration (i.e. there are no fundamental data), so the composite Silurian magnetostratigraphy is still at a much more preliminary stage than that for the Ordovician.
- The majority of Silurian time was characterised by periods of 'mixed' polarity (i.e. intervals of high reversal frequency).
- Wenlock times were mainly characterised by a magnetic field of normal polarity based on single polarity studies from red mudstones of western Ireland (Smethurst & Briden 1988), andesitic volcanics of south-western England (Torsvik et al. 1993), carbonates and mudstones of south-eastern Sweden (Trench & Torsvik 1991) and volcanics of south-eastern Australia (Luck 1973; Klootwijk, unpublished data).

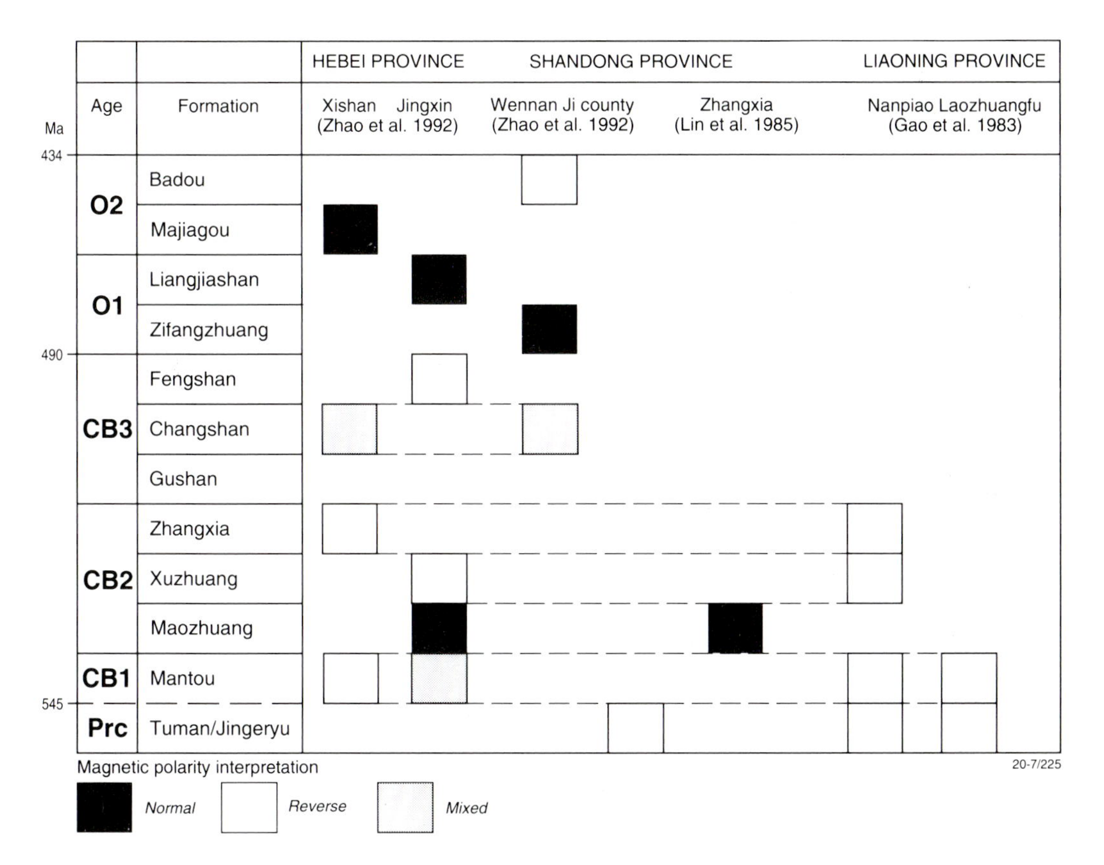

Figure 6: Sketch section showing magnetic polarity zonations observed from Cambrian and Ordovician sediments of the North China Block (after Zhao et al. 1992).

1.5.3 DEVONIAN TO PERMIAN

Chris Klootwijk

Introduction

Data quality for the Devonian, Carboniferous and Permian periods is variable. Considerable progress has been made over the past 5 years in defining the base and top of the Permo–Carboniferous Reversed Superchron (PCRS). Some high-quality magnetostratigraphic profiles for the mid-Carboniferous and the latest Permian are now available, but the boundaries of the PCRS are as yet not defined within zone level. The reversal stratigraphy within the PCRS has not yet been studied at the high-resolution standards that are the current norm. The few earlier reported observations of normal polarity intervals within the PCRS remain to be confirmed and detailed on their primary origin and stratigraphic level. The magnetostratigraphy of the pre-PCRS Early Carboniferous and Devonian is poorly known. A few isolated studies have provided some magnetostratigraphic detail for successions covering limited timespans. Some magnetostratigraphic compilations have been carried out on palaeomagnetic data that were mainly acquired for other than magnetostratigraphic purposes. Many of these data were obtained prior to the recognition of the pervasiveness and wide areal extent of 'Kiaman' reverse polarity overprints in Permo–Carboniferous fold belts and adjacent basins. The primary origin of many of these results, the reversed

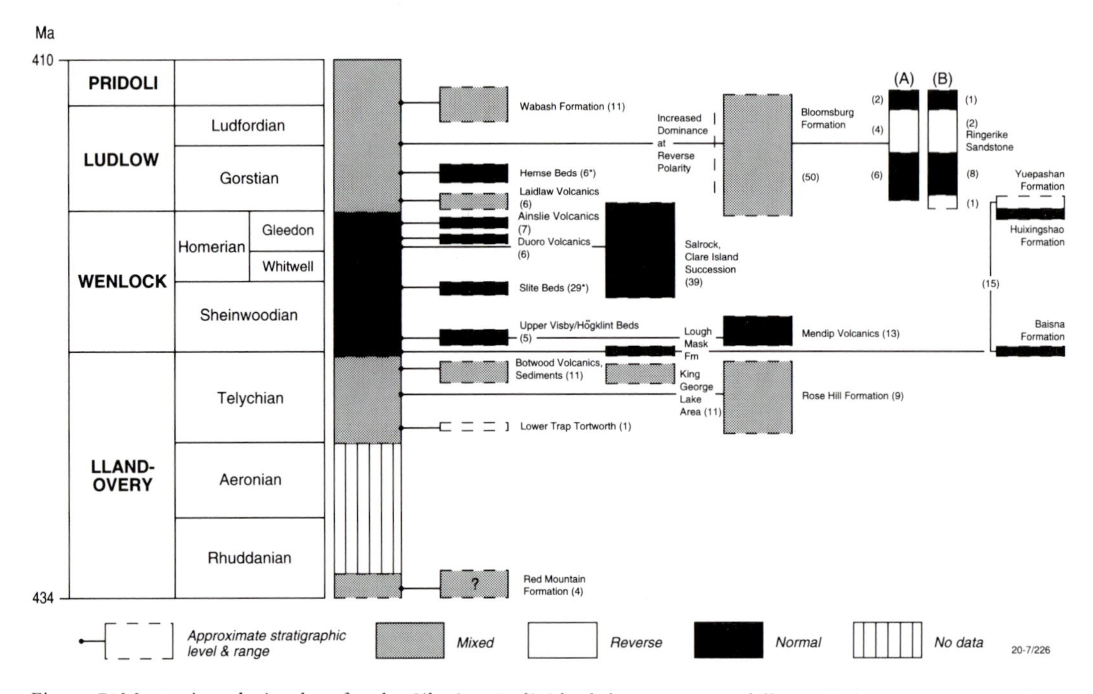

Figure 7: Magnetic polarity data for the Silurian. Individual datasets are as follows: Wabash Formation, USA (McCabe et al. 1985); Hemse Beds, Sweden (Claesson 1979); Laidlaw Volcanics, Australia (Luck 1973); Ainslie Volcanics, Australia (Luck 1973); Douro Volcanics, Australia (Luck 1973); Slite Beds, Sweden (Claesson 1979), Upper Visby/Höglint Beds, Sweden (Trench & Torsvik 1991); Botwood Volcanics and sediments, Canada (Gales et al. 1989); lower lava, Tortworth, UK (Piper 1975); Red Mountain Formation, USA (Morrison 1983); Salrock, Clare Island and Lough Mask Formations, Ireland (Smethurst & Briden 1988); King George IV Lake area, Canada (Buchan & Hodych 1989); Bloomsburg Formation, USA (Kent 1988); Mendip Volcanics, England (Torsvik et al. 1993); Rose Hill Formation, USA (French & Van der Voo 1977, 1979); Ringerike Sandstone, Norway (Douglass 1988); Yuejiashan, Huixingshao and Baisha formations, southern China (Opdyke et al. 1987). Numbers in parentheses next to each set indicate the number of palaeomagnetic sample sites; those with an asterisk refer to number of samples.

polarity interpretations in particular, remains to be confirmed, and the compilations may serve as no more than a guide to further detailed studies. Extensive and important magnetostratigraphic studies were carried out by Russian workers on Palaeozoic stratotype sections in Russia and Tatarstan. Most of these studies were originally published in the 1970s and early 1980s in Russian literature. Details of the studies and results are not easily accessible, although subsequent compilations were published more widely. A further impediment to scrutiny of the results is their acquisition through methods that are largely unfamiliar to workers outside the former USSR. The studied stratotype sections no doubt have great potential for development of a magnetostratigraphic timescale, but the extent to which available results can be used needs further demonstration in the light of widespread Late Palaeozoic remagnetisation.

A long interval of reversed polarity, from about the Middle Carboniferous to the latest Permian, was first defined by Irving & Parry (1963) in a rather informal way as the 'Kiaman Magnetic Interval', with the 'Kiaman Magnetic Division' as the corresponding 'time-rock unit'. They defined the 'Kiaman' on the basis of earlier observations of a prolonged interval of reversed polarity in Europe, North America, and in the Permian stratotype sections in Russia and

Tatarstan (see Irving & Pullaiah 1976), and in particular on the basis of their own observations in the Hunter Valley and Sydney Basin. The lower boundary of the 'Kiaman Magnetic Interval' of reversed polarity was defined as the 'Paterson Reversal' (Irving 1966) at the change from normal polarity in the underlying Paterson Volcanics (Paterson toscanite) to reversed polarity in the overlying Seaham Formation (upper Kuttung glacial sediments). The upper boundary was defined as the 'Illawarra Reversal', then not better located than above the Gerringong Volcanics (upper marine latites) and below the Narrabeen Group (Narrabeen chocolate shales; Irving 1963), which left its position undefined within the uppermost Permian coal measures of the Sydney Basin.

Two problems soon became apparent to Irving and co-workers. First, they realised that the early practice of coining new specific names for magnetic intervals would soon clutter the stratigraphical nomenclature with terms of unclear linkage (Irving 1971). Linkage of such intervals to the existing timescale nomenclature was judged preferable. The second problem related to correlation in time of the 'Paterson Reversal' as the base of the 'Kiaman' in eastern Australia, with observations on the base of the 'Kiaman' in North America, South America and Eurasia (Irving & Pullaiah 1976). It appeared to them that the 'Paterson Reversal' might well represent a short normal polarity event within the 'Kiaman' and could not be used, therefore, to define its base. For these reasons, Irving (1971) withdrew the term 'Kiaman' and suggested 'Late Palaeozoic Reversed Interval' as an easier identifiable term. The term 'Permo–Carboniferous quiet interval' or PCR was introduced subsequently (Irving & Pullaiah 1976).

Terminology for the 'Kiaman' has evolved since then, unfortunately without arriving at a single commonly agreed identifier. So different workers and groups use different terms such as: 'Kiaman Hyperzone (R)' (Khramov & Rodionov 1981; Khramov 1987; DiVenere & Opdyke 1991*a*; Haag & Heller 1991), 'Kiaman Superchron' (Molina-Garza et al. 1989; Magnus & Opdyke 1991), 'Kiaman Reversed Superchron' (Ma et al. 1993), 'Permo–Carboniferous Reversed Polarity Superchron' (McFadden et al. 1988), 'Permo–Carboniferous Reversed Superchron' (PCRS) (DiVenere & Opdyke 1991*a*; Magnus & Opdyke 1991; Opdyke et al. 1993) or 'PC-R superchron' (Cox in Harland et al. 1990). Following the recommendations of the International Subcommission on Stratigraphic Classification (Anonymous 1979), the term 'Permo–Carboniferous Reversed Superchron' (PCRS) is used in the following.

Early Carboniferous and Devonian constraints

The magnetostratigraphy of the Early Carboniferous and the Devonian has not been detailed systematically, other than in Russian studies that are mostly of some vintage. Results from only a few modern studies of limited stratigraphic coverage have been reported:

- Studies on the uppermost Mississippian Mauch Chunk Formation in Pennsylvania (DiVenere & Opdyke 1991*b*) have provided a succession of mixed polarity. The upper part of the succession is tentatively correlated with the basal part of the polarity succession in the Maringouin Formation (Fig. 8).
- Hurley & Van der Voo (1987, 1990) have interpreted a series of five polarity intervals within a thin *Frutexites* stromatolite bed from the Virgin Hills Formation of the Canning Basin. This formation is of Frasnian–Famennian age; the *Frutexites* beds are entirely within the *Palmatolepsis triangularis* conodont subzone of Druce (1976).
- Smethurst & Khramov (1992) studied Middle Palaeozoic successions from the Russian Platform in the Ukraine and observed very pervasive 'Kiaman'-type reversed polarity overprinting, which put doubt as to the interpretation of the earlier Russian results. Primary magnetisation results were established in Lower Devonian successions for which unfortunately only limited stratigraphic detail is available. This showed the occurrence of

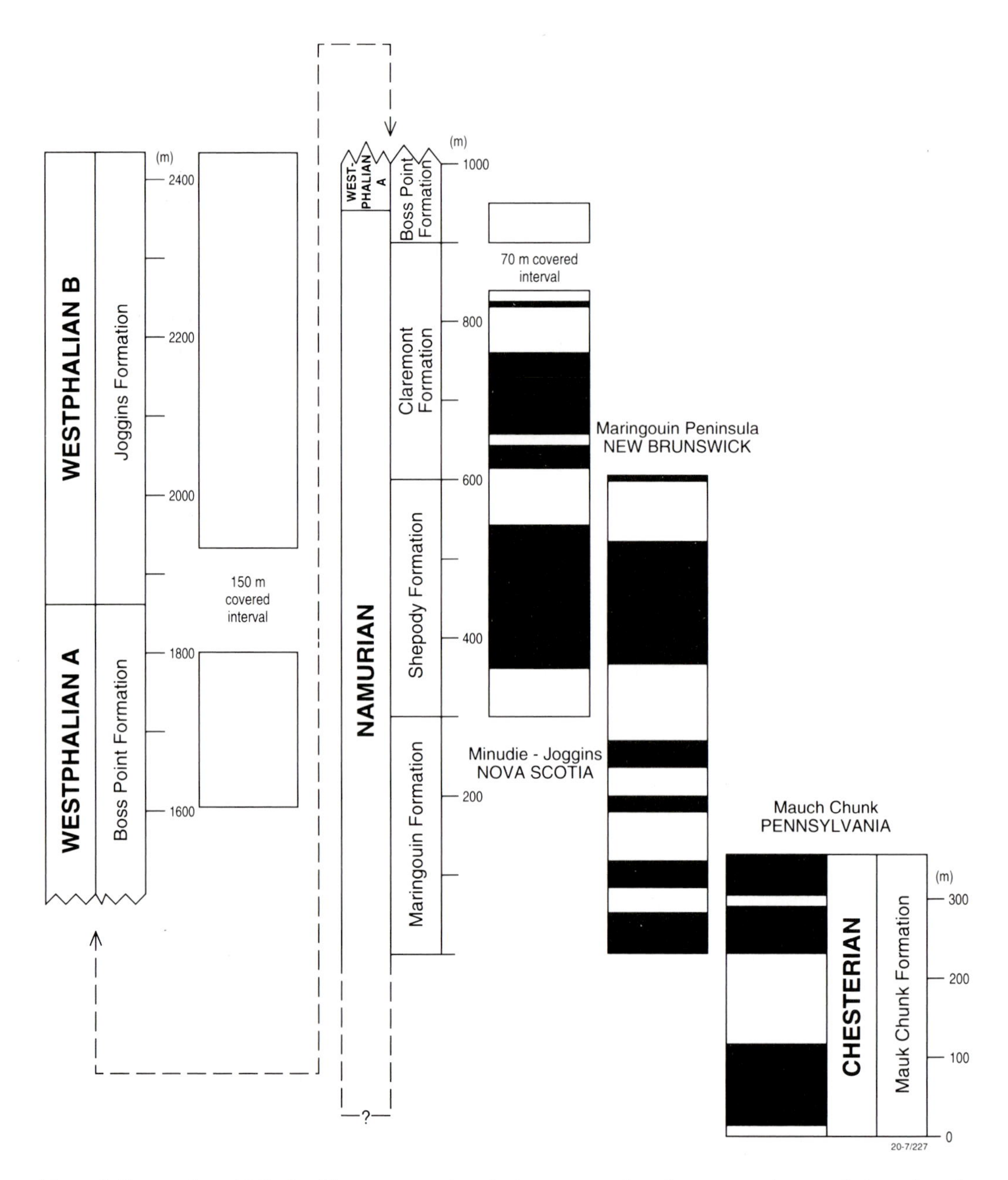

Figure 8: Composite mid-Carboniferous magnetic polarity sequence based on a tentative correlation of results from the Mauch Chunk Formation (Pennsylvania) and from the Cumberland Basin (Nova Scotia, New Brunswick), redrawn after DiVenere & Opdyke (1991*a*, Fig. 10; 1991*b*, Fig. 6).

reversed polarity in Gedinnian–Pridolian sections, normal polarity for the Gedinnian, and reversed polarity for the Gedinnian–Siegenian.

Given the absence of recent detailed magnetostratigraphy studies on Early Carboniferous and Devonian successions, use of the polarity

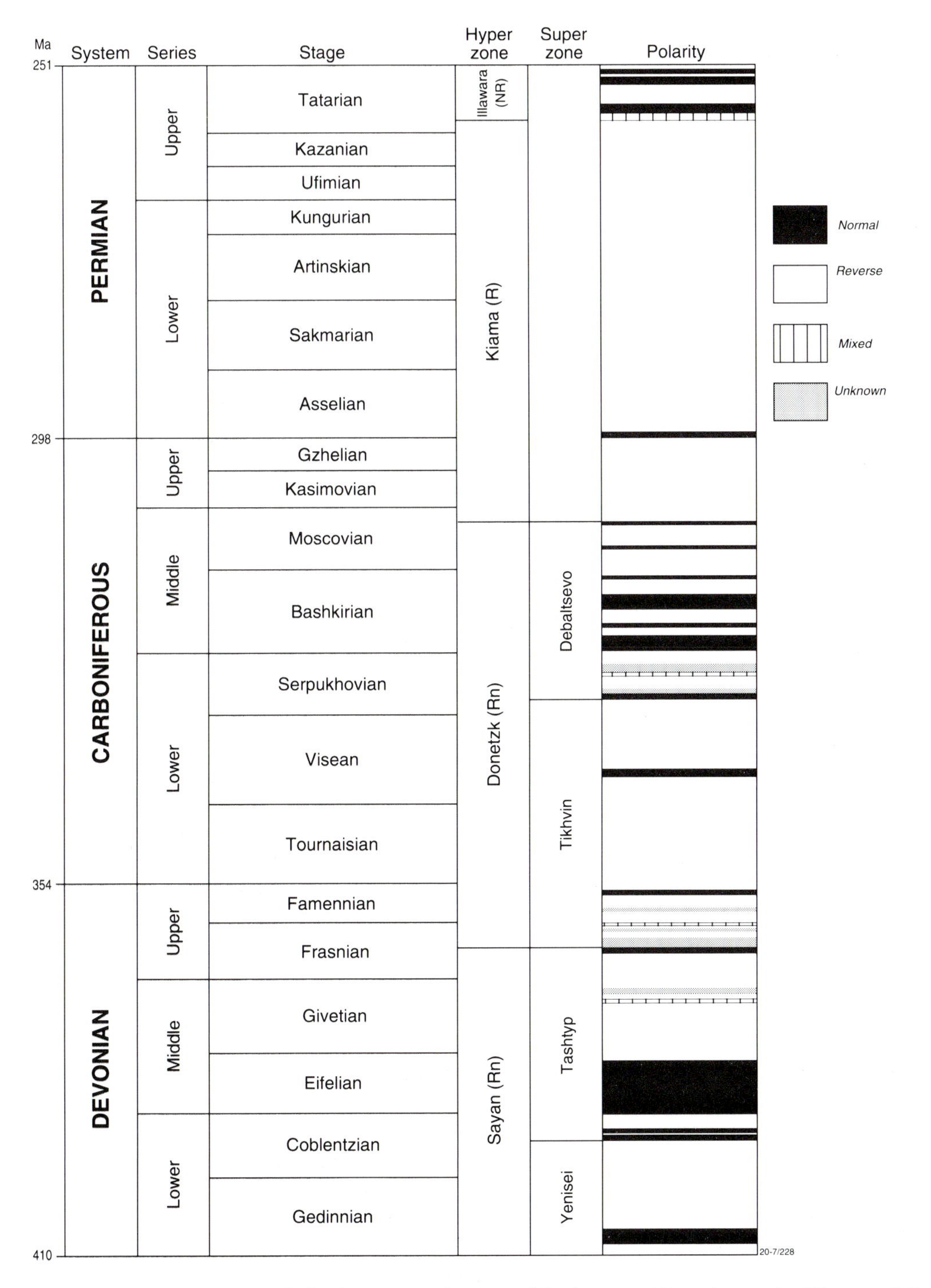

Figure 9: Magnetic polarity scale for the Upper Palaeozoic of the former USSR, redrawn after Khramov & Rodionov (1981, Fig. 1).

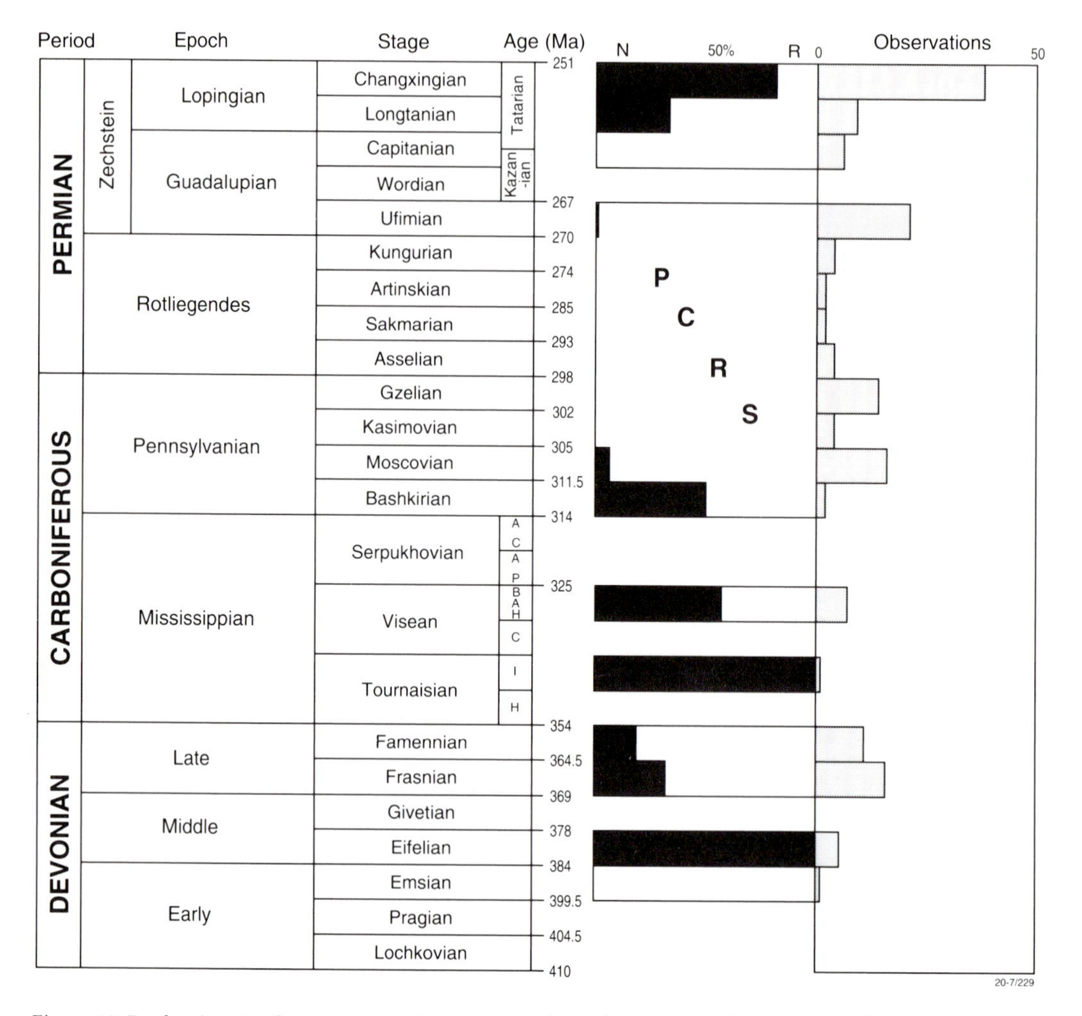

Figure 10: Predominant polarity sequence for the Late Palaeozoic, based on the frequency of normal and reversed polarity results at stage or substage level, as obtained from interrogation of the Global Palaeomagnetic Database (version 2.2). The database was interrogated using ages according to Harland et al. (1990), corresponding ages according to the Timescales Project (herein) are shown.

timescale compiled from earlier Russian studies (Khramov & Rodionov 1981; Khramov 1987; Fig. 9) is still appropriate. Care must be taken, however, with its interpretation as the polarity timescale was compiled before the realisation that 'Kiaman'-type overprinting is very widespread within Late Palaeozoic fold belts and adjacent cratonic basins.

As a corollary to the Russian polarity timescale (Fig. 9) the Global Palaeomagnetic Database (Lock & McElhinny 1991; McElhinny & Lock 1993) was interrogated for polarity bias during the Early Carboniferous and Devonian. Such an interrogation can provide an indication of the polarity structure, but such compilations are fraught with problems related to imprecisions on the age of the rocks and uncertainties about the primary origin of the results. The database was interrogated for directional results from rocks dated within an epoch or occasionally stage, with age limits according to the Harland et al. (1990) timescale. The directional results were interpreted subsequently concerning the polarity and primary origin of the magnetisation. The information that passed such a filtering process is presented in Figure 10 as a polarity bias profile (not a magnetostratigraphic profile).

Middle Carboniferous magnetostratigraphy and base of the PCRS

Determination of the base of the PCRS has come a long way since original definition of the base of the 'Kiaman' (Irving & Parry 1963; Irving 1966) between the Paterson toscanite (Paterson Volcanics) of normal polarity and the overlying main glacial stage (Seaham Formation) for which reversed polarities were observed. Recent U/Pb (SHRIMP) datings on zircons (Roberts et al. 1991*a*,*b*, in press *b*) have shown a much older age for the Paterson Volcanics (328 ± 1.7 Ma; late Visean) than originally envisaged, and preliminary findings of an ongoing magnetostratigraphic study (Théveniaut et al. in prep.) to locate the base of the PCRS in the Hunter Valley have shown the presence of mixed polarities above the Paterson Volcanics, at least up to the Mirannie Volcanic Member which are provisionally dated at 321.3 ± 4.4 Ma (Roberts et al. 1995).

The most detailed control available at the moment comes from North American studies, namely the earlier work of Jean Roy and co-workers (Roy 1977; Roy & Morris 1983) in Nova Scotia and the more recent work of Opdyke and co-workers in Nova Scotia–New Brunswick (DiVenere & Opdyke 1988, 1990, 1991*a*), in Pennsylvania (DiVenere & Opdyke 1991*b*), and in Colorado (Miller & Opdyke 1985; Opdyke 1986; Magnus & Opdyke 1991). Roy's (1977) study at Minudie Point located the base of the PCRS in the top of the Riversdale Group directly below the Pictou Group, corresponding with the upper Westphalian A. Subsequent work by DiVenere & Opdyke (1988, 1990, 1991*a*) in nearby and correlatable successions on the Maringouin Peninsula of New Brunswick provided a detailed series of polarity reversals in the Maringouin Formation, Shepody Formation, Clare Formation and in the base of the Boss Point Formation, with more isolated observations of exclusively reversed polarity in the remainder of the Boss Point Formation and the overlying Joggins Formation (Fig. 8). These observations locate the base of the PCRS somewhere within the Westphalian A. The biostratigraphic control for such an age has been questioned by P.J. Jones (pers. comm., 1993; Jones 1995) because palynological studies by G. Dolby (in Ryan & Boehner 1994) have demonstrated that the entire Claremont Formation and the lowermost Boss Point Formation of the Joggins section are late Namurian in age. Thus, the base of the PCRS is probably within the late Namurian (Namurian C?).

Early Russian studies (for compilation see Khramov & Rodionov 1981; Khramov 1987) identified the base of the PCRS with the occurrence of a single normal polarity subzone in the uppermost Moscovian (Fig. 9). This interpretation disagrees with our revision of the positive findings by Opdyke and co-workers and Roy and co-workers in Nova Scotia–New Brunswick for a Namurian C age. Location of the base of the PCRS at the normal polarity subzone in the middle Bashkirian (Fig. 9) seems a far more reasonable proposition. Preliminary results from magnetostratigraphic studies by Opdyke et al. (1993) on sections in the Russian Donets Basin that are taken as time-equivalent with the early and middle Morrowan have not been able to verify earlier Russian studies in the basin. The previously determined base of the PCRS in the latest Moscovian is thus in doubt.

Permo–Carboniferous Reversed Superchron (PCRS)

The extent of the PCRS is mainly determined from compilation studies (Irving & Pullaiah 1976; Khramov & Rodionov 1981; Khramov 1987; Molostovskiy 1992; Solodukho et al.

1993). Only a few earlier studies have attempted to cover the full length of the superzone (e.g. McMahon & Strangway 1968). Naturally, there has been some emphasis on the establishment of possible normal polarity subchrons as global correlation datums. A few claims for their occurrence have been made, but none has been corroborated by independent studies of modern reliability standards. The 'Quebrada del Pimiento event' (Valencio & Mitchell 1972; Valencio & Vilas 1972; Valencio et al. 1977), dated at 263 ± 5 Ma, may post-date the PCRS and form part of the succeeding Illawarra Superchron. Normal polarity observations in middle Leonardian formations of south-western North America (Peterson & Nairn 1971) are based on thermal cleaning procedures that are less detailed than current standards would require. The observations have not been confirmed. Russian compilations (McElhinny 1969; Khramov & Rodionov 1981; Khramov 1987; Molostovskiy 1992; Solodukho et al. 1993) showed a normal polarity subchron in the earliest Asselian (Fig. 9), directly above the Permo–Carboniferous boundary. This normal polarity event is actually just below the P–C boundary, because recent studies (Davydov et al. 1992) show that it is confined to the *Daixina bosbytauensis–D. robusta* Zone, the latest fusulinid zone of the Carboniferous (see Jones, this volume), and is recognised in the southern Urals, Donets Basin and the northern Caucasus region. It is latest Gzhelian in age, and is also recognised in the terrestrial Manebach Formation in north-eastern Germany (Menning et al. 1988). A suggested correlation (McElhinny 1969; Burek 1970; McElhinny & Burek 1971) with two observations of normal polarity in beds near to the Permo–Carboniferous boundary, namely the Supai Formation at Oak Creek Canyon of Arizona (Graham 1955) and Dunkard Series in the Appalachians of West Virginia (Helsley 1965), is tenuous as the observations were based on results that had been obtained without magnetic cleaning. In any case, coining of the term 'Oak Creek Event' (McElhinny 1969; McElhinny & Burek 1971) or its alias 'Graham Event' (Burek 1970) may not properly reflect these more creditable observations of this event within the Permian stratotype sections.

Late Permian magnetostratigraphy and top of the PCRS

Major defining studies include the detailed magnetostratigraphic study on the Permo–Triassic boundary succession at the Nammal Gorge in the Salt Range, Pakistan (Haag & Heller 1991), which has documented various normal and reverse polarity intervals in the uppermost Permian Chhidru and Wargal Formations. Thus the mixed polarity regime of the Illawarra Superchron had already been established in the Murgabian of the central Tethys (Fig. 11), which

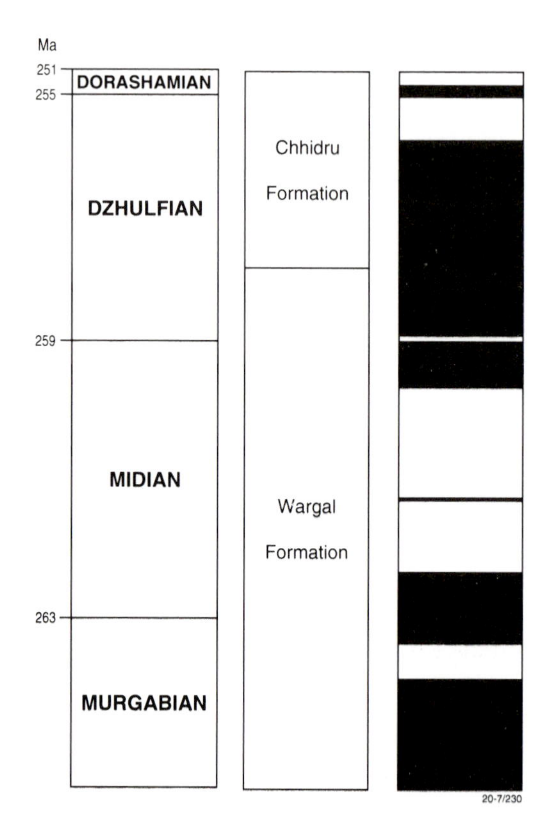

Figure 11: Magnetostratigraphic polarity profile for the Upper Permian of the Nammal Gorge, Salt Range, Pakistan, redrawn after Haag & Heller (1991, Fig. 8).

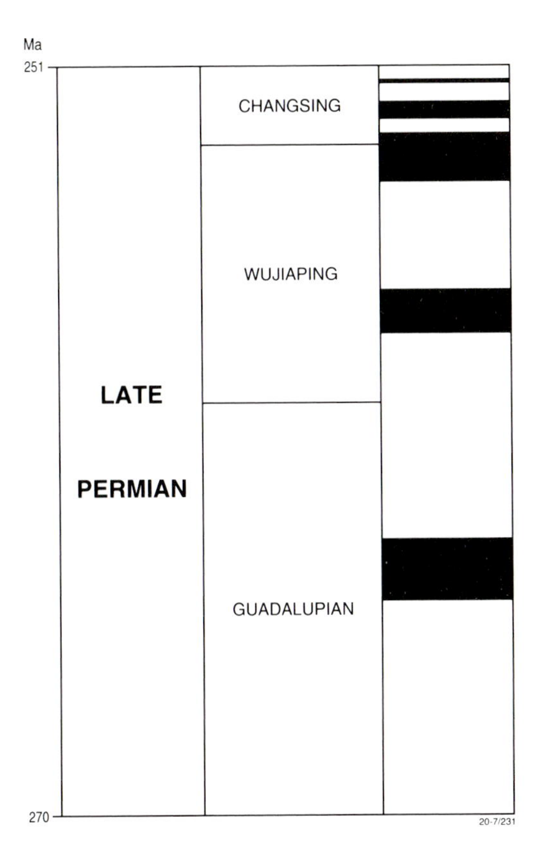

Figure 12: Late Permian magnetic polarity sequence with Chinese Stages from Sichuan, China, after Steiner et al. (1989, Fig. 13).

may be equated with the Guadalupian of North America and the Kazanian of the International Scale. The end of the PCRS thus predates the Tatarian, taken by most earlier workers, Russian in particular, to contain the changeover from the PCRS into the Illawarra Superchron. Magnetostratigraphic studies on Permo–Triassic boundary successions in the Sichuan province of south-central China (Heller et al. 1988; Steiner et al. 1989) do not provide the same level of detail as the Nammal Gorge study (Haag & Heller 1991). The latter supersedes the compilation by Steiner et al. (1989) of then available Permo–Triassic data (Fig. 12).

Russian studies on Permian stratotype sections of Russia and Tatarstan (Khramov & Rodionov 1981; Khramov 1987; Molostovskiy 1992; Solodukho et al. 1993) have located the PCRS–Illawarra Superchron boundary at the boundary between the early and late Tatarian (Fig. 9). This is in conflict with Haag & Heller's (1991) findings of a Murgabian or earlier ending of the PCRS.

A cursory study by Facer (1981) on the magnetostratigraphy of drill-cores from the Illawarra Coal Measures of the southern Sydney Basin claimed location of the PCRS–Illawarra Superchron boundary in the lower part of the coal measures, near the Appin Formation and Tongara coal seam. Limited thermal cleaning, however, may not have been fully successful in separating the primary polarity record and a pervasive normal polarity overprint of mid-Cretaceous origin that is rampant in the Tasman seaboard. Pending further substantiation of Facer's claim, this has left the stratigraphic location of the 'Illawarra Reversal' in its type region no better established than between the top of the Gerringong Volcanics (Irving & Parry 1963) and the base of the Narrabeen Group (Embleton & McDonnell 1980). A study in progress on drill-cores through the coal measures of the Sydney Basin (Illawarra Coal Measures, Newcastle Coal Measures, Wittingham Coal Measures) and Gunnedah Basin, backed up with outcrop samples from tuffs in the Newcastle Coal Measures and volcanics and sediments from the Gerringong Volcanics (Théveniaut et al. 1994), indicates that the top of the PCRS has apparently not yet been reached in any of the cores. This locates the top of the PCRS somewhere between the top of the Gerringong Volcanic Facies and the Jerrys Plain Subgroup of the Wittingham Coal Measures from the north-western Sydney Basin (Fig. 13). Palynological control indicates a probable Kazanian or possibly Ufimian age for the studied lower part of the Wittingham Coal Measures of mixed polarity. If sustained by further work, such observations date the end of the PCRS as no younger than Kazanian.

The early end of the PCRS indicated by the preliminary findings of the 'Illawarra Reversal' study appears to be supported by the well established normal polarity observations of Murgabian age in the basal part of the Nammal Gorge magnetostratigraphic profile (Fig. 13). These results highlight an emerging contradiction concerning the age of the younger end of the PCRS. On the one hand, we now have evidence for a Kazanian or older age from the sections of the continental eastern Australian coal measures and a Murgabian or older age for the marine Nammal Gorge section of the Salt Range. On the other hand, a younger early to late Tatarian age is concluded from the Late Permian stratotypes in Tatarstan and Russia (Fig. 13). Palaeontological control on the base of the Tatarian is reasonably well established, and a correlation with Australia through the marine Permian of Greenland, Austria and the Salt Range of Pakistan has been made (see Foster & Jones 1994). This correlation is not likely to be a source of discrepancy in the ages for the younger boundary of the PCRS. We cannot exclude with confidence that the conflict may arise from unrecognised normal polarity overprints of Late Cretaceous age in the Australian data, whose effect would be to make an otherwise reverse polarity interval appear mixed. Such overprints are of prevalent and pervasive occurrence in rocks along the Tasman seaboard, but their effect is less severe inland. The mutually supportive interpretations for an older age from both continental sections (Sydney and Gunnedah basins) and a marine section (Nammal Gorge, Salt Range) make it more likely that the cause of the discrepancy may have to be sought in the interpretation of palaeomagnetic results from the Late Permian stratotypes of Tatarstan and Russia.

Supporting data come from the magnetostratigraphic study of a Permo–Triassic type section at Biyulopaokutze in the northern Tarim Basin of north-western China (McFadden et al. 1988), which identified the top of the PCRS within level 35 of the sequence. However, no stratigraphic detail has been proffered, other than location of the Permo–Triassic boundary at the overlying level 41. The observed polarity reversal could represent the Illawarra Reversal, but the correctness of the claim remains to be tested pending provision of further stratigraphic detail. Mixed polarities observed in the Dewey Lake Formation, Texas, of Kazanian to Tatarian (Ochoan) age (Molina-Garza et al. 1989) likewise represent the Illawarra Superchron. The lowermost observed reversed polarity zone has been suggested to represent the top of the PCRS, but this remains to be tested through magnetostratigraphic study of the underlying formations.

Some earlier observations of normal polarity magnetozones may provide additional constraints on occurrence of the PCRS–Illawarra Superchron changeover. Peterson & Nairn (1971) observed normal polarities in red beds from south-western North America of late Guadalupian and middle to late Ochoan age. Although a primary origin has not been established beyond doubt, these normal polarity results have generally been interpreted in terms of possible normal polarity magnetozones within the PCRS. Following Haag & Heller's (1991) findings for a Murgabian or earlier onset of the Illawarra Superchron they could possibly represent the latter. Valencio & Vilas (1972) have documented normal polarity results for igneous rocks of the Quebrada del Pimiento Formation in the Paganzo Group of Argentina with a K/Ar date of 263 ± 5 Ma (old constants) (Valencio & Mitchell 1972; Valencio et al. 1977; Valencio 1981). This observation has generally been interpreted as a normal polarity magnetozone within the PCRS. Haag & Heller's (1991) results suggest that this normal polarity magnetozone could post-date the PCRS and thus constrain its upper boundary.

Extensive magnetostratigraphic studies have been carried out on Permo–Triassic successions within the Germanic facies realm (Burek 1970; Dachroth 1976, 1988; Mauritsch & Rother 1983; Beres & Soffel 1985; Menning 1986; Menning et al. 1988; Turner et al. 1989). These studies generally have no detailed biostratigraphic control and will rely on proper stratigraphic

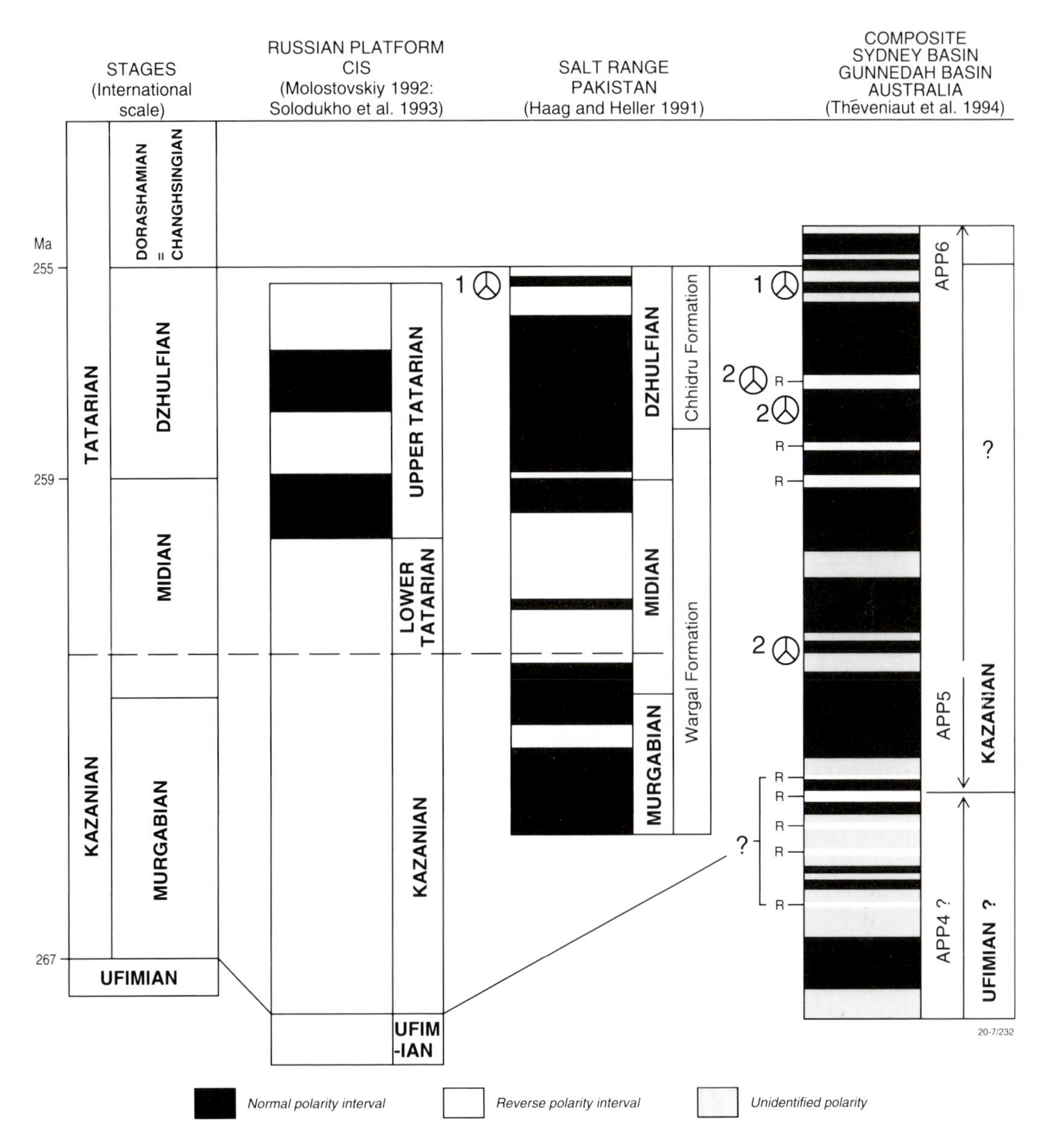

Figure 13: Tentative global correlation of Late Permian magnetostratigraphic profiles. 1 = *Protohaploxypinus microcorpus* Zone; 2 = *Dulhuntyispora parvithola* Zone (= AAP5). Reference timescale after Archbold & Dickins (1991), Kapoor (1992) and Kozur (1992). Salt Range timescale adapted after Pakistani–Japanese Research Group (1985).

location of the PCRS–Illawarra Superchron boundary for regional correlation.

A magnetostratigraphic study on drillcore and outcrop material of the Late Permian Bellerophon Formation and the Early Permian Werfen Formation of the Carnian Alps (Zeissl & Mauritsch 1991) showed mixed polarities throughout the studied succession. The authors, however, do not judge the record a reliable reflection of the Permo–Triassic magnetostratigraphy, as it is plagued by pervasive overprints of both Recent origin and those acquired during the Cretaceous Normal Polarity Superchron.

1.5.4 TRIASSIC TO MIDDLE JURASSIC

Hervé Théveniaut

Introduction

The establishment of a detailed magnetostratigraphy for the Triassic to Middle Jurassic has made good progress over the past decade. The Late Jurassic M25–M38 polarity sequence, identified by Handschumacher et al. (1988) from the north-western Pacific, is generally taken as the oldest sequence that can be established from marine magnetic anomalies. Thus the magnetic polarity sequence for times prior to the Callovian– Bathonian time boundary has to be defined from land sections. In order to provide adequate biostratigraphic control, such studies have generally been undertaken on marine sedimentary successions with well defined ammonite or conodont zonations.

Good polarity control can be obtained from land sections, but there are a number of pitfalls that have to be considered in order to arrive at a reliable magnetostratigraphy:

- Many studies show polarity zones that are based on only a single sample. These are generally taken as indicative for the polarity pattern, even though single observations are not very reliable.
- The magnetostratigraphic detail that can be deciphered depends, among other things, on the sedimentation rate. Variations in sedimentation rate may lead to apparent fluctuations in reversal rate or to erroneous correlations between successions of equivalent age, particularly for periods of medium to high reversal frequency.
- Biostratigraphic control for continental sedimentary successions generally has far less detail than for marine sediments. Thus hiatuses may not be identified and sampling densities may be inadequate for proper resolution of magnetostratigraphic detail.

Selected Early Triassic to Bathonian magnetostratigraphic profiles are presented in Figures 14–16. Their applicability as reference profiles is discussed below with respect to precision of results, precision of biostratigraphic control, and completeness of coverage per stage, period or biozonation. Many Triassic magnetostratigraphic studies have been carried out on continental successions in fluvial or lacustrine environments (e.g. Helsley 1969; Reeve & Helsley 1972; Baag & Helsley 1974*a,b*; Helsley & Steiner 1974; Steiner & Helsley 1974; Larson et al. 1982; Shive et al. 1984; McIntosh et al. 1985; Molina-Garza et al. 1991; Witte et al. 1991). Most of these studies were undertaken on the Colorado Plateau primarily for correlation purposes and to add magnetostratigraphic control to meagre biostratigraphic control. Recent magnetostratigraphic studies, however, of marine sediments from widely different regions in Canada, China, and Turkey have provided magnetically more precise results with more accurate biostratigraphic control. These marine sections are the more promising for establishment of a magnetostratigraphic timescale, despite current pitfalls with incomplete records, absence of stability tests, and sometimes considerable dispersion within the data.

Early Triassic

Magnetostratigraphic data for the Early Triassic have been obtained mostly as part of studies on the Permo–Triassic boundary (Heller et al. 1988; Steiner et al. 1989; Haag & Heller 1991). Recently, however, a nearly complete magnetic polarity pattern was obtained from a continuous Lower Triassic succession in the Canadian Arctic (Ogg & Steiner 1991). The magnetostratigraphy could be tied in with ammonite zonations from stratotype sections. The various studies show the occurrence of a normal polarity interval during the earliest Triassic, but gaps in all current stud-

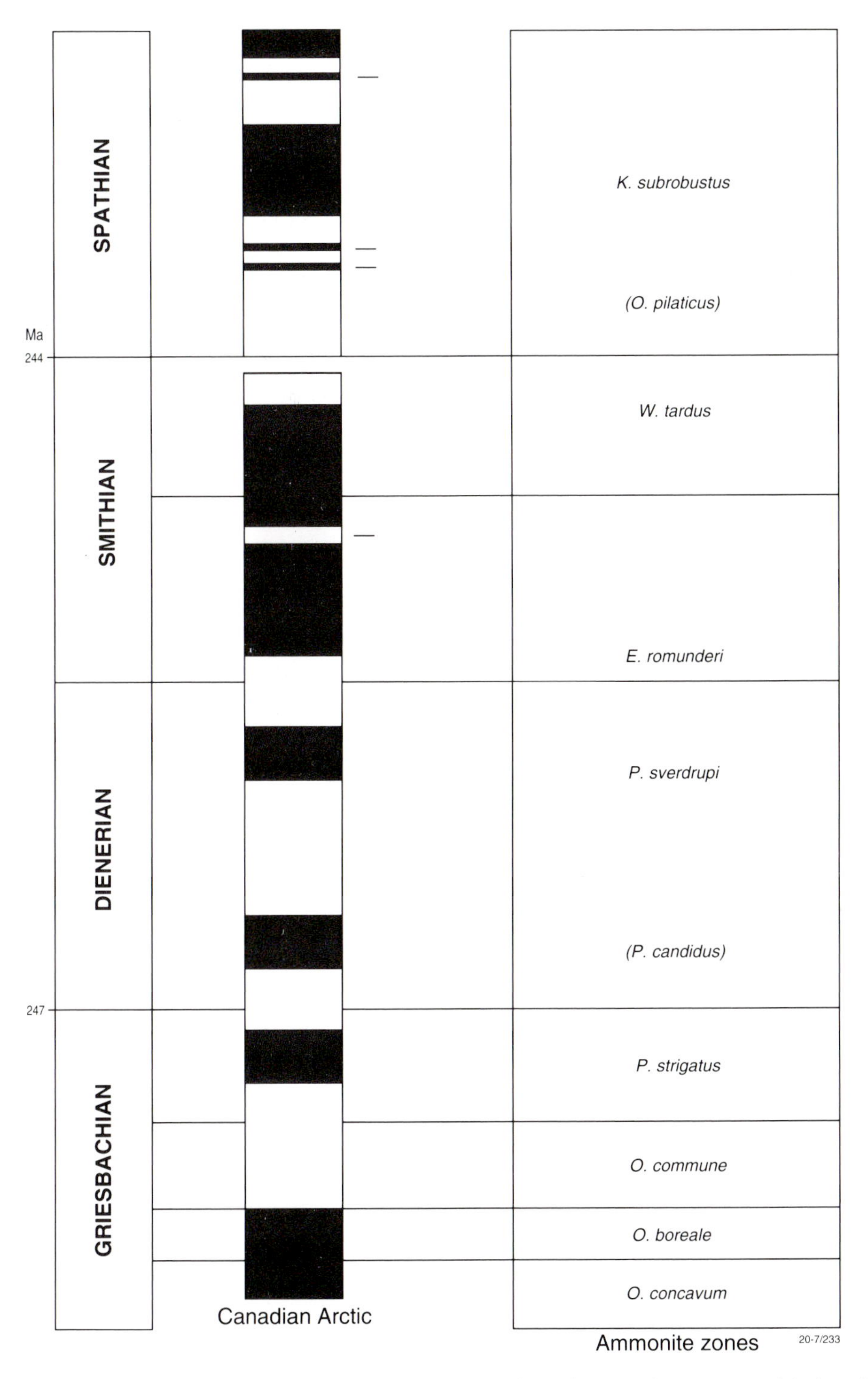

Figure 14: Early Triassic magnetostratigraphy obtained from the Canadian Arctic Archipelago (Ogg & Steiner 1991). Figure redrawn from the original composite diagram with stages taken to be of uniform duration.

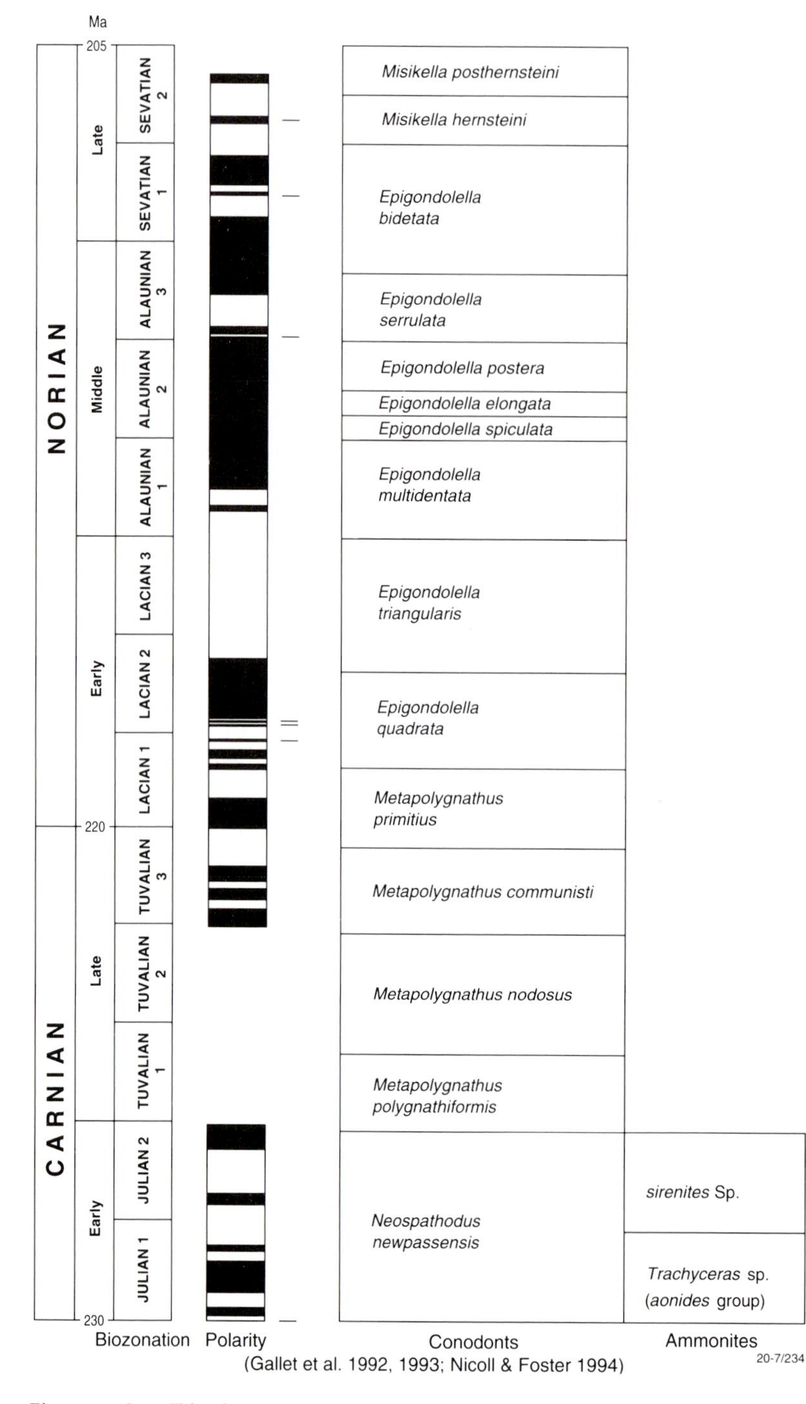

Figure 15: Late Triassic magnetostratigraphy obtained from south-western Turkey (Gallet et al. 1992, 1993). Figure redrawn from the original papers, with ammonite zones taken to be of uniform duration. Conodont zonation follows Nicoll & Foster (1994).

ies prevent identification of the magnetic polarity at the Permo–Triassic boundary. The magnetostratigraphic studies undertaken on Lower Triassic sections in China (Heller et al. 1988; Steiner et al. 1989) are hampered by absence of well defined stage boundaries. For instance, the Feixianguan Formation has a Griesbachian succession that is well calibrated in time but is considerably condensed, and a Dienerian succession cannot be clearly identified. The Smithian and Spathian stages are evident in the Jialingjiang Formation from conodont zonations, but without precision on their timespan.

The stratotype succession of the Canadian Arctic could be taken as the reference magnetostratigraphic succession for the Early Triassic (Fig. 14). Ogg & Steiner's (1991) results for this succession appear similar to the composite magnetostratigraphy compiled by Steiner et al. (1989) from world-wide data, with more accurate biostratigraphic control, and with a more adequate time resolution around 0.04 myr/sample.

Middle Triassic

Detailed magnetostratigraphic results are not available for the Middle Triassic. Occurrence of normal polarity has been suggested for the earliest Anisian and the latest Ladinian (Ogg & Steiner 1991; Gallet et al. 1992). Interrogation of the Global Palaeomagnetic Database (Lock & McElhinny 1991; McElhinny & Lock 1990, 1993) shows a mixed polarity record for the Middle Triassic.

Late Triassic

Two recent studies by Gallet et al. (1992, 1993) gave detailed magnetostratigraphic results for the Late Triassic. The results come from pelagic carbonates from South Tethyan successions in the Antalya region of southern Turkey. The two studied sections have precise biostratigraphic control from ammonites and conodonts and define a detailed magnetostratigraphic succession that is nearly complete for the Norian and the Carnian (Fig. 15), although two biostratigraphic zones seem to be missing from the latter succession. The Norian succession is established from 315 samples with a time resolution of 0.04 myr/sample, and the Carnian succession has a time resolution of 0.05 myr/sample. The two studies detail a total of forty-eight polarity intervals for the Late Triassic. This large number of reversals initially was perceived to conflict with magnetostratigraphic results obtained by Witte et al. (1991) from sequences in the Newark Basin of North America, which detail a lower number of reversals. Problems with resolution through time and biostratigraphic control of the latter study were invoked by Gallet et al. (1993) in an attempt to explain these apparent discrepancies. Recently, however, Kent et al. (1993) reported the presence of forty-three polarity intervals in the Late Triassic succession of the Newark Basin, thus confirming the large number of reversals for this period despite their lack of biostratigraphic control. The magnetostratigraphic results from Gallet et al. (1992, 1993) are supported further, in terms of the number of polarity intervals, by ODP Leg 122 results (Galbrun 1992*a*) from the Wombat Plateau off north-western Australia. Currently they represent the most complete and precise magnetostratigraphic record for the Late Triassic.

Early Jurassic

The magnetostratigraphy of the earliest stage in the Jurassic, the Hettangian, is presently undefined. The Sinemurian is poorly documented from two sections in the Umbrian Apennines, studied by Channel et al. (1984). Six polarity intervals were identified in the Fonte Avellana section and four intervals in the Cingoli section. There is no biostratigraphic control available for these two sections. A magnetostratigraphic correlation between the two sections is, therefore, tentative at best and the sections certainly do not qualify as reference sections.

The Pliensbachian is nearly completely present in the Breggia Gorge section (Fig. 16); only the very base is missing. The overlying Toarcian

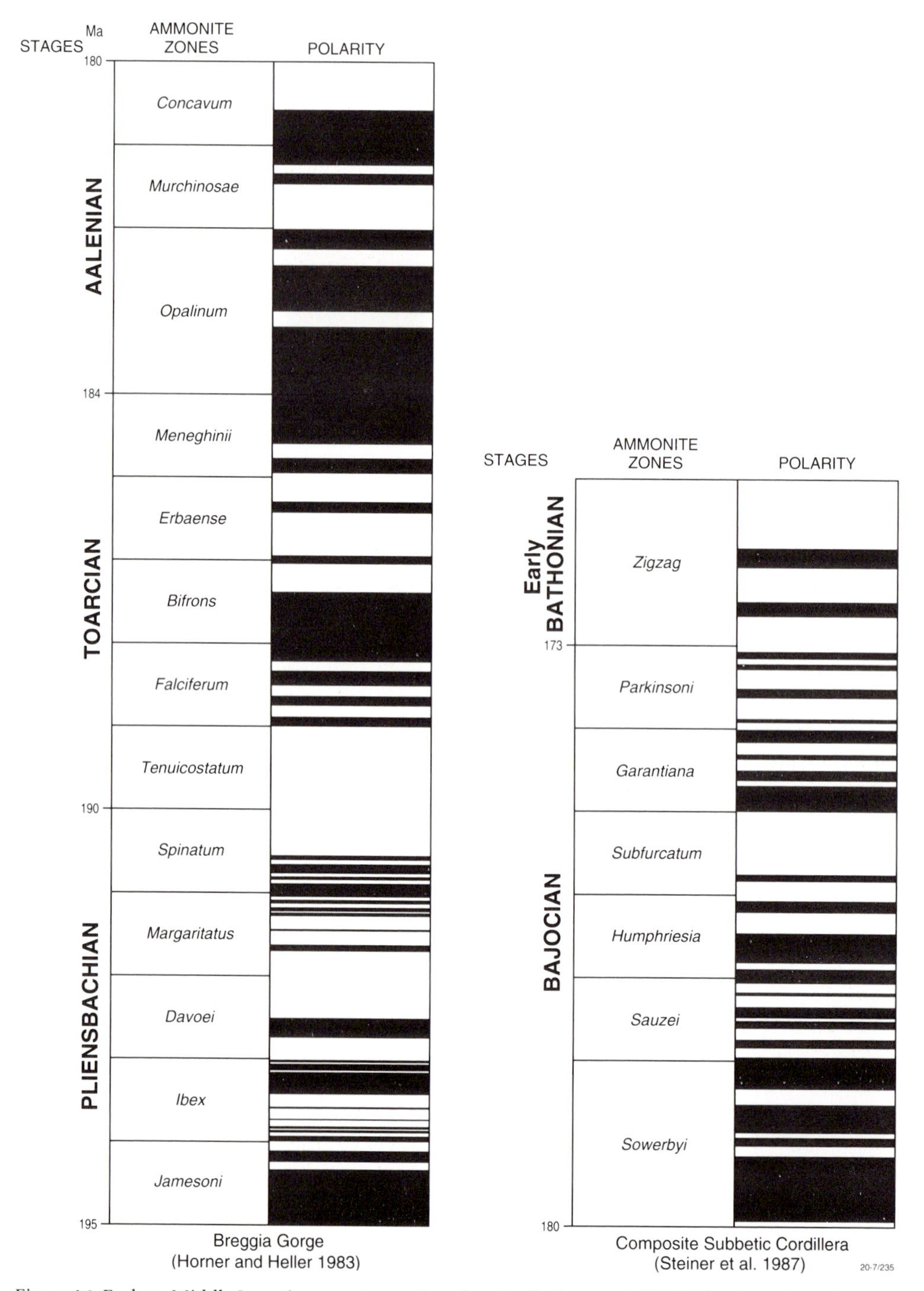

Figure 16: Early to Middle Jurassic magnetostratigraphy after Steiner et al. (1987) for the early Bathonian and the Bajocian, and after Horner & Heller (1983) for the Aalenian to Pliensbachian. Results are redrawn from original papers with transformation of the original depthscale to a linear timescale following Burger (1990*a*) and Burger & Shafik (herein).

succession is less well developed. Horner & Heller's (1983) study of the section provided a continuous magnetostratigraphic record. Results from the upper part of the section show a positive correlation with a latest Pliensbachian section from the Alpe Turati in the Italian Southern Alps, also detailed in the same study. Results from other Pliensbachian sections, namely the Fonte Avellana and Cingoli sections in the Umbrian Apennines (Channel et al. 1984) and the Baconycsernye section from the Transdanubian Central Mountains in Hungary (Marton et al. 1980), cannot be reliably correlated with the Breggia Gorge section. This is probably due to the condensed nature of the former sections, resulting in incomplete and less precisely defined reversal patterns.

The various studies that have been carried out on Toarcian successions exemplify the problems to be faced in magnetostratigraphy. Study of Toarcian type sections at Thouars and Airvault in western France by Galbrun et al. (1988) showed an incomplete, although biostratigraphically precisely controlled, magnetostratigraphy, reflecting the condensed nature of the sections. Channel et al. (1984) interpreted fifteen polarity intervals from their study of the middle and upper part of the Toarcian section in the Valdorbia section of the central Apennines, but a third of the polarity intervals appear poorly defined magnetically. The study by Galbrun et al. (1990) of the Iznalloz section in the Betic Cordillera of Spain showed a total of twenty-eight polarity intervals for the entire Toarcian succession. The polarity intervals are, however, poorly defined and the sampling is not continuous. This section has potential as a reference section, if higher quality and more detailed results can be obtained. The best defined magnetostratigraphic results for the Toarcian available so far come from the Horner & Heller (1983) study of the Breggia Gorge section in the Southern Alps. The section was sampled in detail, with a mean time resolution of 0.07 myr/sample, has precise biostratigraphic control, and shows well defined polarity intervals. This study may well be used as the best established polarity sequence for this stage (Fig. 16), although a few reversals may have been missed because of the low sedimentation rate.

Middle Jurassic

Magnetostratigraphic data for the Bathonian are available from the Subbetic Cordillera in southern Spain. Although the whole of the Bathonian is represented, the three sections that were studied by Steiner et al. (1987) appear generally too condensed to provide a detailed and reliable reference magnetostratigraphy for the whole of this period. Only that part of the sections containing the lowermost *Zigzagiceras zigzag* ammonite zone has provided a well defined reference magnetostratigraphy.

Steiner's study provided a more complete and more detailed magnetostratigraphy for the Bajocian, based on the Carcabuey section from the Subbetic Cordillera. Although this represents the most complete study of the Bajocian currently available, with good biostratigraphic control and a high sample density, about half of the interpreted polarity intervals are based on the result of a single sample only and need further confirmation.

The earliest ammonite zone of the Bajocian (*Sonninia sowerbyi*) has been documented also from the Breggia Gorge section in the Southern Alps (Horner & Heller 1983). The Breggia and Carcabuey sections show good correlation for the basal part of the *S. sowerbyi* Zone, which correlates with the end of a reversed polarity interval and a succeeding normal polarity interval. The magnetostratigraphy for the underlying Aalenian is also documented in studies by Steiner et al. (1987) and Horner & Heller (1983), as well as in a study on the Valdorbia secion from the Umbrian Apenines in central Italy (Channel et al. 1984). The good correlation that was observed between these studies for the earliest Bajocian zonation holds also for the latest Aalenian zonation. Whereas the magnetostratigraphy and biostratigraphy for the whole of the Aalenian was established in the Breggia Gorge section (Horner & Heller 1983), biostratigraphic control is available only for the lower part of the

Aalenian succession in the Valdorbia section (Channel et al. 1984). Thus, although only about half of the polarity intervals defined in the Breggia Gorge section are identified in the Valdorbia section, correlation for the base of the Aalenian stage is possible between the two sections.

With the above limitations in mind, the magnetostratigraphy for the early Bathonian and Bajocian ammonite zones in the Betic Cordillera (Steiner et al. 1987) and for the Aalenian zones of the Breggia Gorge of the Southern Alps (Horner & Heller 1983) are the most complete and precise reference magnetostratigraphies currently available (Fig. 16). Their time resolution is around 0.04–0.06 myr/sample.

1.5.5 MIDDLE JURASSIC TO RECENT

Mart Idnurm

Introduction

Except for the last 5 myr, for which abundant data are available from terrestial sources, the reversal timescale for the Middle Jurassic to Recent is based on oceanic magnetic anomalies. As new sea-floor is extruded at the mid-ocean ridge it becomes magnetised and adds the latest feature to a continuously evolving pattern of normal and reversed polarity lineations. If, as seems likely, the generation and outward spreading of the sea-floor have been continuous, the lineations would represent a complete record of the geomagnetic reversals. In addition, if changes in the rate of spreading have been gradual, linear interpolation between a few well-dated reversals would provide the ages of the other reversals. Combined with a brevity that is well beyond the resolution of isotopic methods, and its global synchroneity, the reversal potentially provides a very precise timescale.

The main difficulty in establishing the reversal timescale has been its time calibration. The first calibration for the entire Cainozoic reversal sequence was obtained from the South Atlantic sea-floor anomalies by Heirtzler et al. (1968), who assumed in their calculation that the rate of sea-floor spreading had been constant. This rate was estimated from a single lineation, dated at 3.4 Ma. Since 1968 the number and quality of age controls have improved greatly, giving better estimates for the sea-floor spreading rates. Nevertheless the calibration is still based on only few data, especially for the Jurassic and Early Cretaceous.

The most direct calibration would be by dating the basalts that cause the oceanic magnetic anomalies. However, the K/Ar ages of oceanic basalts are not considered sufficiently reliable (developments of the $^{40}Ar/^{39}Ar$ technique appear promising; McWilliams 1993). The volcanics are usually altered and have therefore lost or gained radiogenic argon, and because of the hydrostatic pressures at the ocean floor it is not certain that the K/Ar system had been reset during solidification of the magma by removal of all argon (McDougall 1974), as required for the age calculations. Therefore indirect methods are used. Currently these are based on sedimentary reference sections (oceanic or continental) where a dated biozone or stage boundary has been located in a specific part of the reversal pattern. Correlation of the pattern with sea-floor lineations provides a calibration point. For example, Cande & Kent (1992*a*) obtained one of their calibration points from a deep-sea drill-core where the lower boundary of the N9–N10 planktonic foraminiferal zone is linked to the younger part of chron C5Bn (Miller et al. 1985). This boundary had been dated elsewhere by $^{40}Ar/^{39}Ar$ as 14.8 Ma (Andreieff et al. 1976; Tsuchi et al. 1981), giving the calibration point. This, together with dates of several other magnetic lineations, defines an age–distance plot for the sea-floor, and the ages of the remaining lineations are obtained by interpolation along the curve of best fit to the plot.

The above procedure contains several uncertainties. First, the data that define the stratigraphic boundaries may not have been located accurately within the reference section. Second, the biozones may be time-transgressive over long distances. Third, the isotopic ages may not be secure, as seen for example in the revision of the Brunhes–Matuyama boundary from 0.73 to 0.78 Ma (Baksi et al. 1992; Spell & McDougall 1992) or of the Oligocene–Eocene boundary from 36.4 Ma (Harland et al. 1990) to 33.7 Ma (Odin et al. 1991). Fourth, and probably the largest uncertainty, is in the assignment of stage boundary ages by date bracketing (e.g. chronogram technique). Therefore, further significant revisions are likely. The following summarises briefly the current state of work on the Middle Jurassic to Recent sequence of reversals.

Middle Jurassic to Early Cretaceous

The timescales of Kent & Gradstein (1985) and Harland et al. (1990) extend back to the Callovian–Oxfordian boundary, into what had been regarded as the younger part of the Jurassic Quiet Zone. Handschumacher et al. (1988) have identified from aeromagnetic data of the East Mariana Basin still older, low-amplitude anomalies (M38–M30), which they estimate to extend the timescale back another 8 myr, that is, to the Bathonian–Callovian boundary (Fig. 17). It is not completely certain if these anomalies represent reversals rather than, for example, variations in the geomagnetic field (Cande & Kent 1992*b*). However, the existence of reversals in the 'Jurassic Quiet Zone' has been confirmed independently in Polish successions (Ogg et al. 1992).

The timescale from Oxfordian to Barremian (Kent & Gradstein 1985; Harland et al. 1990) is based on the Hawaiian M (Mesozoic) lineations, which span chrons M29 to M0 (Fig. 18). The Hawaiian set has been selected because of its completeness and relatively high resolution, especially in comparison with the Keathley set of the North Atlantic, where low spreading rates give a poorer resolution. However, some of the younger lineations in the Hawaiian set have not been identified in the Keathley set. This raises the possibility that parts of the Hawaiian set may be duplicates due to transfer faulting (Channell et al. 1987; Harland et al. 1990). Recent measurements on sedimentary successions in the Umbrian Apennines of Italy support the Hawaiian set (Channell & Erba 1992).

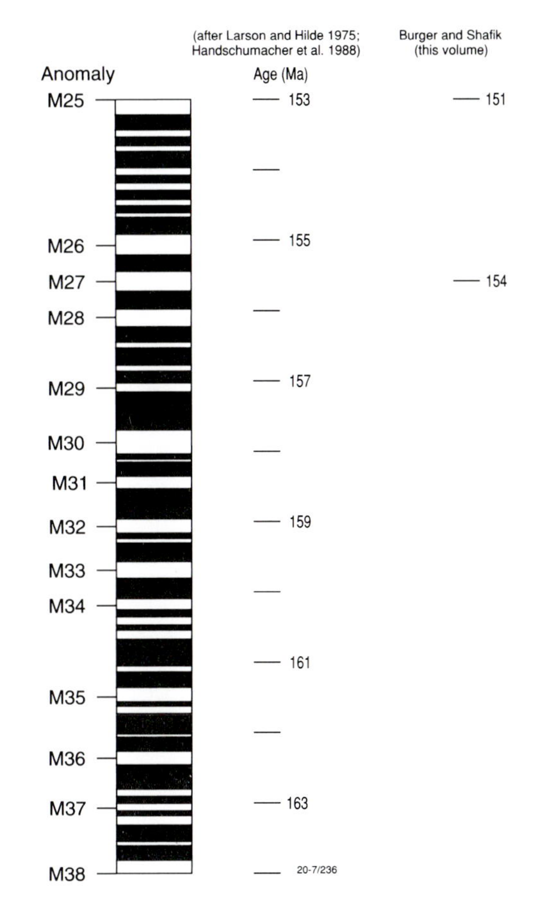

Figure 17: Older part of the Jurassic marine anomaly record, (after Handschumacher et al. 1988). Anomalies M38–M30 from the western Pacific Ocean have been tentatively interpreted by Handschumacher et al. as reversed polarity magnetisations. For anomalies M29-M25, the left-hand timescale is based on Larson & Hilde (1975), and for anomalies M38-M30 on extrapolation beyond 157 Ma of Larson & Hilde's timescale on the assumption that the sea-floor spreading rate had been constant. The right-hand timescale shows some corresponding ages according to Burger and Shafik (herein)

For time calibration, absolute age estimates are available for all Oxfordian to Barremian stage boundaries; however, only those at the

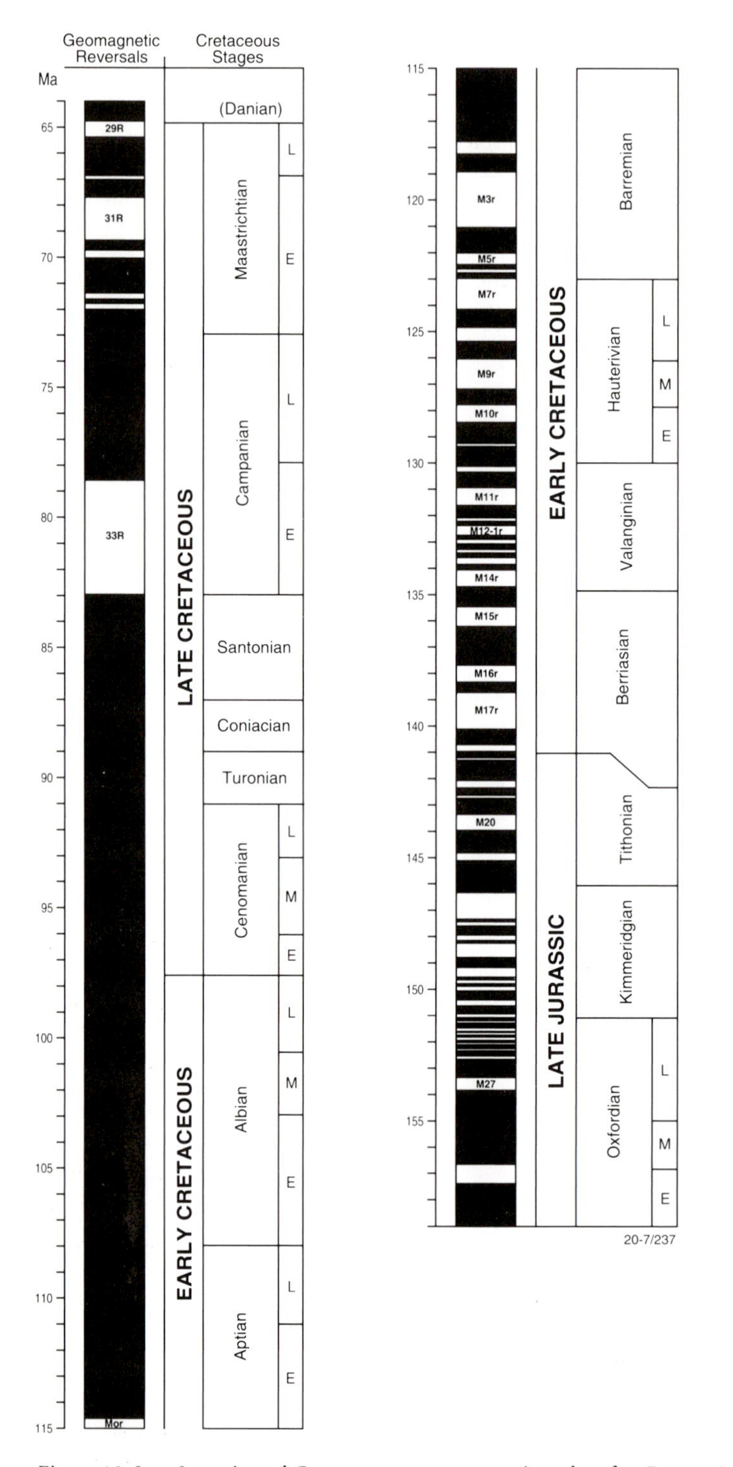

Figure 18: Late Jurassic and Cretaceous magnetostratigraphy after Burger & Shafik (herein).

ends of this period are considered reasonably well determined (Kent & Gradstein 1985). Nevertheless, the latest calibration by Harland et al. (1990) uses the ages of all stage boundaries irrespective of reliability and assumes that the rate of sea-floor spreading was constant throughout the period. (Support for a constant rate has since come from Early Cretaceous successions in the Umbrian Apennines; see Channell & Erba 1992.) If the intermediate stage boundaries with poorly defined chronogram ages are excluded, as by Kent & Gradstein (1985), the calibration points change by up to 8 myr. In the latest time calibration, Bralower et al. (1990) have obtained tight age estimates for the Jurassic–Cretaceous and Berriasian–Valanginian boundaries.

Although consolidation of the Late Jurassic to Early Cretaceous timescale requires additional dates, few have been reported in recent years. Instead, most recent advances have been in linking reversals to biostratigraphy. The latter studies have been principally on uplifted pelagic successions in southern Europe, especially Italy. The reversals have been correlated with ammonite zonations in the Jurassic and Cretaceous, and with calpionellid, nannofossil and planktic foraminiferal zones in younger successions. References to the studies are listed in Channell & Erba (1992) and Ogg et al. (1991).

Cretaceous Normal Polarity Superchron

Only a few, brief intervals of reverse polarity have been reported for the 45 myr period (chron 34N) that extends from the base of the Aptian to the Santonian–Campanian boundary (Fig. 18) (e.g. Van den Berg et al. 1978; Tarduno 1990). Until the early 1990s magnetostratigraphy seemed therefore not generally applicable to this period. However, a mixed polarity interval has now been found in an Albian succession from the Umbrian Apennines, Italy (Tarduno et al. 1992). Should the polarities prove to be original, rather than an artefact of remagnetisation, this would considerably extend the applicable time-range of magnetostratigraphy in the Middle Cretaceous.

Late Cretaceous to Recent

For this period (chrons 34–1) the most recent timescale revision is by Cande & Kent (1992*a*). This timescale, with revised Cainozoic stage boundaries of Berggren et al. (in press *a,b*), is shown in Figure 19. It is based on the South Atlantic lineations and involves a revision of their relative spacing. Cande & Kent's timescale differs from the preceding versions (Berggren et al. 1985*c*; Harland et al. 1990) principally for the Palaeogene, where the revision reduces the chron ages by up to 3 myr. This difference is mainly due to reassessment of the stage boundaries by Berggren et al. (1992) in response to mounting evidence for younger ages (e.g. Glass et al. 1986; Odin et al. 1991).

The Cande & Kent calibration uses the dates of nine stage boundaries selected at roughly equal intervals, and assumes that the rate of seafloor spreading varied smoothly. Not all of the available age constraints were used and it is not clear how the inclusion of other ages would affect the calibration. While the main features of the reversal sequence appear to be reliably established the existence of brief reversal events remains uncertain. This is reflected in the omission by Cande & Kent (1992*a*) of some features in the oceanic magnetic profiles that had been interpreted in the earlier timescales as brief events.

1.5.6 AUSTRALIAN CONSTRAINTS

In Australia, magnetostratigraphic studies for the Jurassic to Recent period have been restricted to the latest Neogene and the Quaternary. These include both age dating and correlation, and have been on beach ridges (Idnurm & Cook 1980), ephemeral and dry lakes (Singh et al. 1981; An Zhisheng et al. 1986; Chivas et al. 1986; McEwan Mason 1991; Cheng & Barton 1991) and vertebrate fossil sites (MacFadden et al. 1987; Whitelaw 1991*a,b*, 1992). Notably lacking are attempts to link the reversal timescale to

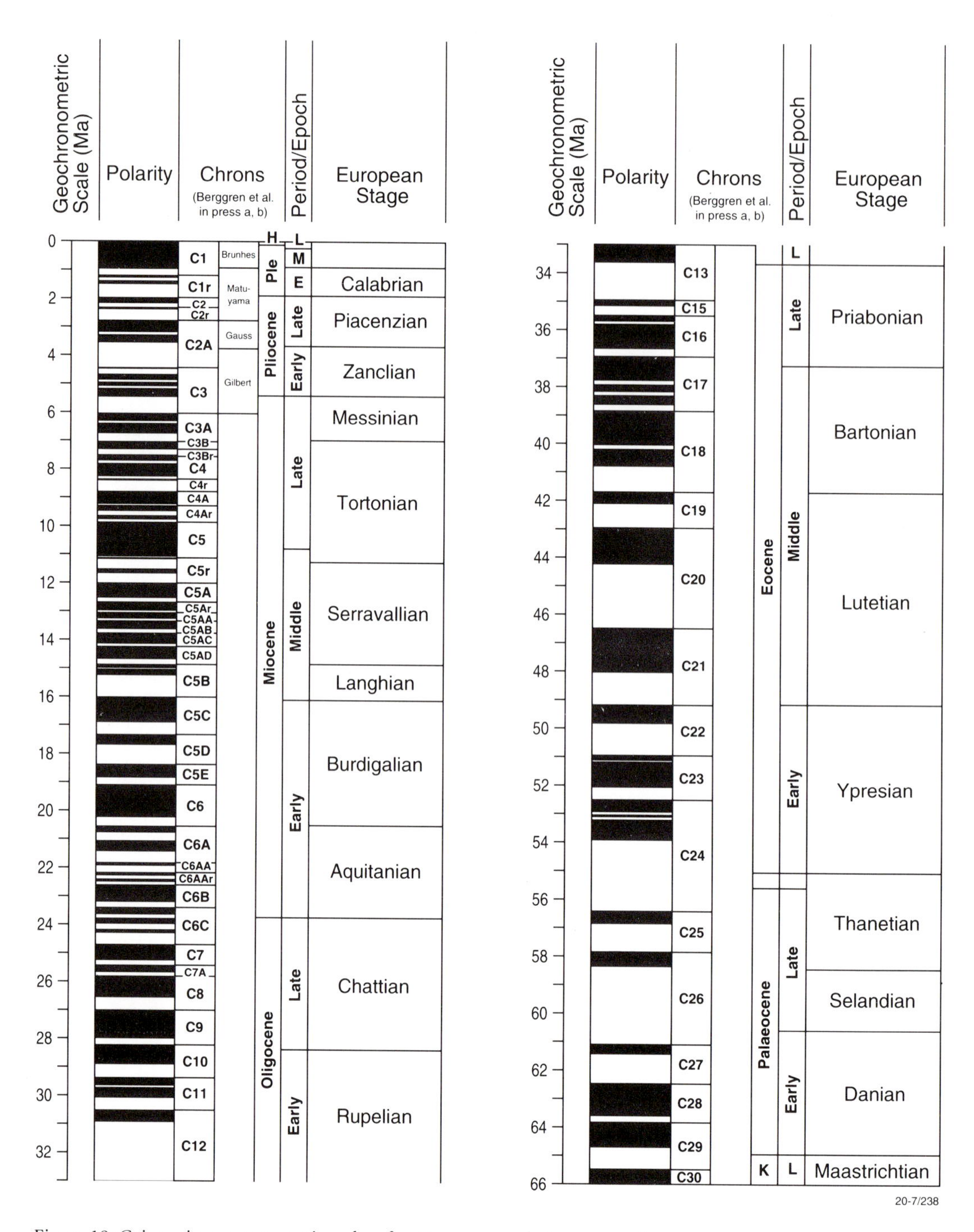

Figure 19: Cainozoic magnetostratigraphy after Berggren et al. (in press *a*, *b*).

Australian biostratigraphy. Such links would show if the Australian biozones correspond to their inferred international equivalents and would provide checks on possible time-transgressiveness of species data.

1.6

Numerical Calibration of Major Phanerozoic Boundaries

Compiled by G.C. Young and J.R. Laurie

1.6.1 PROTEROZOIC–CAMBRIAN BOUNDARY

The Cambrian timescale is currently in a state of flux. New age measurements in biostratigraphically well established horizons, by variants of the zircon U/Pb dating technique, are superseding the previous very large database of less confident or less relevant ages on which previous Cambrian scales have been based. For example, recent estimates of the age of the Proterozoic–Cambrian boundary have ranged between 530 Ma and 590 Ma (Cowie & Harland 1989). This wide range of uncertainty has now been narrowed to an age close to 544 Ma from the conventional zircon dating of a volcanic breccia in northern Yakutia. The breccia has an age of 543.9 ± 0.24 Ma and lies near the base of the *Anabarites trisulcata* Zone, of earliest Nemakit-Daldynian age (Bowring et al. 1993).

Subdivisions of the Cambrian

Nemakit-Daldynian–Tommotian boundary Constraints higher in the Cambrian are sparse, but Bowring et al. (1993) report a zircon age of 534.6 ± 0.5 Ma from porphyry clasts in a proximal conglomerate overlain by earliest Tommotian carbonates in Siberia. This provides a maximum age for the base of the Tommotian.

Atdabanian–Botoman boundary A SHRIMP zircon age of 526 ± 4 Ma has been measured at Sellicks Hill on the Fleurieu Peninsula, South Australia, for a tuff of the Truro Volcanics in the upper part of the Heatherdale Shale (Cooper et al. 1992), which is considered to be early or mid Botoman in age. Further SHRIMP measurements of a complex zircon population in bentonites of the Meischucun section in southern China give an interpreted age of 525 ± 7 Ma for Bed 5 of the early Meishucunian (Compston et al. 1992). These zircon dates contradict previous (much older) age measurements in the same section, which were based on dating sedimentary rocks. However, they also conflict with the dates from Sellicks Hill and Siberia that suggest that the Nemakit-Daldynian–Tommotian boundary lies near 534 Ma and the Atdabanian–Botoman boundary below 526 Ma.

Early–Middle Cambrian boundary No data are available, and an estimate of 509 Ma is used herein.

Middle–Late Cambrian boundary Recent SHRIMP zircon ages from the Mount Read Volcanics of Tasmania (Perkins & Walshe 1993), which contain Undillan to Boomerangian (late Middle Cambrian) trilobites, give four dated samples with an age of 502.6 ± 3.5 Ma, and one interpreted at 494.4 ± 3.8 Ma. Accepting the majority agreement near 503 Ma suggests an age for the Middle–Late Cambrian boundary of younger than 500 Ma.

1.6.2 CAMBRIAN–ORDOVICIAN BOUNDARY

Reliable isotopic age information is lacking in the higher stages of the Cambrian and in the Lower Ordovician. Most previous compilations placed the top of the Cambrian at 500–510 Ma, but with acceptance of 545 Ma as its base, the Cambrian Period becomes unacceptably short. Taking account of the results of Perkins & Walshe (1993) regarding a Middle–Late Cambrian boundary age of less than 500 Ma (see above), and allowing for the accumulation of Upper Cambrian strata, the top of the Cambrian must be considerably younger than 500 Ma, and is here arbitarily placed at 490 Ma, with a caution that the real age may be substantially younger.

Subdivisions of the Ordovician

As with the Cambrian, the database underpinning previous Ordovician compilations is being superseded by new, precise zircon dating. However, the new rèsults are subject to conflicting interpretations. The primary set of new data is that of Tucker et al. (1990), who used conventional methods to date multigrain samples of between six and thirty zircons from volcanic rocks interbedded with the Ordovician graptolite sequences in the United Kingdom. These ages range from 465.7 ± 2.1 Ma in the Early Llanvirn, to 438.7 ± 2.0 Ma in the Early Llandovery. Tucker et al. (1990) proposed a base for the Ordovician near 513 Ma, and a top near 441 Ma, immediately below their dated Llandovery sample, with subdivisions scaled appropriately between the precise zircon ages.

However, SHRIMP reanalysis of probed zones within individual grains of the same zircon concentrates, and other Ordovician samples (Compston & Williams 1992), have given significantly younger ages (by up to 10 Ma) than the conventionally measured zircon data. This might be due to inter-laboratory bias, and/or may indicate that inherited zircon grains in several of the samples (and detected by both laboratories) were not successfully excluded from the multigrain conventional analyses, but were resolved by the microbeam technique.

The younger age interpretations suggested by Compston & Williams (1992) are provisionally followed here. This is consistent with the Cambrian–Ordovician boundary being about 490 Ma, or younger, as required by Cambrian zircon ages, and with an Ordovician–Silurian boundary near 434 Ma, as is suggested by available Silurian evidence.

Tremadoc–Arenig boundary Zircon ages with which to constrain the Tremadoc are lacking. The Tremadoc and the base of the Ordovician are therefore scaled between the constraints of the SHRIMP dates within the Arenig discussed next, and the Middle Cambrian Mount Read Volcanics zircon ages, with a caution that calibrations in this interval are based on little information.

Arenig–Llanvirn boundary The top of the Arenig is constrained near 465 Ma by SHRIMP zircon ages of 471 ± 3 Ma within the Arenig Llynfant Flags, and 462 ± 3 Ma within the Llanvirn Llanrin Volcanics (conventional measurement of the same sample gives 465.7 ± 2.1 Ma and would require an older top to the Arenig).

Early–Late Ordovician boundary Higher parts of the Ordovician are scaled by a single SHRIMP age determination of 448 ± 4 Ma for the Caradoc Pont-y-ceunant Ash (conventional measurement 457.2 ± 2.2 Ma).

1.6.3 ORDOVICIAN–SILURIAN BOUNDARY

This scale largely follows the compilation by McKerrow et al. (1985), which comprehensively reviews the available data, but which uses the base of the *Glyptograptus persculptus* Zone as the base of the Silurian. This has now been superseded by international ratification of a boundary at the base of the overlying *Parakidograptus acuminatus* Zone (Cocks 1985). Appropriately adjusted, the 435 Ma age suggested by McKerrow et al. (1985) for the base of the Silurian becomes 434 Ma herein, and this is in accord with a SHRIMP zircon date of 430 ± 3 Ma for the Rhuddanian (*Coronograptus cyphus* Zone) Birkhill Shale in Scotland (Compston & Williams 1992).

Subdivisions of the Silurian

With the exception of the Llandovery Birkhill Shale, the Silurian has not yet been subjected to age study using the new zircon dating techniques, and its timescale is based on earlier data. This scale follows the compilation by McKerrow et al. (1985), which comprehensively reviews the available data. The later compilation of Harland et al. (1990) uses a similar database but does not take account of relative age information such as sediment thicknesses and relative durations of biozones, which, in the absence of precise radiometric data, offer some of the more useful constraints.

Llandovery–Wenlock boundary The Wenlock lacks confident constraints, but extrapolation suggests a base close to 425 Ma.

Wenlock–Ludlow boundary Within the Silurian, the most tightly constrained radiometric age is that of 421 ± 2 Ma for the early Gorstian Laidlaw Volcanics, near Canberra, Australia (Wyborn et al. 1982). This is consistent with an age for the basal Ludlow close to 420 Ma.

1.6.4 SILURIAN–DEVONIAN BOUNDARY

Owen & Wyborn's (1979) suggested numerical age of 410 Ma is adopted, following Cowie & Bassett (1989). This date was obtained from granites intruding the Mountain Creek Volcanics, which underlie fossiliferous limestones including strata at least as old as the latest Pragian *pireneae* Zone (Mawson et al. 1992). A maximum age of earliest Devonian is provided by the early Lochkovian conodont *Icriodus woschmidti* in the Elmside Formation of the Bowning Group (Link & Druce 1972), which is older than the Bowning event. The granites have Rb/Sr dates in the 400–406 Ma range, and are inferred on similar chemical characteristics to be comagmatic with the Mountain Creek Volcanics (Wyborn et al. 1987), and therefore younger than the Bowning event, but pre-late Pragian. Odin (1985*a*, p.96) quoted three other radiometric studies indicating a boundary clearly older than 400 Ma, and McKerrow et al. (1985) and Kirchgasser et al. (1985) cited other evidence (e.g. the Katahdin Batholith in Maine intruding Oriskany [Pragian] rocks) also indicating an older age for the boundary than the Harland et al. (1982, 1990) date of 408 Ma. This tie-point specified by Harland et al. (1990, Fig. 1.7), and followed by Fordham (1992), is based on a less than compelling 'chronogram' (A4.88).

Subdivisions of the Devonian

Isotopic constraints are poor within the Devonian and consideration of individual stage boundaries is little more than arbitrary. The series durations may be compared with earlier compilations. The **Early Devonian** has a duration of about 25 Ma, compared with the 22 Ma of Harland et al. (1990), but their stage durations are completely different, with the Emsian the shortest, and the Lochkovian longer than the other two together (Fig. 20). On this chart the Emsian is the longest, consistent with previous palaeontological assessment (Boucot 1975; Ziegler 1978)

Stage Boundary	This chart	Dur	GTS89	Dur	Diff in b'dary ages	% total (this chart)	% total (A)	% total (B)
Famennian/Tournaisian	354	11	363	4	+9	19.6	18.9	20.3
Frasnian/Famennian	365	5	367	10	-2	8.9	16.2	17.6
Givetian/Frasnian	370	8	377	5	-7	14.3	16.2	14.9
Eifelian/Givetian	378	7	381	5	-3	12.5	13.5	12.2
Emsian/Eifelian	385	15	386	4	-1	26.8	13.5	17.6
Pragian/Emsian	400	5	390	6	+10	8.9	10.8	8.1
Lochkovian/Pragian	405	5	396	13	+9	8.9	10.8	9.5
Pridolian/Lochkovian	410		409		+1			20-1/49

Figure 20: Comparison of numerical ages for Devonian stage boundaries and stage durations between the current chart and GTS89 (Harland et al. 1990). Stage duration (dur.) refers to the first named in the left column. In the three right-hand columns, the stage duration is expressed as a percentage of the duration of the Devonian Period for this compilation and two earlier estimates (col. A, Boucot 1975; col. B, Ziegler 1978).

that the Emsian and pre-Emsian were approximately of equal length. The **Middle Devonian** on the chart lasts about 15 Ma, which is almost twice the 8.6 Ma of Harland et al. (1990), but their duration is in serious conflict with previous palaeontological assessment (Boucot 1975; Ziegler 1978), and with McKerrow et al. (1985), who by extrapolation estimated the Middle Devonian at 17 Ma duration, with the Givetian (9 Ma) slightly longer than the Eifelian. Other authors also agree that the Givetian is the longer stage. The **Late Devonian** on the new chart reflects the widely held view that the Frasnian is shorter than the Famennian, to give a total duration of about 16 Ma (cf. 15 Ma in Harland et al. 1990).

1.6.5 DEVONIAN–CARBONIFEROUS BOUNDARY

The 354 Ma date previously proposed (Jones 1988, 1991; Young 1989*a*) based on consideration of biostratigraphic evidence relating to isotopic dates from eastern Victoria (Richards & Singleton 1981; Williams et al. 1982; Odin 1985*b*) has been confirmed. The reference datum for the base of the Carboniferous is now the zircon age of 353.7 ± 4.2 Ma for bentonite Bed 79 in the Hasselbachtal auxiliary stratotype section in Germany, 35 cm above the boundary and within the lower part of the *Siphonodella sulcata* conodont zone (Claoué-Long et al. 1992; renormalised by Claoué-Long et al., in press). In eastern Australia, the age for a volcanic constrained by the same conodonts is 355.8 ± 5.6 Ma for the lower part of the Kingsfield Formation (Claoué-Long et al. 1992). This is within measurement uncertainty of the Hasselbachtal date.

Subdivisions of the Carboniferous

Tournaisian–Visean boundary This boundary is now placed at 343 Ma, based on new isotopic evidence from the Curra Keith Tongue (342.0 ± 3.6 Ma) of the Isismurra Formation (NSW), which provides the best constraint so far on the age of this boundary in Australia (Roberts et al. 1993*b*). The previously used age of 342 Ma (Jones 1988, 1991; also followed by Cowie & Bassett 1989 for the IUGS Global Stratigraphic Chart) was based on the K/Ar age of an andesite at Foybrook in the Waverley Formation (NSW), but revisions to the stratigraphy now suggest that this bed is somewhat older than the boundary (Roberts et al. 1993*a*).

Within the Visean, the SHRIMP zircon age for the Martins Creek Ignimbrite of the Hunter Valley (NSW) is now revised to 332.3 ± 2.2 Ma (Roberts et al. 1993*b*). This ignimbrite lies between the *Linoprotonia tenuirugosus* Subzone, the upper subzone of the *Delepinea aspinosa* Zone and the *Rhipidomella fortimuscula* Zone, and recent conodont work (Jenkins et al. 1993) demonstrates that the *D. aspinosa–R. fortimuscula* zone boundary lies within the Holkerian. This gives a mid-Holkerian date at 332 Ma, and an estimated 335 Ma for the base of the Holkerian.

A new zircon age of 328.0 ± 1.7 Ma (Visean) for the normally magnetised Paterson Volcanics of the Hunter Valley, NSW (Claoué-Long et al., in press), previously correlated with the zone of mixed polarities within British Westphalian C coals (Noltimier & Ellwood 1977), shows that the magnetostratigraphic synthesis for the Carboniferous of Palmer et al. (1985) needs extensive improvement before it can be applied to the solution of stratigraphic problems.

Visean–Namurian boundary The most useful age constraints in the Upper Carboniferous are in Europe, where Hess & Lippolt (1986) have reported accurately measured $^{40}Ar/^{39}Ar$ plateau ages for sanidine, which supersede pre-existing data for the Upper Carboniferous in both measurement accuracy and biostratigraphic constraints. Claoué-Long et al. (in press) have shown that these $^{40}Ar/^{39}Ar$ ages are directly comparable with SHRIMP zircon ages being measured in the Australian sequences. The base of the Namurian is constrained by two $^{40}Ar/^{39}Ar$ ages within the Namurian A: one at 319 ± 8 Ma and another at 325 ± 8 Ma. This scale is constructed on the basis that the older of these constrains the base of the Namurian, whose base must be close to 325 Ma. In contrast, the 332.9 Ma date suggested by Harland et al. (1990) for this boundary is too old and, on our scale, lies within the Holkerian.

Namurian–Westphalian boundary A recent determination of SHRIMP zircon ages of 314.4 ± 4.6 Ma and 314.5 ± 4.6 Ma for the $E2_{a3}$ and $E2_{b2}$ Arnsbergian ammonoid zones in England (Riley et al. 1993) indicates that the 315 Ma age suggested by Hess & Lippolt (1986) for the Namurian–Westphalian A boundary is probably too old. We suggest a 312.5 Ma age for this boundary based on interpolation from the Arnsbergian tie-point (314.5 Ma) below, and the Westphalian B–C boundary tie-point (311 Ma) above, which is dated by both $^{40}Ar/^{39}Ar$ (Lippolt & Hess 1985) and SHRIMP zircon techniques (Claoué-Long et al., in press). The 309 Ma for the Westphalian C–D boundary is taken from the scale of Hess & Lippolt (1986).

Westphalian–Stephanian boundary The date of 306 Ma for the base of the Stephanian is again based on the Hess & Lippolt (1986) scale, which also gives a 303 Ma date for the Stephanian A–B boundary.

1.6.6 CARBONIFEROUS–PERMIAN BOUNDARY

The previous date of 295 Ma adopted for this boundary by Jones (1988, 1991) was a compromise between the 290 Ma date of De Souza (1982) and Forster & Warrington (1985), and the 300 Ma date of Hess & Lippolt (1986). The most important Carboniferous constraint on this boundary comes from the 300.3 Ma $^{40}Ar/^{39}Ar$ age reported by Hess et al. (1983) and Lippolt & Hess (1985) for a Stephanian B or C tuff (159/71S) from Baden Baden, Germany. Correlation with the Autun Basin in France implies that this dated sample lies stratigraphically below the uppermost Carboniferous Igornay Formation (Stephanian D of Bouroz & Doubinger 1977). The most important Permian constraint comes from the 297.8 Ma and 298.7 Ma $^{40}Ar/^{39}Ar$ ages reported by Lippolt & Hess (1983) respectively for tuffs in the Grenzlager Formation (basal Upper Rotliegend) at Lohmühle and Hohlbusch from the Saar–Nahe Basin, Germany. Although precise correlation from these non-marine beds to the marine succession in the southern Urals is still uncertain, the base of the Permian is taken here at about 298 Ma, to accommodate the isotopic evidence. This is considerably older than the 290 Ma date of Harland et al. (1990).

Subdivisions of the Permian

The Permian is the least well constrained period in the timescale, with no reliable ages in areas with satisfactory stratigraphic control. The ages of even major subdivisions are therefore largely conjectural. A program of zircon dating is now in progress, pending the outcome of which the scale here is constructed arbitarily between the reasonably closely measured ages for the top and base. In this we follow Harland et al. (1990), rather than Forster & Warrington (1985), who attempted a subdivision of the Permian from the poor available evidence.

1.6.7 PERMIAN–TRIASSIC BOUNDARY

The top of the Permian has been dated at the boundary bentonite in the Chinese stratotype as 251.1 ± 3.6 Ma by SHRIMP zircon U/Pb dating (Claoué-Long et al. 1991, renormalised by Claoué-Long et al., in press). The bentonite is at the lithological change from the stratotype of the uppermost Permian Changxing Formation limestone, to shales containing mixed faunas, which include probable Triassic *Otoceras* ammonoids and, a few centimetres higher, definite Triassic *Hindeodus parvus* conodonts. Tozer (1986) suggested that the bentonite occupies the time interval of a local unconformity, and thus represents a position within the uppermost Permian, rather than its top. Despite this biostratigraphic ambiguity, this age is presently the most direct constraint on the age of the boundary; it is adopted here pending the outcome of new dating studies in the uppermost Permian.

Subdivisions of the Triassic

Absolute age control within the Triassic is poor, with many of the so-called control points not adequately constrained biostratigraphically. Because of uncertainty of the absolute ages the timespan of any one stage can vary by 1–5 Ma, depending on the authority used. As an example, the Norian of Harland et al. spans 6 Ma; as shown here, the stage occupies 10.5 Ma years. Users of this chart must be aware that until accurate dating of the type stages is achieved, the length of time for any one stage will vary from authority to authority.

1.6.8 TRIASSIC–JURASSIC BOUNDARY

Armstrong (1982) dated the Triassic–Jurassic boundary as 208 Ma, based on analysis of many isotopic analyses of Upper Triassic (Takla Group, 215–223 Ma) and Lower Jurassic (Hazelton Group, 182–191 Ma) volcanics, and Upper Triassic? crystalline rocks (200–209 Ma), all in British Columbia, Canada. However, Odin & Letolle (1982) and Kennedy & Odin (1982) commented on the poor stratigraphic control of several of the Canadian samples analysed. Quoting additional results from Late Triassic intrusives in Thailand (209–211 Ma), the Early Jurassic of the USA (190–206 Ma), and the Hettangian of France (194 Ma), Odin & Odin (1990) deduced 205 Ma for the Triassic–Jurassic boundary, a value accepted here.

Subdivisions of the Jurassic

Time frameworks published for the Jurassic are based on extrapolation from isotopic key data, assuming equal duration of stages (from comparable thicknesses of sequences) or ammonite zones (assuming a constant rate of evolution), or constant rates of sea-floor spreading. Very little reliable isotopic control has so far been obtained from the ocean floor. More useful ages (K/Ar, Rb/Sr, $^{40}Ar/^{39}Ar$) have been measured in onshore sedimentary and crystalline rocks linked with the fossil record. The data discussed by Armstrong (1978, 1982) and Odin (1982, pp.659–948) offer some indications between which age limits the Jurassic stages may lie.

Hallam et al. (1985), who discussed the Jurassic timescale, criticised Kennedy & Odin's (1982) choice of Late Jurassic boundary ages based on glauconies. They suggested that systematic discrepancies with the ages calculated by Van Hinte (1976) indicated underestimations due to argon loss. Van Hinte based his ages on equal duration of Jurassic ammonite subzones (on average about 1 Ma), assuming a constant rate of ammonite evolution. In the absence of firmer evidence, stage boundaries are here dated on the same principle, and the ages for the Early and Middle Jurassic boundaries approximate those suggested in recent literature. However, it is obvious that such a procedure can only be a temporary vehicle to fill the data gap. It may in practice yield plausible time estimates for timespans of parts of the Jurassic (Geological Society of London 1964; Harland et al. 1982; Westermann 1984; Hallam et al. 1985), but the technique of 'averaging out' durations of ammonite zones are guesses based on isotopic dates. The basic assumption is an unproven one since ammonite evolution, being tied to the environment, was equally likely not to be uniform (Hallam 1984).

1.6.9 JURASSIC–CRETACEOUS BOUNDARY

There are few reliable age indicators for the base of the *Berriasella grandis* Zone. Kennedy & Odin (1982) cited glaucony datings of 122 and 134 Ma for the Early Cretaceous in England and the USSR. Harland et al. (1982) calculated 135 Ma for the Berriasian–Valanginian boundary on the basis of minimum error functions from dated glauconies, but by the same method Lowrie & Ogg (1986) obtained 135 Ma for the Jurassic–Cretaceous boundary (base of *B. grandis* Zone). This value is accepted by many geoscientists, but this paper accepts an age for the Jurassic–Cretaceous boundary of 141 Ma, following the estimate of Bralower et al. (1990).

Subdivisions of the Cretaceous

There are generally accepted analytical results only for a few, mainly Late, Cretaceous stage boundaries. Ages in different parts of the sequence move continually as new isotopic data become available. On the basis of radiometrically dated Early Tertiary and Late Cretaceous ocean-floor anomalies, Tarling & Mitchell (1976), Lowrie & Alvarez (1981), Harland et al. (1982) and Berggren et al. (1985*a*) have extrapolated closely comparable radiometric ages for Late Cretaceous anomalies 29 to 34. Absolute ages of the Aptian–Maastrichtian geomagnetic reversals are those of Harland et al. (1982).

At present, no accurate ages have been obtained for the Early Cretaceous sea-floor anomalies. As a result, age calibrations by Larson & Hilde (1975), Vogt & Einwich (1979), Harland et al. (1982) and Lowrie & Ogg (1986) have yielded timespans for the Neocomian varying between 21 and 29 Ma. The most recent data have not yet been collated into an integrated chronology. As a compromise, this timescale follows that of Harland et al. (1982) for the Late Cretaceous, and Haq & Van Eysinga (1987) for the Early Cretaceous, slightly modified to accept 141 Ma for the Jurassic–Cretaceous boundary, as interpreted by Bralower et al. (1990).

1.6.10 CRETACEOUS–TERTIARY BOUNDARY

The age of the Cretaceous–Tertiary boundary is now taken at 65 Ma, principally from $^{40}Ar/^{39}Ar$ dating of tektites in the boundary event layer, thought to be associated with the impact represented by the Chicxulub crater (Swisher et al. 1992; Berggren et al., in press *a*). Berggren et al. (in press *a*) discuss how variations on this age in the literature are mainly a function of intercalibration problems in the $^{40}Ar/^{39}Ar$ dating method, and show that the weighted mean of identically normalised $^{40}Ar/^{39}Ar$ ages is 65.06 ± 0.02 Ma.

Subdivisions of the Tertiary

The latest available age information for the Tertiary is discussed by Berggren et al. (in press *a,b*), whose reviews are followed in constructing this scale, and to which the reader is referred for detailed discussions of the age calibration. This section of the timescale is becoming very accurately constrained from a combination of precise $^{40}Ar/^{39}Ar$ dating and biostratigraphic interpolations securely tied to the magnetostratigraphic record. In the youngest parts of the scale, astronomical calibrations are helping to provide an independent check on radiometric age measurements.

The Palaeocene–Eocene boundary is constrained at 55 Ma by $^{40}Ar/^{39}Ar$ dating of ashes at this level in Denmark, and correlatives in England. The calibration of the Eocene–Oligocene boundary at 33.7 Ma is thoroughly discussed by Berggren et al. (1992), and supported by new data discussed in Berggren et al. (in press *a*). The Oligocene–Miocene boundary is taken at 23.8 Ma, an age agreed by both Berggren et al. (in press *b*) and Harland et al. (1990). The top of the Miocene is placed at 5.32 Ma, significantly older than the age used in previous compilations by Cande & Kent (1992*a*: 5.16 Ma) and Berggren et al. (1985*c*: 4.86 Ma), partly on the basis of astronomical calibration. This has also been establishing the congruence of $^{40}Ar/^{39}Ar$ age estimates for the Pliocene–Pleistocene boundary at 1.78 Ma (Bergren et al., in press *b*).

SECTION TWO

Explanatory notes on biostratigraphic charts

2.1

Cambrian (Chart 1)

J.H. Shergold

INTRODUCTION

Rocks of assumed Cambrian age were first reported in Australia by Burr (1846) during the early stratigraphic investigations of South Australia (*fide* Cooper 1984). Cambrian rocks at Ardrossan, on Yorke Peninsula, were confirmed on their fossil content of archaeocyaths and trilobites by Etheridge (1890*a*), and Foord (1890) determined the Cambrian age of rocks found in the north-east of Western Australia by Hardman (1884, 1885). Brown (1895) collected Cambrian fossils on the Barkly Tableland, Northern Territory, which were described by Etheridge (1897, 1902, 1905). Selwyn (in Fairfax 1859) recorded the possibility of Cambrian rocks in central Victoria, which were documented by their trilobites by Etheridge (1896). In north-western Tasmania, Gould (1867) described rocks now known to be Cambrian, but not demonstrated as such until the work of Thomas & Henderson (1945). In western Queensland, Saint-Smith (1924) discovered Middle Cambrian trilobites in the Mount Isa region, which were described by Chapman (1929). Cambrian rocks were not known in New South Wales until 1960 when fossils were found in the Mootwingee area (Warner & Harrison 1961; described by Öpik 1975*b*).

Today rocks of Cambrian age are known to occur in all onshore Australian Lower Palaeozoic sedimentary basins except the Canning Basin, where their existence is suspected, possibly in metamorphosed condition. Offshore, only the Arafura Basin is presently known to contain dated Cambrian rocks (Bradshaw et al. 1990).

COLUMN 1: GEOCHRONOLOGY

The base of the Cambrian is herein taken at 545 Ma based on recent radiometric dating from Siberia (Bowring et al. 1993). Few reliable dates are available for the Cambrian–Ordovician boundary; so an arbitrary figure of 490 Ma is selected, with the acknowledgement that the boundary may be still younger.

COLUMN 2: MAGNETOSTRATIGRAPHY

The rudimentary magnetic polarity scale is based on the limited investigations of Kirschvink (1976, 1978*a,b*) and Klootwijk (1980) in central and South Australia and the work of Ripperdan & Kirschvink (1992) on the Cambrian–Ordovician transition at Black Mountain in the eastern Georgina Basin.

COLUMN 3: STANDARD BIOCHRONOLOGICAL SCALES

As noted above, the Global Stratotype and Point for the base of the Cambrian is now internationally agreed. The biostratigraphical datum, based on the first occurrence of the ichnofossil *Phycodes pedum*, is not yet precisely correlated in Australian sequences, although the index fossil is known in the Arrowie Basin of South Australia, the Broken Hill district of western New South Wales, and the Amadeus and western Georgina Basins of central Northern Territory (see Walter et al. 1989).

The Cambrian–Ordovician boundary is undecided, but there is an international preference for its placement at the base of the Tremadoc Series of north-western Europe, which is defined by the incoming of nematophorus graptolites. Since these are rarely found in platform carbonate sequences, which commonly occur at the Cambrian–Ordovician boundary around the world, a boundary datum point at an appropriate conodont zone at or near the base of the Tremadoc is being sought. Currently the base of the *Cordylodus lindstromi* Zone is considered to approximate the first occurrence of nematophorus graptolites, and in Australian sections *C. lindstromi* is considered an appropriate index fossil for defining the base of the Ordovician (Shergold & Nicoll 1992).

Although several series names are available (see historical discussion in Öpik et al. 1957), none is currently applied to the Cambrian of Australia, where the traditional tripartite division of Lower, Middle and Upper Cambrian provides an informal series base. The right-hand side of Column 3 shows twelve local stage subdivisions for the Cambrian of Australia, with suggested global correlations through northern China, Siberia and Kazakhstan to Scandinavia and North America.

The research of Whitehouse (1927–39) led to a biostratigraphic classification for the Cambrian, which was adopted by David (1950), but Öpik (1956) was the first to apply the classical European Middle Cambrian agnostoid zonation developed by Westergård (1946) to Australian successions. Öpik (1956–1982) demonstrated the potential of these organisms in international correlation, established the correct stratigraphic order of Whitehouse's (1939) Cambrian stages, and introduced various subdivisions for the Australian Middle and early Upper Cambrian, viz. Ordian, Templetonian, Floran, Undillan, Boomerangian, Mindyallan and Idamean. The Payntonian and Datsonian Stages were introduced for the latest Cambrian (Jones et al. 1971), and Iverian has been proposed (Shergold 1993) for the stratigraphic interval previously known as post-Idamean-pre-Payntonian (e.g. Shergold 1989). Biostratigraphic subdivision of the pre-Ordian Cambrian is rudimentary, with no local stage nomenclature for the Early Cambrian of Australia. Archaeocyathans, small shelly fossil assemblages, and trilobite correlations (see Bengtson et al. 1990) show the stage nomenclature of the Siberian Platform and adjacent Sayan–Altai Fold belt to be generally applicable, with 'Atdabanian–Toyonian' faunal assemblages recognised throughout southern and central Australia, and Tommotian and Nemakit-Daldynian faunas perhaps represented in South Australia by ichnocoenoses in the basal Lower Cambrian Parachilna and Uratanna Formations respectively.

Australian Cambrian biochronological schemata are currently based on three groups of organisms: archaeocyathans in the Early Cambrian, trilobites throughout the period, and conodonts in the latest Cambrian.

Other groups shown on the 1989 chart as having lesser biostratigraphic resolution, are not included in the present one. Instead, upgraded information on the ranges of selected inarticulate brachiopods, taken from Rowell & Henderson (1978) and Henderson & McKinnon (1981), and bradoriid and archaeostracan crustaceans (Öpik 1961, 1968*b*; Fleming 1973; Glaessner 1979; Jones & McKenzie 1980; Hinz 1991*a*,*b*; Hinz & Jones 1992) is shown in Figure 21. The relatively little biostratigraphic work done on other fossil groups has been summarised under

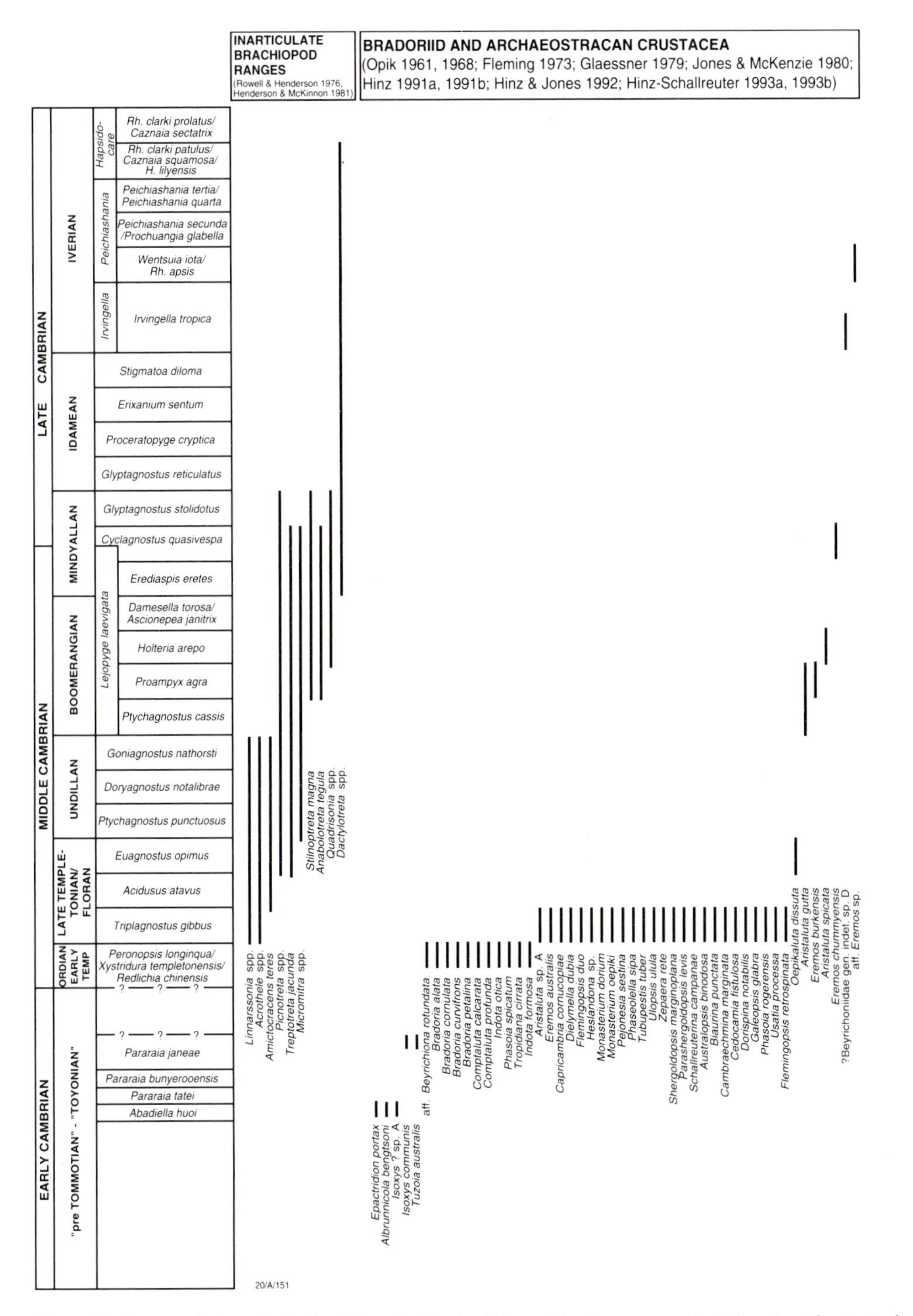

Figure 21: Ranges of phosphatic Brachiopoda, Bradoriida and Archaeostraca plotted against the Australian Early to Middle Late Cambrian biostratigraphic scheme.

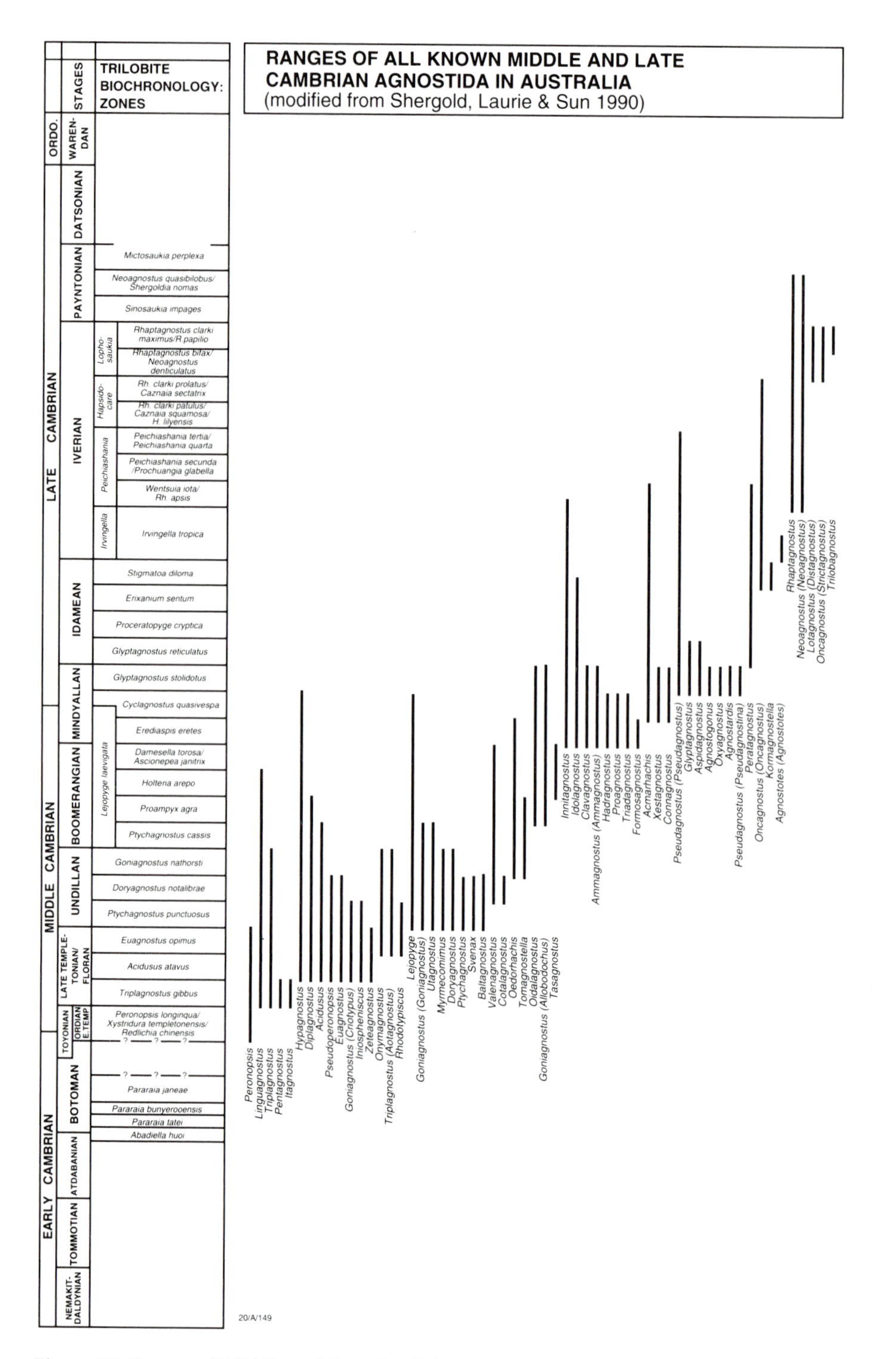

Figure 22: Ranges of Middle and Late Cambrian Agnostida plotted against the Australian trilobite biostratigraphic scheme.

the appropriate time context in the notes to the earlier chart (Shergold 1989). To that should be added recent references to work on a variety of algal and shelly fossil groups from the Early Cambrian of South Australia (Bengtson et al. 1990), and from the early Middle Cambrian of the northern Georgina Basin, eastern Northern Territory (Kruse 1991); on sponge spicules from the Georgina Basin of western Queensland (Bengtson 1986); on Late Cambrian dendroids from north-western Tasmania (Rickards et al. 1990); and organic-walled microfossils from the Early Cambrian of South Australia (Foster et al. 1985), and from the early Middle Cambrian of the Amadeus Basin, Northern Territory (Zang in Shergold 1991*a*; Zang & Walter 1992).

COLUMN 4: ARCHAEOCYATHA

This group has been used in subdivision of the Early Cambrian in South Australia (Gravestock 1984; Debrenne & Gravestock 1990; James & Gravestock 1990; Zhuravlev & Gravestock 1994), New South Wales (Kruse 1978, 1982), and central Australia (Kruse & West 1980). The biostratigraphy shown in Column 4 is based on Gravestock's (1984) Faunal Assemblages I–V, which are formally named in Zhuravlev & Gravestock (1994).

COLUMN 5: TRILOBITES

Early Cambrian

A subdivision of the Early Cambrian by trilobites is based on the work of Jell (in Bengtson et al. 1990), who defined four zones in the Arrowie Basin of South Australia at sections on Yorke Peninsula and in the Flinders and Mount Scott Ranges. The oldest, based on *Abadiella huoi* and *Pararaia tatei*, correlates directly with the Chinese *Parabadiella* and *Eoredlichia* Zones (Zhang 1988), of Qiongzhusian age (see Column 3); the *P. tatei* Zone permits correlation with the late Atdabanian of the Siberian Platform. The youngest zone of *Pararaia janeae* suggests a Botoman age in Russian terms.

Above the zone of *P. janeae*, in the Wirrealpa Limestone and Moodlatana Formation, occur sequentially the species *Redlichia guizhouensis* and *Onaraspis rubra*, which Jell (Bengtson et al. 1990) would correlate with the Longwangmiaoan of China, considered here to equate with the Ordian of northern Australia. There is apparently no locality yet identified where there is a faunal passage from confirmed Early into Middle Cambrian.

In the Georgina Basin, which contains the most complete Middle and Late Cambrian sequences known in Australia, the Early Cambrian is poorly developed, seemingly only represented by archaeocyathan and shelly fossil faunas, lacking trilobites, of 'late Atdabanian–early Botoman' ages (Kruse & West 1980; Laurie & Shergold 1985; Shergold et al. 1985).

The Middle Cambrian sedimentary sequences of the Georgina Basin contain the faunal assemblages used by Öpik (1968*a*, 1979) to define his Ordian, Templetonian, Floran, Undillan and Boomerangian Stages. Trilobites were mainly used in the original definition of these, and particularly agnostoid trilobites whose species evolved rapidly and had a wide geographical distribution resulting from an apparent pelagic mode of life. Ranges of Australian agnostoid genera, after Shergold et al. (1990), are plotted in Figure 22. Middle and Late Cambrian trilobite biochronology has been reviewed and discussed at length previously (Shergold 1989) and, where repeated here, it is for the sake of completeness or where important revisions that require explanation have been made.

Middle Cambrian

Ordian–early Templetonian As foreshadowed by Shergold et al. (1989), discussed by Shergold (1989), and suggested by the sequence stratigraphic analysis of Southgate & Shergold (1991), the *Redlichia chinensis* and *Xystridura templeto-*

nensis Zones are considered to represent lateral biofacies. Initially, Shergold et al. (1989) considered the *Peronopsis longinqua* Zone to be a third lateral biofacies in the south-western Georgina Basin because Öpik (1979) intimated it to predate the occurrence of species of *Pentagnostus* that might indicate the *Triplagnostus gibbus* Zone there. Recent examination of core material from this region suggests that the *P. longinqua* Zone is perhaps a lateral biofacies of the *T. gibbus* Zone (Southgate & Shergold 1991). Accordingly, the Ordian Stage of Öpik (1968*a*), characterised by the occurrence of the *Redlichia chinensis* assemblage (Öpik 1970*b*), is combined with that part of the Templetonian, in turn characterised by the occurrence of *Xystridura templetonensis,* and predating that of *Triplagnostus gibbus*, to form the initial Middle Cambrian Stage in the Georgina and related basins.

A Middle Cambrian age is retained for this stage although it appears to correlate with the combined Longwangmiaoan and Maozhuangian Stages of China, and thence with the Toyonian of the Siberian Platform (see Column 3), all regarded as having terminal Early Cambrian ages. Arguments on the age of the Ordian Stage, based on the overlap of species of *Redlichia* and *Xystridura,* the taxonomic affinity of the latter *vis-à-vis* the Paradoxididae, and historical concepts of Early and Middle Cambrian, posed by Öpik (1968*a*) and discussed in Shergold (1989), are therefore maintained. The position of Xystriduridae versus Paradoxididae put forward by Jell (in Bengtson et al. 1990) is noteworthy, but in need of further elaboration at this time.

Palaeontologically, it is difficult to characterise the early Templetonian because four of the diagnostic xystridurine generic groups recognised by Öpik (1975*a*) have their origins in the Ordian. Moreover, similar eodiscoid and ptychoparioid trilobites, some bradoriid ostracods, like *Zepaera*, several micromolluscs, like *Mellopegma, Protowenella,* and *Pelagiella,* and Problematica, like *Chancelloria,* are present in rocks of both Ordian and early Templetonian ages. These observations have been persuasive in the recognition of the Ordian–early Templetonian as a single biochronological unit.

Late Templetonian–Floran Rocks of late Templetonian and early Floran ages in the eastern Georgina Basin belong to the second Middle Cambrian stratigraphic sequence recognised by Southgate & Shergold (1991), and their faunas represent a biostratigraphic continuum. In the Burke River and Thorntonia areas, Öpik (1968*a*) expanded earlier concepts of the Templetonian Stage by recognising the *Triplagnostus gibbus* Zone as its youngest division. The index species may have a long stratigraphic range and overlap that of the *Acidusus atavus* Zone, which follows. Species of *Pentagnostus* are integral to the *T. gibbus* Zone, often predating and ranging coeval with the index species. The overlap of *T. gibbus* and *A. atavus* suggests that these zones and their biofacies equivalents in the south-west of the Georgina Basin, together with the late Floran Zone of *Euagnostus opimus*, should form a single unified stage. This stage has a global distribution, and is a most significant datum for international correlation (Robison et al. 1977).

The Floran Stage as originally defined contains two agnostoid trilobite zones: *Acidusus* [*Ptychagnostus*] *atavus* (early) and *Euagnostus opimus* (late). Öpik (1979) recorded some twenty-three agnostoid species in the *A. atavus* Zone, which is characterised by the earliest diplagnostids, the ascendency of the *Goniagnostus* lineage, the first *Hypagnostus*, and a diagnostic association of the genera *Triplagnostus, Iniospheniscus, Rhodotypiscus, Criotypus* and *Zeteagnostus.* Particularly important are *Zeteagnostus incautus* and *Triplagnostus gibbus posterus.*

Euagnostus opimus occurs at the stratigraphic level occupied by *Hypagnostus parvifrons* on the European agnostoid biochronological scale. Since *H. parvifrons* is so rarely reported in Australia, Öpik (1970*a*) designated the commonly occurring *Euagnostus opimus* as the index species for this interval. Some authorities (e.g. Jell & Robison 1978), disputing the taxonomy of *Euagnostus*, have regarded it as a subjective junior synonym of *Peronopsis*, and accordingly refer this biostratigraphical interval to the zone of *Peronopsis opimus*. Occurring also in the *Euagnostus opimus* Zone are species of *Onymagnostus*, *Ptychagnostus*, *Triplagnostus*, *Criotypus* and *Pseudoperonopsis*, some of which have their origins in the earlier *A. atavus* Zone. While these agnostoid taxa commonly occur in the eastern Georgina Basin of western Queensland, their distribution elsewhere in Australia is quite limited. Relatively few polymeroid trilobites are associated: of those that are, the ptychoparioids (Jell 1978) are localised, but the nepeiids (Öpik 1970*a*; Jell 1977), dolichometopids (Öpik 1982), anomocarids and damesellids (Jell in Jell & Robison 1978) offer potential for wider correlation both in Australia and elsewhere.

Undillan No major revision has been made on the Undillan Stage, which embraces the *Ptychagnostus punctuosus* Zone (early) and *Goniagnostus nathorsti* Zone (late). Prior to 1979, Öpik (e.g. 1956) had recognised an interval of overlap between these zones that contains some fifteen agnostoid trilobite taxa. This interval of overlap was subsequently designated as the *Doryagnostus notalibrae* Zone (Öpik 1979). Restricted to the Undillan Stage in Australia are the genera *Svenax*, *Baltagnostus*, *Doryagnostus*, *Myrmecomimus* and *Oedorhachis* (*sensu* Öpik). Species of *Pseudoperonopsis*, *Acidusus*, *Aristarius*, *Onymagnostus*, *Euagnostus*, *Aotagnostus* and *Rhodotypiscus* commonly occur but have earlier origins. The agnostoid faunas of the Undillan Stage are cosmopolitan in their distributions. Besides agnostoids, the occurrence of ptychoparioids (Whitehouse 1939; Jell 1978), anomocarids (Whitehouse 1939; Jell in Jell & Robison 1978), mapaniids and damesellids (Öpik 1967), conocoryphids (Shergold 1973), corynexochids (Whitehouse 1945; Öpik 1967), nepeiids (Öpik 1970*a*) and dolichometopids (Öpik 1982) are characteristic and widespread. Undillan trilobites have also been described from north-western Tasmania (Jago 1977, 1979).

Boomerangian The Boomerangian Stage embraces the agnostoid trilobite zone of *Lejopyge laevigata*, which Öpik (1961) divided into three: *Lejopyge laevigata* I, II and III. At the same time these divisions were diagnosed by polymeroid trilobites. *Lejopyge laevigata* II is the zone of *Proampyx agra* and *L. laevigata* III the zone of *Holteria arepo*. *L. laevigata* I is also known as the zone of *Ptychagnostus cassis*. In terms of agnostoid trilobites, only *Delagnostus* is confined to the stage, which is nevertheless characterised by the common occurrence of species of *Lejopyge*, *Hypagnostus*, *Grandagnostus*, *Oidalagnostus* and *Diplagnostus*, most of which originate in earlier stages. *Allobodochus* and *Agnostus* begin their ascendency during the Boomerangian.

A good range of polymeroid trilobites accompanies the agnostoids, most important of which are species of *Centropleura*, dolichometopids, olenids, mapaniids, corynexochids and damesellids, all described in western Queensland by Öpik (1958, 1961, 1967, 1970*a*, 1982). Boomerangian trilobites have also been described from the Dundas Trough, Dial Range Trough and Adamsfield Trough in Tasmania (Jago 1972*a*,*b*, 1974*b*, 1976*a*,*b*, 1981; Jago & Daily 1974; Daily & Jago 1975); and their occurrence is also noted in the Warburton Basin of north-eastern South Australia by Daily (1966) and Gatehouse (1986).

According to Öpik (1966, 1967), a Zone of Passage between the Middle and Late Cambrian is interposed between the late Middle Cambrian Boomerangian and early Late Cambrian Mindyallan stages, but its stratigraphic position is ambiguous. In 1966, Öpik quite clearly regarded it classifiable with the Mindyallan, but by 1967 the Zone of Passage was attributed to neither stage. Daily & Jago (1975), however, show it (1975, Table 3) as Middle Cambrian. Öpik (1967, p.8) has stated that the relationship between the Zone of Passage and the underlying Boomerangian *Lejopyge laevigata* Zone is palaeontologically inconclusive. Nevertheless, the interval contains a fauna characterised by the occurrence of *Damesella torosa* and *Ascionepea janitrix* together with species of *Ptychagnostus*, *Hypagnostus*, and *Lejopyge*, dorypygid, damesellid, solenopleurid and rhyssometopid trilobites, all decidedly Middle Cambrian. Öpik's faunal lists (1967, pp.41–43) show the presence of quite conclusive Boomerangian trilobites. Accordingly, the chart follows Daily & Jago (1975) and shows the zone of passage below the top of the Middle Cambrian.

Late Cambrian

The traditional Middle–Late Cambrian boundary has been taken at the base of the *Agnostus pisiformis* Zone, which in northern Europe overlies the *Lejopyge laevigata* Zone or its correlatives. In northern Australia *Lejopyge laevigata* characterises the Boomerangian Stage of the late Middle Cambrian, and a subspecies extends into the overlying Mindyallan Stage (Daily & Jago 1975; *cf.* Öpik 1967). The base of the *Agnostus pisiformis* Zone is placed within the second Mindyallan zone of *Acmarhachis quasivespa*, at a level between faunas based on *Lejopyge cos* and *Blackwelderia sabulosa*. Thus, they advocate with good argument a Middle–Upper Cambrian boundary lying within the Mindyallan Stage as conceived by Öpik.

Mindyallan In the original simplified form that it was introduced by Öpik (1963), the Mindyallan Stage was conceived as containing two zones: a zone of *Glyptagnostus stolidotus* (above), and a 'pre-*stolidotus*' zone (below).

Following subsequent description of the Mindyallan trilobite faunas (Öpik 1967), which contain an estimated 170 species (Öpik 1966), the early Mindyallan was divided into two biostratigraphical zones (Öpik 1966, 1967). The late Mindyallan, constituting the *Glyptagnostus stolidotus* Zone, with eighty-one species, is the most geographically widespread, having been identified also in the eastern Amadeus Basin (Öpik 1967; Shergold 1986, 1991*a*), Bonaparte Basin (Öpik 1969), western New South Wales (Öpik 1975*b*; Wang et al. 1989), central Victoria (Thomas & Singleton 1956), and Tasmania (Jago 1972*a*, 1986). Probably, it also occurs in the Warburton Basin (Gatehouse 1986).

The initial zone of the Mindyallan Stage is that of *Erediaspis eretes*, a tricrepicephalid trilobite, which occurs in western Queensland and in Tasmania (Öpik 1967). It contains some forty-five trilobites including eighteen agnostoid genera. The latter include species whose genera range up from the Middle Cambrian (*Agnostus*, *Ptychagnostus*, *Hypagnostus*, *Grandagnostus*), associated with taxa occurring for the first time (*Proagnostus* [=*Agnostascus*], *Agnostoglossa*, *Hadragnostus*, *Idolagnostus*, *Clavagnostus* and *Triadagnostus*), several of which continue into younger zones. Some fifteen polymeroid trilobites are confined to this zone (Öpik 1967, p.9). They belong to a wide variety of families that typically occur in the early Mindyallan: anomocarid, asaphiscid, catillicephalid, damesellid, leiostegiid?, lonchocephalid, menomoniid, nepeiid, norwoodiid and rhyssometopid. *Erediaspis eretes*, *Cermataspis abundans*, *Aedotus instans* and *Rhyssometopus* (*Rostrifinis*) *rostrifinis* are typical (Öpik 1967).

The youngest zone of the early Mindyallan is that of *Acmarhachis quasivespa* (formerly *Cyclagnostus*) (Öpik 1966, 1967), which has eighteen species of trilobites restricted to it. According to Öpik (1967, p.10), the most important components of this zone are: *Blackwelderia sabulosa*, *Griphasaphus griphus*, *Bergeronites dissidens*, *Rhyssometopus* (*R.*) *rhyssometopus*, *Stephanocare richthofeni* and *Acmarhachis quasivespa*. Many other species, however, range upwards from older zones, but very few range into that of *Glyptagnostus stolidotus*, which succeeds. In fact, only eight of the many species considered by Öpik (1967) range from the *A. quasivespa* or older zones into that of *Glyptagnostus stolidotus*. The last, introduced by Öpik in 1961 (p.39) and subsequently developed by him (1963, 1966, 1967), contains an estimated seventy-five trilobite species. Öpik (1967, p.11) has cited the following as diagnostic of the zone: *Aulacodigma quasispinale*, *Auritama aurita*, *A. trilunata*, *Biaverta biaverta*, *Blackwelderia gibberina*, *Meteoraspis bidens*, *Mindycrusta mindy-*

crusta, *Bergeronites dissidens*, *Rhodonaspis longula*, and *Rhyssometopus princeps* among the polymeroids; and *Agnostardis amplinatis*, *Aspidagnostus inquilinus*, *Glyptagnostus stolidotus* and *Xestagnostus legirupa* among the agnostoids. Daily & Jago (1975) considered that the *quasivespa* Zone could be divided into two assemblages characterised by *Lejopyge cos* and *Blackwelderia sabulosa*. As indicated above, the faunas of this zone have a very wide distribution in Australia and can also be correlated elsewhere.

Idamean Öpik (1963) conceived the Idamean as being composed of five successive assemblage zones: *G. reticulatus* with *Olenus ogilviei*, *Glyptagnostus reticulatus* with *Proceratopyge nectans*, *Corynexochus plumula*, *Erixanium sentum*, and *Irvingella tropica* with *Agnostotes inconstans*. Henderson (1976) combined the assemblages with *G. reticulatus* into a single *G. reticulatus* Zone, and recognised a series of three zones in the *Corynexochus plumula–Erixanium sentum* interval: that is, the zones of *Proceratopyge cryptica*, *Erixanium sentum* and *Stigmatoa diloma*. The *I. tropica–A. inconstans* assemblage was renamed the *I. tropica* Zone.

There is a major faunal reorganisation, a faunal crisis (Öpik 1966), at the incoming of the *Glyptagnostus reticulatus* assemblage. As documented by Öpik (1966) none of the eighty plus trilobite species described from the *G. stolidotus* Zone, and very few of the genera, persist into the early Idamean. There is also a major reorganisation of trilobite families as the endemic shallow shelf carbonate communities listed above are virtually instantaneously replaced by cosmopolitan outer shelf assemblages dominated by agnostoids, olenids, pterocephaliids, leiostegiids, eulomids and ceratopygids. These incoming faunas lack the Mindyallan diversity, and Idamean species total only about a hundred. Thus the beginning of the Idamean Stage is readily recognised biostratigraphically.

Both Öpik (1963, 1966, 1967) and Henderson (1976, 1977) regarded the *Irvingella tropica* assemblage as the youngest zone of the Idamean Stage. It has been demonstrated subsequently (Shergold 1982) that in the Burke River area of the eastern Georgina Basin a sharp faunal change exists between the *Stigmatoa diloma* and *I. tropica* zones, and that the latter shows palaeontologically more in common with succeeding post-Idamean (Iverian) trilobite assemblages than with those predating the *Stigmatoa diloma* Zone. Hence, the Iverian Stage is considered to begin with the incoming of *Irvingella tropica* (see Shergold 1982, pp.15–16 for justification; and Shergold 1993). On the accompanying chart, therefore, the Idamean Stage terminates with the *Stigmatoa diloma* Zone.

The early Idamean zones of *Glyptagnostus reticulatus* and *Proceratopyge cryptica* are very readily identifiable. The former, a cosmopolitan species of limited duration (Kobayashi 1949 regarded its range as a 'world instant'), is associated with equally wide-ranging species of *Olenus*, *Aphelaspis* and *Eugonocare* among polymeroid trilobites, and the rapid rise to ascendency of the agnostoids *Pseudagnostus* and *Oncagnostus* (*sensu stricto*), all of which permit the diagnosis of an accurately and globally correlatable biostratigraphic unit.

The later Idamean is similarly identifiable. Together the *Erixanium sentum* and *Stigmatoa diloma* Zones can be equally widely correlated. They have more faunal variation, however, and represent mainly an admixture of cosmopolitan and Australo–Sinian genera, for example *Pagodia*, *Prismenaspis*, *Pseudoyuepingia* [*Iwayaspis*], *Yuepingia*, *Eugonocare*, *Proceratopyge*, *Corynexochus* and the inevitable Pseudagnostinae.

The faunas of the Idamean Stage yield a highly resolved biochronology that permits very accurate global correlations. In Australia, Idamean trilobite faunas have been described to date from the Georgina Basin (Whitehouse 1936, 1939; Öpik 1963, 1967; Henderson 1977; Shergold 1982), western New South Wales (Jell in Powell et al. 1982), and western and south-central Tasmania (Jago 1974*a*, 1978, 1979, 1987; Jago & Brown 1989, 1992), and an Idamean fauna has been noted by Gatehouse (1986) in the Warburton Basin. Appropriate faunas have yet to be identified in more cratonic settings, for example the Amadeus and Bonaparte Basins, unless they are represented by trilobite biofacies not commonly occurring in Australia (e.g. Parabolinoidid Assemblage of Öpik 1969 in the Bonaparte Basin).

Iverian This stage has recently been proposed (Shergold 1993) for the stratigraphic interval previously referred to (Shergold 1989) as post-Idamean-pre-Payntonian. It has been zoned on the basis of trilobite faunas from the Burke River Structural Belt, eastern Georgina Basin, the only region where a probable complete sequence has so far been described (Shergold 1972, 1975, 1980, 1982, 1993). Faunas of Iverian age do occur elsewhere in Australia, principally in Tasmania where Jago (1978, 1979) and Jago & Brown (1992) have described trilobites from the Climie Formation, and Jell et al. (1991) those from the Upper Huskisson Group. Additionally, there is inference by the latter that the faunas of the Singing Creek Formation may have the same age, as does poorly preserved material from the Newton Creek Sandstone (Corbett 1975; Jago & Brown 1989). Possibly contemporaneous basinal Iverian biofacies are reported from western New South Wales (Watties Bore) by Webby et al. (1988).

Probably also material from the Wagonga Beds on the New South Wales coast (Bischoff & Prendergast 1987) has an Iverian age, but could be Idamean. In other areas, on the Australian craton, one or more stratigraphic hiatuses occur within the Iverian, for example in the western Georgina, Amadeus, Warburton, Wiso and Ngalia Basins (but see Column 7 for comments on the Bonaparte Basin).

The Iverian zonation starts with *Irvingella tropica*, as discussed above. Justification for the recognition of the zone, and its exclusion from the Idamean Stage where it had been previously classified (by Öpik 1963; Henderson 1976, 1977), has been given by Shergold (1982). The trilobite fauna comprises globally wide ranging correlatable elviniid, eulomid, leiostegiid, olenid and ceratopygid genera such as *Irvingella*, *Olenus*, *Proceratopyge*, *Eugonocare*, *Stigmatoa*, *Protemnites*, and *Chalfontia* mingled with Australo–Sinian pagodiid leiostegiaceans (*Pagodia* (*Idamea*) and *Prochuangia*), and agnostoids, of which *Agnostotes* is particularly diagnostic. Despite the cosmopolitan nature of the trilobites, the assemblage is so far only recorded from the Georgina Basin.

It is succeeded at Mount Murray in the Burke River Structural Belt by an assemblage previously separated as the post-*Irvingella* Zone. This contains a limited fauna consisting of elviniid, pterocephaliid, ceratopygid and leiostegiid genera related to those of the *I. tropica* Zone, and now included in it (Shergold 1993).

Four younger assemblages, *Wentsuia iota–Rhaptagnostus apsis*, *Peichiashania secunda–Prochuangia glabella*, *Peichiashania tertia–P. quarta* and *Hapsidocare lilyensis*, occur in the vicinity of the type section of the Chatsworth Limestone at Lily Creek, near Chatsworth Homestead in the Burke River area (Shergold 1980). These form a group of biostratigraphical entities dominated by leiostegiid trilobite genera (particularly the pagodiids *Prochuangia* and *Lotosoides*, and the mansuyiinids *Peichiashania*, whose species form a lineage, and *Hapsidocare*) associated with ceratopygids and the first true asaphids. Olenids, pterocephaliids, catillicephalids, eulomids and the first shumardiids and saukiids occur, but not so commonly. Agnostoids of the subfamilies Agnostinae and Pseudagnostinae are significant. The latter include associated species of *Pseudagnostus*, *Rhaptagnostus* and *Neoagnostus*, which appear to have separated morphologically during the *Irvingella* Zone (Shergold 1977, 1981). On the eastern New South Wales coast, near Batemans Bay, trilobites recovered by Bischoff & Prendergast (1987) include catillicephalids and agnostoids, which may be correlated at the *iota–apsis* level.

At Black Mountain, a further 54 km to the south, four more Iverian assemblages occur in the Chatsworth Limestone (Shergold 1975), and are diagnosed on the basis of their saukiid and pseudagnostinid trilobites: *Rhaptagnostus clarki patulus–Caznaia squamosa*, *R. clarki prolatus–C. sectatrix*, *R. bifax–Neoagnostus denticulatus*, and *R. clarki maximus–R. papilio*. The first two of these assemblages correlate with the *Hapsidocare lilyensis* Assemblage-Zone at Lily Creek.

These assemblages, *patulus–squamosa* to *maximus–papilio*, are characterised by prosaukioid and saukioid dikelocephaceans (*Caznaia*, *Prosaukia*, *Lophosaukia*), pagodiid (*Pagodia*, *Oreadella*, *Lotosoides*) and kaolishaniid (*Mansuyia*, *Mansuyites*, *Hapsidocare*, *Ceronocare*, *Palacorona*) leiostegiaceans, asaphids (*Golasaphus*, *Atopasaphus*), and the first kainelloid remopleuridaceans (*Sigmakainella*, *Richardsonella*, *Elkanaspis*) among other polymeroids. Of the agnostoids, species of the pseudagnostinid genera *Rhaptagnostus* and *Neoagnostus* are characteristic and diagnostic. They are associated with species of *Oncagnostus* and *Distagnostus* representing the Agnostinae.

These trilobite assemblages, dominated by leiostegiaceans, and increasingly by dikelocephalaceans, are quite distinct from those of immediate post-Idamean age at Mount Murray, and at Lily Creek. It is apparent that outer shelf family groups became replaced in the biostratigraphical sense by carbonate bank dwelling associations of American–Asian aspect.

It has been recently suggested (Shergold 1993) that this detailed Iverian zonation might be simplified by the overlay of four generic range zones on the existing assemblage-zones. As shown on Column 5, in ascending order, these are the zones of *Irvingella*, *Peichiashania*, *Hapsidocare* and *Lophosaukia*.

Payntonian The Payntonian Stage was originally defined on the basis of its trilobite faunas (Shergold 1975) at the datum on the type section where the comingled American–Asian assemblages of the Iverian are replaced by others of totally Asian affinity. Few previously occurring Iverian species at Black Mountain pass into the Payntonian but several existing genera extend their ranges. The Payntonian is diagnosed palaeontologically by the appearance of tsinaniid leiostegiaceans, and the diversification of dikelocephalaceans (saukiids and ptychaspidids) and remopleuridaceans. Early and late Payntonian assemblages were originally recognised in the southern Burke River Structural Belt. Trilobites of the early Payntonian Assemblage-Zone of *Neoagnostus quasibilobus* with *Shergoldia nomas* (Shergold 1975) are characteristically an association of tsinaniid, saukiid, shumardiid, leiostegiid and kaolishaniid genera. Those of the late Payntonian *Mictosaukia perplexa* Assemblage-Zone are dominated by Saukiidae (Shergold 1975).

Subsequently, following deliberations on the position of the Cambrian–Ordovician boundary, and misleading conodont determinations, Payntonian biostratigraphy has been revised and reassessed (Shergold & Nicoll 1992). Nicoll & Shergold (1991) and Nicoll (1990, 1991) have published new conodont information that refutes earlier statements (such as discussed in Shergold 1989), and the Payntonian Stage redefined on their basis. A modified trilobite biostratigraphy has resulted, in which the base of the Payntonian Stage is now drawn at the appearance of the *Sinosaukia impages* Assemblage-Zone, and accordingly a tripartite zonation is now recognised. This inclusion does not alter the original definition of the Payntonian Stage in terms of trilobites.

Correlatable Payntonian trilobite faunas were, until recently, described only in the eastern Georgina Basin of western Queensland. Since 1989, they have been described from the Pacoota Sandstone of the Amadeus Basin, Northern Territory (Shergold 1991*b*), and from an unnamed sequence at Misery Hill, western Tasmania (Jago & Corbett 1990). Documented Payntonian faunas on the Gnalta Shelf, western New South Wales (Shergold et al. 1985), and in the Bonaparte Basin (Öpik 1969) remain undescribed.

The Payntonian Stage is succeeded by the Datsonian (Jones et al. 1971), which has been widely regarded as the initial stage of the Ordovician in northern Australia. The Datsonian Stage contains few documented trilobites in Australia, and is defined solely on the basis of conodonts (see discussion on Column 6).

Since there seems to be an acknowledgement that the Cambrian–Ordovician boundary has been correlated too low in Australia, at the incoming of the *Cordylodus proavus* Zone, and that a more appropriate correlation might be at the level of *Cordylodus lindstromi*, the Datsonian Stage is here, following Shergold & Nicoll (1992), regarded as the terminal Cambrian Stage. Its internal zonation is discussed more comprehensively under the notes on Column 6.

COLUMN 6: CONODONT BIOCHRONOLOGY

A conodont biostratigraphy has not yet been developed for the Early and Middle Cambrian in Australia, although the presence of conodonts in the Middle Cambrian phosphate deposits of western Queensland has been known for many years. Similarly, although the existence of conodonts in the pre-late Iverian has been known since the 1960s, they have not been biostratigraphically exploited, nor has their value been assessed. Müller & Hinz (1991) actually dispute the biostratigraphic utility of Cambrian conodonts from their experience working with Scandinavian material, especially if samples have low yield. Here, however, the stratigraphic scheme developed by Miller (1969, 1980, 1988) in North America, and Druce & Jones (1971) and Jones et al. (1971) in Australia has been further developed in combination with the taxonomic philosophy and conclusions of Nicoll (1990, 1991, 1992) and Nicoll & Shergold (1991).

First identified by Jones (1961*b*), and later described by Druce & Jones (1968, 1971) and Druce (1978*a,b*), the late Iverian to earliest Ordovician conodonts of the Burke River Structural Belt, particularly from Black Mountain, in western Queensland, have provided the basis for a highly resolved, globally correlatable, biostratigraphy. Following recent revisions (Nicoll & Shergold 1991; Shergold & Nicoll 1992), they have assumed prime importance in the redefinition of the Payntonian, Datsonian and Warendan Stages (Jones et al. 1971), and the Cambrian–Ordovician boundary (Fig. 23).

Iverian Conodonts have first demonstrated biostratigraphic value from late Iverian time onwards, in both North America and Australia. In the latter, a single latest Iverian assemblage based on *Teridontus nakamurai* has been documented by Nicoll & Shergold (1991), and correlated to the *Proconodontus posterocostatus* and early *P. muelleri* Zones of western USA, as used, for example, by Miller (1988).

Payntonian Following Shergold et al. (1991*b*) and Shergold & Nicoll (1992), the Payntonian Stage is redefined on the basis of successive species of the genus *Hispidodontus* that constitute the *H. resimus*, *H. appressus* and *H. discretus* assemblages. The base of the stage is now considered to coincide with the first appearance of the *Hispidodontus resimus* assemblage, as defined by Nicoll & Shergold (1991). This level coincides with the incoming of the *Sinosaukia impages* Assemblage-Zone on the trilobite biochronological scale, and correlates within the *Proconodontius muelleri* Zone of North America.

The succeeding *Hispidodontus appressus* assemblage contains such species as *Eoconodontus notchpeakensis* and *Eoconodontus* [*Cambrooistodus*] *minutus* and appears to correlate with the middle part of the *Eoconodontus* Zone of North America, while the *Hispidodontus discretus* assemblage represents the latest part of that zone. The first appearance of *Hirsutodontus*, *H. nodus*, occurs in the *Hispidodontus appressus* assemblage. This is the species misidentified by Miller in 1976 as *Hirsutodontus hirsutus*, which gave rise to the discrepant notions of the age of the Payntonian–Datsonian boundary on the conodont versus trilobite biochronological scales discussed in the previous version of the Cambrian Timescales Chart (Shergold 1989). On the present chart, the *Mictosaukia perplexa*

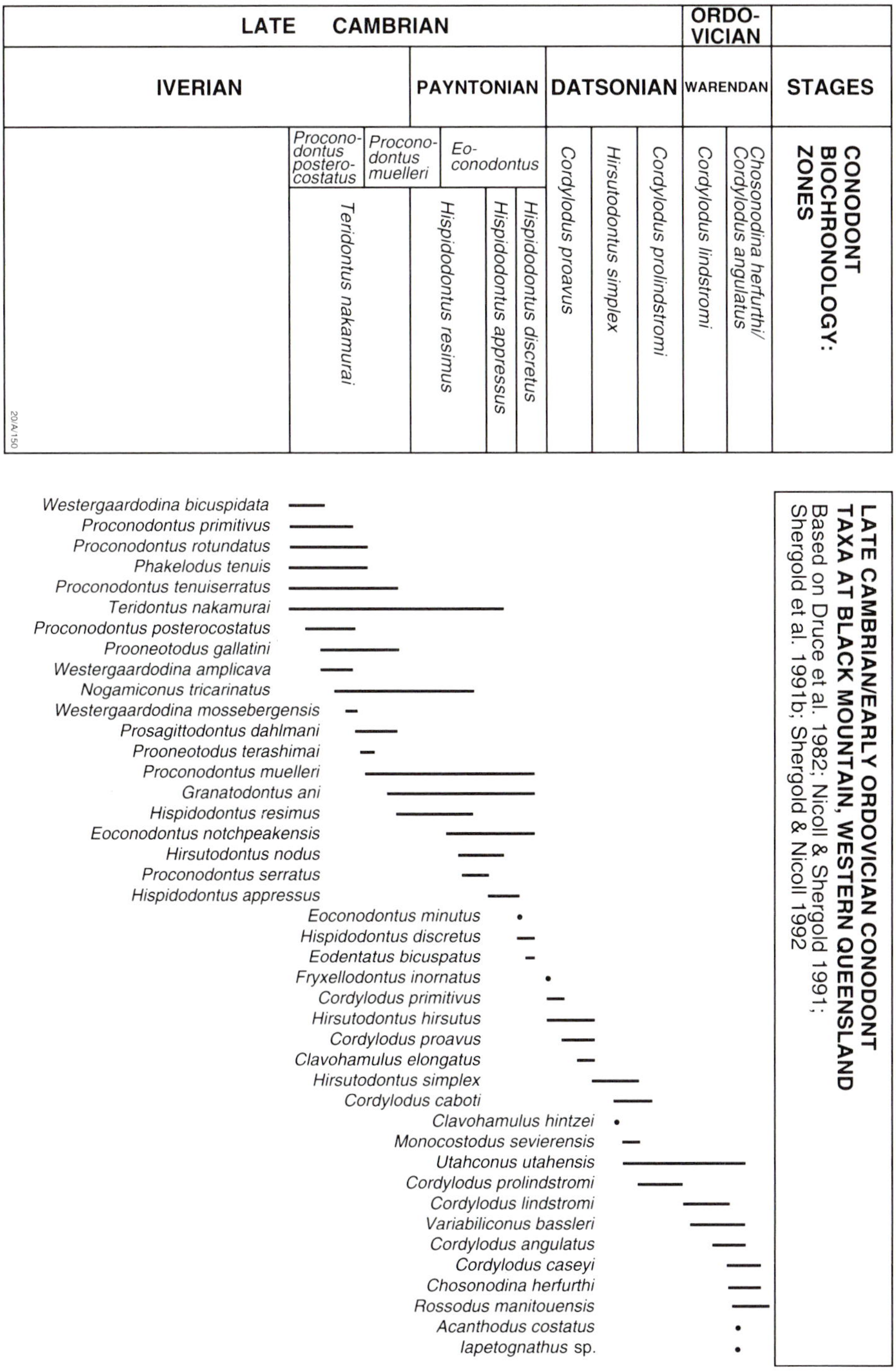

LATE CAMBRIAN | ORDOVICIAN

STAGES: IVERIAN | PAYNTONIAN | DATSONIAN | WARENDAN

CONODONT BIOCHRONOLOGY: ZONES

Proconodontus posterocostatus
Proconodontus muelleri
Eoconodontus
Teridontus nakamurai
Hispidodontus resimus
Hispidodontus appressus
Hispidodontus discretus
Cordylodus proavus
Hirsutodontus simplex
Cordylodus prolindstromi
Cordylodus lindstromi
Chosonodina herfurthi/Cordylodus angulatus

20/A/150

LATE CAMBRIAN/EARLY ORDOVICIAN CONODONT TAXA AT BLACK MOUNTAIN, WESTERN QUEENSLAND
Based on Druce et al. 1982; Nicoll & Shergold 1991; Shergold et al. 1991b; Shergold & Nicoll 1992

Westergaardodina bicuspidata
Proconodontus primitivus
Proconodontus rotundatus
Phakelodus tenuis
Proconodontus tenuiserratus
Teridontus nakamurai
Proconodontus posterocostatus
Prooneotodus gallatini
Westergaardodina amplicava
Nogamiconus tricarinatus
Westergaardodina mossebergensis
Prosagittodontus dahlmani
Prooneotodus terashimai
Proconodontus muelleri
Granatodontus ani
Hispidodontus resimus
Eoconodontus notchpeakensis
Hirsutodontus nodus
Proconodontus serratus
Hispidodontus appressus
Eoconodontus minutus
Hispidodontus discretus
Eodentatus bicuspatus
Fryxellodontus inornatus
Cordylodus primitivus
Hirsutodontus hirsutus
Cordylodus proavus
Clavohamulus elongatus
Hirsutodontus simplex
Cordylodus caboti
Clavohamulus hintzei
Monocostodus sevierensis
Utahconus utahensis
Cordylodus prolindstromi
Cordylodus lindstromi
Variabiliconus bassleri
Cordylodus angulatus
Cordylodus caseyi
Chosonodina herfurthi
Rossodus manitouensis
Acanthodus costatus
Iapetognathus sp.

Figure 23: Latest Cambrian and earliest Ordovician conodont distributions and biostratigraphy.

trilobite assemblage-zone is rightfully returned to the Payntonian, its single occurrence at Black Mountain, in western Queensland, falling within the uppermost range of the *Hispidodontus discretus* assemblage.

Datsonian The base of the Datsonian Stage remains as originally conceived by Jones et al. (1971) at the incoming of the *Cordylodus proavus* assemblage, now known to include besides the index species, *Cordylodus primitivus, Eodentatus bicuspatus, Fryxellodontus inornatus* and *Hirsutodontus hirsutus*. This occurs at the globally recognised Lange Ranch Eustatic Event (Miller 1984; Nicoll et al. 1992; Ripperdan et al. 1993).

The Datsonian Stage currently also embraces two further conodont assemblages, based on *Hirsutodontus simplex* and *Cordylodus prolindstromi*, which previously formed the *Cordylodus oklahomensis* Zone of Jones et al. (1971). The *C. prolindstromi* assemblage has only recently been recognised as a biostratigraphic entity following Nicoll's (1990, 1991) analysis of the element composition of species of *Cordylodus*; his review and validation of *Cordylodus lindstromi*, conceptualisation of *C. prion*, and suggested evolution of the cordylodid lineages in western Queensland are followed here. While no question remains about the separation of *C. prolindstromi* and *C. lindstromi* there, elsewhere in more condensed sequences, the differentiation is more difficult to demonstrate.

As indicated in the comments on Column 3, the first appearance of *C. lindstromi* is gaining acceptance as defining the base of the Ordovician System. Nicoll (1990) regards *C. prion* Lindström *sensu* Druce & Jones (1971) as a synonym of *C. lindstromi*, hence the *C. prion–Scolopodus* Assemblage-Zone recognised by those authors falls within the span of the *C. lindstromi* Assemblage-Zone, and the base of the Warendan Stage, distinguished by Jones et al. (1971) on the occurrence of *C. prion*, can be correlated to the base of the *C. lindstromi* Assemblage-Zone.

COLUMN 7: ALTERNATIVE SCHEME FOR THE BONAPARTE BASIN

This column shows a preliminary trilobite biostratigraphic scheme for the Bonaparte Basin prepared by A. A. Öpik in 1969. Twelve informal biostratigraphic units are recognised but remain unsupported by taxonomy. It is possible, however, to suggest varying degrees of correlation with the more highly resolved centralian trilobite biostratigraphy, particularly in the Late Cambrian. Little can be offered at this stage for the Middle Cambrian: Units I and II are not stratigraphically associated; Unit III is poorly fossiliferous, constrained by underlying and overlying data; Unit IV contains agnostoids and damesellids indicative of a biostratigraphic level close to the Middle–Late Cambrian boundary.

In the Late Cambrian, Unit V contains elements of the Mindyallan assemblages of western Queensland, and aphelaspidinid trilobites occur in Unit VI, which may be indicative of the Idamean Stage. Iverian, Payntonian and possibly Datsonian trilobite assemblages are relatively well represented (Shergold 1993). Units VII–IX contain elements of the *Peichiashania, Hapsidocare* and *Lophosaukia* Zones of western Queensland. Saukiid trilobites characterise Units X–XI, Öpik's cf. *Tellerina* apparently representing *Mictosaukia* and thereby indicating a late Payntonian age. Unit XII contains kainellid and leiostegiid trilobites, which Öpik (1969) considered to be late Tremadoc–early Arenig in age, but may be older. Jones (1971) recognised both Datsonian and Warendan conodont assemblages, but did not identify the *Cordylodus proavus* Assemblage-Zone. If there is a correlation of his taxonomy to that of Nicoll (cited under Column 6), then equivalents of the *Hirsutodontus simplex, Cordylodus prolindstromi* and *Cordylodus lindstromi* are likely to be represented, but the material on which these assemblages are based is poorly preserved, and the samples low yielding.

2.2

Ordovician (Chart 2)

R.S. Nicoll and B.D. Webby

INTRODUCTION

Ordovician rocks were identified in Australia with the discovery of graptolites in auriferous slates in Victoria by A.R.C. Selwyn in June 1856 (McCoy 1875). Etheridge (1874) and McCoy (1874) provided the first illustrations and fossil descriptions. Ordovician fossils were recorded in Tasmania from about 1860, with the stromatoporoid *Stromatocerium* listed in Bigsby (1868), and the first Ordovician trilobites were described by Etheridge (1883). In central Australia Etheridge (1891*a*, 1892, 1893, 1894) and Tate (1896) described some of the Ordovician shelly faunas (molluscs, brachiopods and trilobites) from the Amadeus Basin. Papers by Teichert (1939, 1947) and Teichert & Glenister (1952, 1953, 1954) described nautiloids, Hill (1955, 1957) dealt with corals, and Ross (1961) worked with bryozoans.

Ordovician sediments are now known from all states and mainland territories of Australia (Webby et al. 1981) where they were deposited both in intracratonic basins, and on the platforms and shelves of the margin of the Australian Block. Since the late 1960s descriptive work on Ordovician fossils has increased greatly, with many groups being studied for the first time, thereby providing the systematic database for more varied biostratigraphic schemes to provide age control for a wider range of facies. Correlation of these sediments is essential for a detailed and accurate interpretation of the depositional relationships of sediments in areas that are now geographically isolated.

The level of the Cambrian–Ordovician boundary has not yet been ratified internationally; so there is no formal agreement on the precise lower limit of the Ordovician System. However, the conodont zone closest to the first appearance of the nematophorous graptolite *Rhabdinopora flabelliforme* is generally accepted as the most likely boundary level (Norford 1988), and this provisional boundary is adopted here. As presently understood, this is at the base of the *Cordylodus lindstromi* conodont zone, which can be identified in the Georgina Basin (Nicoll 1991; Shergold & Nicoll 1992). However, the boundary cannot be located in the Victorian graptolitic successions because it lies below the earliest records of planktic graptolites currently recognised in the Lancefieldian Stage (Cooper 1979), and conodont recovery in this interval in Victoria is insufficient for biostratigraphic zonation. Another Cambrian–Ordovician boundary section in western New South Wales is developed in deeper, off-shelf, silty trilobite-bearing beds with a diverse latest Cambrian fauna of *Rhaptagnostus, Pseudoyuepingia, Proceratopyge, Hedinaspis* and ?*Prosaukia* succeeded by an earliest Ordovician (basal Tremadoc) occurrence of index fossil *Hysterolenus*. Similar faunal components are seen in Cambrian–Ordovician boundary sections in China (Webby et al. 1988).

The Ordovician–Silurian boundary is not well exposed in Australia. VandenBerg et al. (1984) reported a succession of beds spanning the boundary but representatives of key graptolites of the *persculptus* Zone below and the *acuminatus* Zone above the boundary (Cocks 1985) are not found in any one section. Cas & VandenBerg (1988) extended the latest Ordovician Bolindian Stage upwards to include the *persculptus* Zone and, in Tasmania, there is a graptolitic horizon in the Westfield Sandstone suggesting either the *persculptus* Zone or a position low in the *acuminatus* Zone (Baillie et al. 1978; Banks 1988).

COLUMN 1: GEOCHRONOLOGY

New dates for the Middle Cambrian (Shergold, this volume) indicate that the age of the Cambrian–Ordovician boundary is significantly less than 500 Ma and a figure of 490 Ma is used in this report until new data constrain the boundary more precisely. The Ordovician–Silurian boundary is interpolated at 434 Ma from the work of McKerrow et al. (1985) and Compston & Williams (1992).

COLUMN 2: ORDOVICIAN SERIES AND STAGES

British, American, Baltoscandian and Chinese subdivisions of the Ordovician are shown against the Australian stages. A discussion of subsystemic, series and stage subdivisions and classification of the Ordovician System in Australia and New Zealand, and the applicability of overseas series names in Australasia, was recently presented by Webby et al. (1991). Because there is no international agreement on the limits of the Middle Ordovician (Jaanusson 1960), the term is not used here, with the Darriwilian regarded as the uppermost stage of the Early Ordovician, as recommended by Cas & VandenBerg (1988) and Webby et al. (1988). The base of the Late Ordovician is placed at the base of the *Nemagraptus gracilis* Zone, and coinciding with the base of the Gisbornian, as shown on the chart.

Correlation between the Victorian graptolite scheme and the 'standard' British subdivision has been complicated by the endemism of many British graptolites, including zonal index species, and the problem of relating the Scottish graptolite zones with the Welsh stages based on shelly fossils. Whittington et al. (1984) largely resolved the latter problem, and the correlations adopted here are based on that account.

In the absence of key forms, the **Tremadoc** Series can only be broadly correlated with a substantial portion of the Lancefieldian (La1–La2) on the basis of anisograptids (*Clonograptus* and *Adelograptus*). The **Arenig** Series has traditionally been the most difficult to correlate with Victoria and other regions in the Pacific faunal province, but is now regarded as equivalent to most, if not all, of the Bendigonian, and the entire Chewtonian, Castlemainian and Yapeenian. The occurrence of '*Glyptograptus*' *austrodentatus anglicus* in highest (but not precisely zoned) Arenig strata in Britain (Bulman 1963) suggests that the top of the Arenig lies within the Darriwilian, with the upper Arenig *Didymograptus hirundo* Zone possibly as high as Da1 or Da2 (Williams & Stevens 1988).

The **Llanvirn** and **Llandeilo** series have been combined and placed in a revised **Llanvirn** Series (Fortey et al. 1995). The base of the revised Llanvirn Series is characterised largely by the occurrence of 'tuning-fork' didymograptids (chiefly *Didymograptus artus*), which are unknown from the Pacific faunal province. Other Llanvirn species, such as *Glossograptus acanthus* and *G. ciliatus*, are relatively long ranging within the Darriwilian, suggesting a broad correlation with its middle part. The upper boundary of the revised Llanvirn Series is not precisely defined in the Australian succession because the upper Darriwilian '*Glyptograptus (Hustedograptus)* *teretiusculus* Zone' is very poorly documented. The Llanvirn is provisionally shown extending into the Gisbornian.

The base of the **Caradoc** Series has yet to be defined in its type region; according to Whittington et al. (1984), it lies either within or below the *Nemagraptus gracilis* Zone. On this basis, it correlates with a level in the upper Darriwilian–lower Gisbornian interval. Most important for correlating the Caradoc–**Ashgill** boundary with the Victorian scheme is the occurrence of *Dicellograptus gravis* in a *D. complanatus* Zone assemblage at a level that Whittington et al. (1984) regarded as Pusgillian. This assemblage is typical of Ea4 age (with the exception of *D. complanatus*, which is absent from Australia). On their evidence the base of the Ashgill lies within the *Pleurograptus linearis* Zone; in Victorian terms, this correlates with a level within the Ea3 Zone of *Dicranograptus kirki.*

COLUMN 3: VICTORIAN GRAPTOLITE ZONES

A.H.M. VandenBerg

Ordovician graptolites are abundant in much of the turbidite and black shale facies of the Lachlan Fold Belt in south-eastern Australia, and the Victorian Graptolite Sequence presented in Column 3 is the standard used in Australasia, and for the larger part is the most finely subdivided in the world. It is the result of a long history of research, beginning with McCoy and Etheridge in the 1870s, which by the end of the 1930s had resulted in the publication of over 100 papers on Victorian graptolite taxonomy and biostratigraphy. This research has continued to the present, with studies concentrating on particular problems in the Lancefieldian (Cooper & Stewart 1979; Erdtmann & VandenBerg 1985), in the Castlemainian and Yapeenian (Cooper 1973; Cooper & McLaurin 1974; McLaurin 1976; Cooper & Ni 1986), and at the Ordovician–Silurian boundary (VandenBerg et al. 1984). The most recent work is represented in the papers by Cas & VandenBerg (1988), Cooper & Lindholm (1990), Bergström & Mitchell (1990) and VandenBerg & Cooper (1992).

Elsewhere in Australia, graptolites are known from the Florentine Valley sequence in central Tasmania (Banks & Burrett 1980; Rickards & Stait 1984), and the Mathinna beds in the north-east and Westfield Sandstone in the south-west (Baillie et al. 1978). From the Warburton Basin very sparse, poorly preserved graptolites are recorded by Cooper (1986). Henderson (1983) recorded a Lancefieldian–?Chewtonian graptolite fauna in the Mount Windsor Subprovince of north-eastern Queensland, and the Horn Valley Siltstone in the Amadeus Basin (Northern Territory) contains a small graptolite fauna of Early Ordovician age (Bagas 1988). The Canning Basin sequence in Western Australia contains reasonably diverse and often well-preserved graptolite faunas extending from the Lancefieldian to about the top of the Lower Ordovician. The lower faunas (Bendigonian and Chewtonian) belong entirely to the Pacific faunal province as in Victoria, whereas the upper fauna from the Goldwyer Formation is largely of the Atlantic faunal province, but with several important Pacific elements (Legg 1976). This faunal association suggests a probable correlation of the Llanvirnian *Didymograptus artus* Zone (= '*D. bifidus* Zone') of Wales, with much of the Darriwilian of Victoria.

COLUMNS 4–6: GRAPTOLITE ZONATIONS

A.H.M. VandenBerg

Graptolite zonations are shown for the Canadian Cordillera (Column 5), the Baltic (Column 6) and for north-eastern North America (Column 4). The Canadian Cordillera scheme is based on Larson & Jackson (1966), Lenz & Chen (1985), Lenz & Jackson (1986) and Lenz (1988). The Baltic scheme is derived from Jaanusson (1982) and numerous other contributions summarised by Cooper & Lindholm (1990).

The north-eastern North America scheme is based on Riva (1974), Rickards & Riva (1981) and Williams & Stevens (1988). All three schemes have been summarised by VandenBerg & Cooper (1992).

COLUMNS 7–10: CONODONTS

The earliest report of conodonts in Australia is that of Crespin in 1943 on a small fauna from the Ordovician Larapinta Group, probably the Horn Valley Siltstone, of the Amadeus Basin, Central Australia. This was followed by the mention of conodonts in the Ordovician Prices Creek Group (Emanuel Formation) of the Canning Basin by Guppy & Öpik (1950) and the later recognition of subsurface Ordovician sediments in the Canning Basin through the identification of conodonts by Glenister & Glenister (1958).

The next phase of conodont studies in Australia was also devoted to the Ordovician with reports by Jones (1961*b*) and later by Druce & Jones (1968, 1971), and Jones (1971) on faunas from the Georgina, Daly River and Bonaparte Basins. The establishment by Druce & Jones (1971) of a series of six assemblage-zones in the lowermost Ordovician of the Burke River Structural Belt (Georgina Basin) and a seventh zone added by Jones (1971) in the Bonaparte Basin represents the first Australian Ordovician conodont biostratigraphic scheme. A second zonal scheme was proposed by McTavish & Legg (1976) for the Arenig to Llanvirn interval of the Canning Basin. However this scheme was never backed with the supporting subsurface distribution data. Nicoll et al. (1993) and Nicoll (1992, 1993) have introduced some preliminary revisions of the biostratigraphic zonation of the Canning Basin and these are summarised in Column 8.

COLUMN 7: VICTORIAN AND TASMANIAN CONODONT ZONES

Conodont faunas from the continental margins of eastern Australia are preserved in black shales and cherts, and are not easily extracted for study. However, Stewart (1988) was able to put together a graptolite-controlled conodont range chart, using data from thick section examination of cherts and bedding plane studies for conodonts within the Victorian graptolite sequence. Using these ranges, a series of informal zones is presented in this column, with established zonal names used where applicable.

The *Cordylodus* Zone is based on the occurrence of several species of *Cordylodus* in the Digger Island Formation, the Howqua Shale and the Romsey Group. The *Paracordylodus–Paroistodus* Zone is marked by the appearance of *Paracordylodus gracilis* and *Paroistodus proteus* in with La2 graptolites. The appearance of *Prioniodus elegans* is differentiated as the *Prioniodus elegans* Zone and the first occurrence of *Oepikodus evae* marks the base of the *Oepikodus evae* Zone. A distinctive conodont fauna has not yet been identified in the Ca1–Da1 interval. The *Histiodella–Periodon* Zone is named for the co-occurrence of these genera associated with Da2–Da3 graptolites. The uppermost interval is the *Pygodus serra* Zone found associated with Da4 graptolites.

Tasmanian conodont occurrences are shown at the top of the column.

COLUMN 8: CONODONT ZONATION—AUSTRALIAN CRATONIC BASINS

The Cambrian–Ordovician boundary interval exposed in the Burke River Structural Belt on the eastern margin of the Georgina Basin has been studied in a series of papers by Nicoll (1990, 1991, 1992), Nicoll & Shergold (1991) and Shergold & Nicoll (1992). These revise the biostratigraphy of the Datsonian–Warendan sequence that contains the Cambrian–Ordovician boundary. In addition the Canning Basin subsurface conodont zonation has been re-examined by Nicoll (1992, 1993) and Nicoll et al. (1993) to update the earlier McTavish & Legg (1976) study. Based on this research, Column 8 gives a series of twelve conodont zones extending from the base of the Warendan to the upper part of the Darriwilian (late Tremadoc–Llanvirn).

COLUMNS 9–10: NORTH AMERICAN AND BALTOSCANDIAN CONODONT ZONES

Bergström (1973) and Sweet & Bergström (1974) recognised two major conodont biofacies for the Ordovician. The North American Midcontinent Province (NAMCP) fauna (Column 8) is found predominantly in shallow and warmer waters of the cratonic basins and shelf margins, especially those located in equatorial areas. The North Atlantic Province (NATP) fauna (Column 9) is found in sediments associated with the deeper or cooler waters of the shelf margins and higher latitudes. More recently Sweet & Bergström (1984) recognised that in terms of the world-wide distribution the province names were not entirely appropriate and have used the informal terms 'warm-water and cold-water pelagic faunas'. Miller (1984) used the terms 'warm and cold faunal realms' (WFR and CFR) and these terms are adopted here.

In Australia, conodont faunas from the cratonic basins (Canning, Bonaparte, Amadeus and Georgina Basins) are all indicative of the WFR, as are most of the conodont faunas found in Tasmania, western New South Wales and Queensland. Conodont faunas from Victoria and eastern New South Wales are more often associated with the CFR. Following Sweet & Bergström (1974) the genera *Oulodus, Plectodina, Phragmodus* and *Belodina* are indicative of the WFR and the genera *Eoplacognathus, Pygodus, Prioniodus* and *Periodon* are indicative of the CFR. The presence of these genera has been used to assign Australian faunas to one of the two faunal realms.

Bergström (1971) indicated that some Australian conodont faunas did not closely correspond to either the NAMCP (WFR) or NATP (CFR), and might represent a distinctive Australian province. There is no strong evidence yet that such a province does exist. However, there may be some indications that an Australasian or Gondwanan province may have existed in the Early Ordovician. For example, the genus *Serratognathus* is now found in China, Korea and Australia (Canning and Arafura Basins). Many other forms are found in common at the species level in the three countries. Until detailed studies of more of the Australian faunas have been published the extent and nature of such a province would be difficult to define.

Preliminary analysis of conodont faunas in the Amadeus, Canning and Georgina Basins does indicate that there is a gradation from shallow-water near-shore faunas to deeper-water offshore faunas. In the Pacoota Sandstone of the Amadeus Basin the clastic sequence is dominated by coniform elements, especially hyaline species like *Drepanoistodus*, but in the laterally equivalent deeper-water carbonate-shale lithofacies of the Emanuel Formation, of the Canning Basin, the fauna is dominated by ramiform species of genera like *Prioniodus*. Species of *Bergstroemognathus* are found in both faunas.

COLUMN 11: UTAH ZONATION

Biostratigraphic zonation schemes for trilobites and brachiopods from northern Utah (Ross 1951) and west-central Utah (Hintze 1951, 1953) established a series of faunal zones for the early to mid-Ordovician that served as a reference point for both trilobite and brachiopod biostratigraphy in North America. A parallel conodont zonation was later introduced by Ethington & Clark (1971). Subsequently the zones have been refined (Hintze et al. 1972; Hintze 1979; Ethington & Clark 1981; Ross et al. 1993) and more formal zones have been defined using faunal names. However, the concepts introduced by the original alphabetically designated zones are still useful in understanding the evolution of early Ordovician biostratigraphic schemes.

COLUMN 12: TRILOBITES
J.R. Laurie and J.H. Shergold

Trilobites are relatively common throughout most of the Ordovician sequences of Australia, with the notable exception of the graptolitic facies in Victoria and south-eastern New South Wales.

Trilobites were first described from the Australian Ordovician in Tasmania (Etheridge 1883, 1904*b*) and central Australia (Etheridge 1893; Tate 1896). It was much later that further work was done on isolated faunules from Tasmania (Kobayashi 1940*a,b*). Three more decades elapsed before any more work was published on Ordovician trilobites from New South Wales (Campbell & Durham 1970; Webby et al. 1970; Webby 1971, 1973, 1974; Webby et al. 1988), Queensland and the Northern Territory (Gilbert-Tomlinson in Hill et al. 1969; Shergold 1975; Henderson 1983; Fortey & Shergold 1984; Shergold 1993; Laurie, in prep.), Western Australia (Legg 1976, 1978), Tasmania (Stait & Laurie 1980; Burrett et al. 1983*a,b*; Jell & Stait 1985*a,b*) and Victoria (Jell 1985).

No comprehensive biostratigraphic scheme for Ordovician trilobite faunas has been erected in this country; however, local assemblage successions and zonations have been used by several workers. Webby (1974) recognised four stratigraphically distinct Upper Ordovician trilobite faunules from central New South Wales. Legg (1978) erected a sequence of combined trilobite–graptolite faunas for part of the Lower Ordovician of the Canning Basin; the trilobite faunas are being revised by Laurie & Shergold (in press). Stait & Laurie (1980) used trilobites (later described by Jell & Stait 1985*a*) and brachiopods to define a sequence of seven assemblages for the Lower Ordovician Florentine Valley Formation in south-central Tasmania. Their scheme was later incorporated in the preliminary biostratigraphic subdivision of the Tasmanian Ordovician by Banks & Burrett (1980). More recently, successions of trilobite assemblages have been recognised in the Lower Ordovician sequences of the Amadeus (Shergold 1991*a*; Laurie, in prep.), Canning (Laurie & Shergold, in press) and Georgina Basins (Fortey & Shergold 1984).

COLUMN 13: PALYNOMORPHS
G. Playford

Published occurrences of Australian Ordovician palynofloras, specificially of acritarchs and chitinozoans, are restricted to unweathered and essentially unmetamorphosed marine sedimentary rocks obtained through exploratory drilling in the cratonic Canning and Georgina Basins. Consequently, the biostratigraphic potential of palynomorphs in the Australian Ordovician has yet to be adequately explored, and may well remain so due to the scarcity of suitably palyniferous strata.

Moderately diverse and fairly well preserved acritarch assemblages were described by Combaz & Peniguel (1972) and Playford & Martin (1984) from portions of the Canning Basin's Lower Ordovician sequence, especially the Goldwyer and Nita Formations (probably late Arenig–Llanvirn). In the first-mentioned work, chitinozoans were also reported. Playford & Wicander (1988) detailed a profuse acritarch flora from the early–mid Arenig Coolibah Formation of the Georgina Basin. As noted by Playford and his co-authors, these Early Ordovician acritarch assemblages appear to be largely endemic. Chitinozoans described by Playford & Miller (1988) from the Coolibah and basal Nora Formations in the Georgina Basin bear some similarities, in terms of their general morphological simplicity, with forms recorded from Arenig deposits in Spitsbergen and Quebec.

COLUMN 14: CORALS, STROMATOPOROIDS AND SPONGES

B.D. Webby

Ordovician corals and stromatoporoids are mainly distributed through the shallow-water carbonate successions of the Tasmanian Shelf (at the margin of Gondwana) and offshore settings of the Molong High and Parkes Platform (volcanic 'island-arc' remnants) of central New South Wales. There are also more localised occurrences in the Tamworth Terrane of north-eastern New South Wales, and from the flanks of the Anakie High and the Broken River Embayment of central and northern Queensland (Webby 1985, 1987; Webby et al. 1981; Strusz et al. 1988).

Little was known about Australian Ordovician sponges until recently. Most of the early discoveries were of discrete siliceous spicules. For example, there were species of *Protospongia* and other doubtfully assigned forms from Lower Ordovician 'deeper water' sequences of Victoria (Hall 1888, 1889) and *Hyalostelia australis* (Etheridge 1916) from the Stairway Sandstone of the Amadeus Basin, central Australia (Pickett 1983; Shergold 1986). However, the first calcareous sponges were recognised more recently as occurring in the Upper Ordovician 'shallow-water' successions of the Molong High and Parkes Platform, and these comprise the sphinctozoans *Cliefdenella* (Webby 1969), first described as a stromatoporoid, and the genera *Angullongia* and *Belubulaia* (Webby & Rigby 1985). More remarkable are the beautifully preserved siliceous sponges described by Rigby & Webby (1988) from 'deeper water', allochthonous limestone blocks in the basinal, graptolitic Malongulli Formation of central New South Wales. They include most varied associations of demosponges (thirty-five species), hexactinellids (eight species), and silicified? sphinctozoans (two species). Apart from the sphinctozoans, they are thought to have lived in a deeper slope habitat on the flanks of the offshore island shelf (Molong High).

Only the corals and stromatoporoids have been employed in biostratigraphical work, mainly for local use in correlating limestones of the Parkes Platform and Molong High in central New South Wales (Webby 1969, 1975; Webby et al. 1981; Strusz et al. 1988). The four biostratigraphically distinct faunas (I–IV) range through the Upper Ordovician (upper Gisbornian–middle Bolindian). Only occurrences of *Stromatocerium* and *Foerstephyllum* from the lower part of the Billabong Creek Limestone of the Parkes Platform (Pickett 1985) apparently came from an earlier (?middle Gisbornian) horizon. Pickett has reported the North Atlantic conodont index *Pygodus anserinus* from the underlying beds (i.e. from beds of early Gisbornian *Nemagraptus gracilis* Zone age).

The Fauna I assemblage is of late Gisbornian–early Eastonian age, and is typified by abundant labechiid stromatoporoids (notably *Cystistroma*, *Labechiella* and *Stratodictyon*), a varied tabulate component including many species of *Tetradium* and appearance of the rugosan *Hillophyllum*. It is best represented in the Fossil Hill Limestone of the Cliefden Caves area, eastern side of the Molong High.

Fauna II is of middle Eastonian age and is characterised by the first appearances of clathrodictyid stromatoporoids (*Clathrodictyon* and *Ecclimadictyon*), the sphinctozoan *Cliefdenella*, and rugosan *Palaeophyllum* and a most varied and abundant assemblage of heliolitine corals. It is well developed in the upper part of the Belubula Limestone and in the Vandon Limestone at Cliefden Caves, and in the Quondong Limestone at Bowan Park (to the eastern and western sides of the Molong High, respectively).

Fauna III may be subdivided into IIIa and IIIb depending on certain specific faunal elements (Webby et al. 1981, p.9), but may alternatively be viewed as an undifferentiated assemblage of broadly late Eastonian–earliest Bolindian age. It is typified by first appearances of streptelasmatinid rugosans, *Favistina*, halysitines and favositids. It is best represented in the upper parts of carbonate sequences of the western

(Bowan Park) side of the Molong High and also occurs in shallow-water limestone blocks of the breccia deposits in the graptolitic Malongulli Formation of the eastern (Cliefden Caves) side of the Molong High.

The latest assemblage, Fauna IV, is of about middle Bolindian age, and has a restricted distribution at the top of the Malachi's Hill Beds in the Bowan Park area. It is characterised by its rich rugosan element (*Bowanophyllum, Rhabdelasma* and *Cyathophylloides*) and tabulates (abundant favositids, the first agetolitids, *Catenipora* and *Adaverina*).

In Tasmania, elements of a pre-*Tasmanognathus* (?late Darriwilian or North American Chazyan) age assemblage of stromatoporoids (*Labechia* aff. *prima, Stratodictyon vetus* and *Stromatocerium bigsbyi*) occur in the Cashions Creek Limestone of the Florentine Valley and the Standard Hill Formation of the Mole Creek area (Webby 1979; Banks & Burrett 1980; Burrett & Goede 1987), a part of the Tasmanian OT10 assemblage of Banks & Burrett (1980), and immediately prior to the first appearances of corals. However, the precise stratigraphic relationship between the Florentine Valley and Mole Creek sections is unclear because Burrett & Goede (1987, Table I) have shown the Standard Hill Formation at Mole Creek correlating with the Karmberg Limestone of the Florentine Valley, that is, to a level beneath the Cashions Creek Limestone. Pickett's (1985) association of *Stromatocerium* and *Foerstephyllum* from the Billabong Creek Limestone of central New South Wales is probably younger than the Tasmanian OT10 assemblage because it includes the coral *Foerstephyllum*, which first appears with conodont *Tasmanognathus* and other corals like *Lichenaria* and *Tetradium* (numerous species) in the Lower Limestone Member of the Benjamin Limestone, that is, through the interval of assemblages OT12–14 (Corbett & Banks 1973; Burrett 1979; Banks & Burrett 1980). The *Thamnobeatricea–Tetradium–Lichenaria–Foerstephyllum* assemblages of the Lower Limestone Member of the Benjamin Limestone and equivalents in Tasmania broadly correlate with the Fauna I associations of central New South Wales.

The lower part of the Upper Limestone Member of the Benjamin Limestone exhibits an association of *Palaeophyllum, Bajgolia, Eofletcheria* and various heliolitines that characterise the OT16 assemblages, and this equates with the New South Wales Fauna II association. Towards the top of the Upper Limestone Member, a rich and varied assemblage (OT19) occurs. It includes a large *Aulacera* and the first clathrodictyids, *Favistina*, favositids and halysitines. This correlates with the central New South Wales Fauna III assemblages of late Eastonian–earliest Bolindian age.

Taxonomic descriptions of the Tasmanian coral and stromatoporoid faunas are so far limited to the contributions of Hill (1942, 1955), Webby & Banks (1976), and Webby (1979).

The isolated Tamworth Terrane remnants in north-eastern New South Wales have similar Fauna III type coral assemblages (Hall 1975). They include a varied fauna comprising *Favistina, Palaeophyllum, Cyathophylloides, Crenulites, Calopoecia*, halysitines, heliolitines and favositids. The Fork Lagoons Beds of the Emerald area, on the south-eastern flanks of the Anakie High in central Queensland (Anderson & Palmieri 1977), have an undescribed coral fauna that may be of similar age.

Occurrences of *Agetolites, Catenipora* and *Plasmoporella* have been recorded from the Carriers Well Limestone of the Broken River Embayment in northern Queensland (Hill et al. 1969), and may, on the basis of the presence of agetolitids, represent a younger? middle Bolindian (Fauna IV) age.

COLUMN 15: TASMANIAN OT ZONES

B.A. Stait

Banks and Burrett (1980) subdivided the Tasmanian Ordovician sequences into twenty zones, labelled OT1 to OT20. This zonation was an extension of the nine informal zones established by Stait & Laurie (1980) for the Lower Ordovician clastics of the Florentine Valley Formation. Although the OT zones are established using co-occurring faunas they are not rigorously defined using the total ranges of the constituent species. Therefore, they are of only very limited use in describing the Tasmanian biostratigraphy in general and can have no formal biostratigraphic significance. Once the ranges of the faunas of Tasmania are well established and documented there will be a good case for the establishment of a zonation like the OT zones, which may even prove useful standard for the shallow-water shelly facies of Australia.

COLUMN 16: BRACHIOPODS

J.R. Laurie

Brachiopods are widespread in the Ordovician shallow-water sequences of central and northern Tasmania and in the deeper-water successions of southernmost Tasmania and central New South Wales. They also form a ubiquitous but only occasionally dominant portion of the faunas of the central and northern Australian basinal sequences. Apart from some shell debris contained in turbidites, and occasional phosphatic brachiopods in siliceous siltstones, they are absent from the thick turbidite and black shale sequences of Victoria and south-eastern New South Wales.

First noted and described mostly in papers dealing predominantly with trilobite faunas (e.g. Etheridge 1893, 1904*b*; Kobayashi 1940*b*), it is only more recently that they have been the major topic of publications in their own right (Brown 1948; Percival 1978, 1979*a*,*b*, 1991, 1992; Laurie 1980, 1987*a*,*b*, 1991*a*,*b*).

Brachiopods have been used to erect biostratigraphic schemes only in Tasmania, by Stait & Laurie (1980), who used brachiopods in conjunction with trilobites to characterise a zonation of the Early Ordovician of the Florentine Valley, southern Tasmania. This was followed by a more extensive study (Laurie 1991*a*,*b*) in which a brachiopod biostratigraphic scheme was proposed for the entire Ordovician shallow-water predominantly carbonate platform sequence of Tasmania. Percival (1978, 1979*a*,*b*, 1991, 1992) has related the brachiopod faunas from the central New South Wales sequence to the coral–stromatoporoid biostratigraphic sequence of Webby (1969, 1972, 1975).

COLUMN 17: NAUTILOID FAUNAS

B.A. Stait

Although nautiloids are known from most of the Ordovician platform and slope sequences in Australia, definition of a formal biostratigraphic zonation is currently impossible. The main problem is the rarity of specimens, which does not allow ranges for individual species, and in most cases genera, to be established. This lack of detailed information means that the nautiloids do not provide the contribution to Ordovician biostratigraphy that their rapid evolution in the Lower Ordovician would otherwise allow.

Ordovician nautiloid faunas have been described from Tasmania, central New South Wales, western New South Wales, and the Georgina Basin, Amadeus Basin and Canning Basin. Extremely rare specimens have also been described from the deep-water sequences of Victoria. Unfortunately the collecting localities are often not accurately known and cannot be reliably slotted into subsequent biostratigraphic frameworks.

The Ninmaroo Formation, Georgina Basin, contains an Upper Cambrian fauna very similar to that of northern China. This fauna contains the typical genera *Protactinoceras*, *Acaroceras* and *Plectronoceras*. Higher in the Ninmaroo Formation the oldest Ordovician nautiloids yet recorded from Australia occur, including *Ellesmeroceras*, *Muriceras*, *Quebecoceras*, *Ectenolites* and *Clarkoceras*. The Canning Basin faunas begin in the Middle Ibexian, Emanuel Formation, with *Kyminoceras*, *Eothinoceras* and *Antho-*

ceras. The Upper Ibexian faunas are possibly the most widespread in the Australian Ordovician. In Tasmania this is the oldest assemblage containing *Manchuroceras, Piloceras, Yehlioceras, Alloctoceras* and *Pycnoceras*. A fauna of similar age in the Georgina Basin not only contains endocerids such as *Manchuroceras*, but also has the oldest known actinocerid faunas, including *Actinoceras, Georgina* and *Armenoceras*. Equivalent faunas of the Canning Basin include *Hardmanoceras, Aethinoceras, Thylacoceras* and *Notocycloceras*. The oldest identifiable nautiloids from the Amadeus Basin are Upper Ibexian to Whiterockian faunas of the Horn Valley Formation. Some genera, such as *Ventrolobendoceras* and *Anthoceras*, also occur in the Canning Basin faunas, while others, such as *Madiganella* and *Bathmoceras*, are not known elsewhere in Australia. *Anthoceras* is widespread in the Upper Ibexian, with *A. decorum* occurring in the Canning Basin and Amadeus Basin, and *A. arrowsmithense* known from western New South Wales and possibly the Georgina Basin.

The next, Whiterockian, fauna represents the great expansion of the actinocerids in Australia. In Tasmania it is characterised by the *Wutinoceras–Adamscoceras* assemblage of the Karmberg Limestone and equivalents. Coeval faunas of the Georgina Basin contain *Georgina, Armenoceras, Wutinoceras* and *Williamoceras*. In the Amadeus Basin the Stairway Sandstone contains *Armenoceras* and *Georgina*. Faunas of this age have not been described from the Canning Basin although rocks of appropriate age and lithology exist. Further study should locate cephalopods in these sequences. Above this level no nautiloid faunas are known from the northern Australian basins (Georgina, Canning and Amadeus Basins).

Throughout the remainder of the Ordovician, the only faunas known are those from the platformal sequences of Tasmania and the island-arc of central New South Wales. Although these faunas are contemporaneous they contain only one genus in common—the nearly cosmopolitan *Gorbyoceras*. In Tasmania nautiloid sequences above the Whiterockian are dominated by the Gouldoceratidae and have been divided into four assemblages. These assemblages are not formal biostratigraphic units and are only intended as convenient groupings of broadly coexisting faunas useful when describing the Tasmanian sequences. These assemblages, in ascending order, are: *Discoceras–Gorbyoceras* assemblage; *Tasmanoceras–Hectoceras–Gouldoceras* assemblage; *Gordonoceras* assemblage; and *Westfieldoceras* assemblage.

These assemblages are difficult to correlate outside Tasmania due to the preponderance of endemic, as well as long ranging and widespread, genera. However, reasonable correlations can be made using the coexisting conodont faunas.

The central New South Wales faunas contain endemic and cosmopolitan genera, possibly reflecting the small area and isolation of these island-arc carbonate platforms. The value of these faunas for biostratigraphy is extremely limited.

COLUMN 18: ROSTROCONCHS AND BIVALVES

J.H. Shergold

It is now known that the 'pecular bivalve' illustrated by Etheridge in 1883 (p.158, Pl. 2, Figs 15*a,b*) from the Caroline Creek Sandstone (north-western Tasmania), now referred to *Tolmachovia corbetti* by Pojeta & Gilbert-Tomlinson (1977, p.32), was the first rostroconch mollusc to be described in Australia. A second, described as *Conocardium* sp. indet. by Tate (1896, p.110, Pl. 2, Fig. 13) from what is probably the Stokes Siltstone of the Amadeus Basin, is a rostroconch belonging to the Superfamily Conocardiacea (Pojeta & Gilbert-Tomlinson 1977). Eighty years later, Pojeta & Runnegar (1976, p.62, Pl. 14, Figs. 9–19) described *Tolmachovia? jelli* from the top of the Ninmaroo Formation in the Georgina Basin. This is now refered to the genus *Pauropegma* by Pojeta et al. (1977) (see chart), who monographed the Cambrian and Ordovician rostroconchs of the Amadeus and Georgina Basins. Occurrences in the Ninmaroo Formation of the Georgina Basin were correlated to the Early Ordovician conodont biostratigraphy by Druce et al. (1982, range chart, Fig. 4). Occurrences in the Amadeus Basin are less readily assessed in the absence of conodont control. They are plotted against the trilobite stratigraphy established by Shergold (1993) and Laurie (in prep.).

Ordovician pelecypod molluscs were first described from Tasmania by Johnston (1888), and subsequently from the Amadeus Basin by Etheridge (1894) and Tate (1896). Hill et al. (1969) illustrated pelecypods, mostly under open nomenclature, from the Toko Range, Georgina Basin. Subsequently, all known Australian Ordovician pelecypods were systematically described for the first time by Pojeta & Gilbert-Tomlinson (1977). The ranges established by these authors lack a detailed chronological control and are listed essentially by formations. Mostly, they are from the Pacoota Sandstone, Stairway Sandstone and Stokes Siltstone of the Amadeus Basin and correlative Nora Formation and Carlo Sandstone of the Georgina Basin. Other material, described from the Gordon Limestone of Tasmania by Pojeta & Gilbert-Tomlinson (1977), is not included on the chart because of the difficulty in locating it biostratigraphically.

COLUMN 19: SEA-LEVEL FLUCTUATION AND EUSTATIC EVENTS

Recent examination of sea-level fluctuation in Ordovician cratonic basins (Gorter 1992; Nicoll et al. 1992; Nielsen 1992) reveals a preliminary curve reflecting the transgression and regression of marine seas onto the Australian continental margins and into interior basins. Increased sophistication of Ordovician biostratigraphy, both in Australia and internationally, has made possible the recognition of patterns of sedimentation and sedimentary breaks that now can be placed in context across the continent. Where sea-level fluctuations have been identified in a number of basins within Australia, or intercontinentally, these fluctuations are regarded as broadly eustatic events. Events recorded only in a local area are thought to result from regional tectonic processes.

Column 19 traces the fluctuation of relative sea-level and reflects aspects of an inundation profile and water depth in interior cratonic basins, such as the Amadeus Basin. Significant eustatic events are noted. These include ARE (Acerocare Regressive Event), BMEE (Black Mountain Eustatic Event), KCEE (Kelly Creek Eustatic Event), Bill (Billingen Transgressive Event) (see Nicoll et al. 1992), ECEE (Ellery Creek Eustatic Event), HVEE (Horn Valley Eustatic Event), MCEE (Maloney Creek Eustatic Event), D/GEE (Darrwilian–Gisbornian Eustatic Event), E/BEE (Eastonian–Bolindian Eustatic Event) (see Gorter 1992).

2.3

Silurian (Chart 3)

D.L. Strusz

INTRODUCTION

General summaries for the Silurian in Australia have been given by Talent et al. (1975), Pickett (1982), Jell & Talent (1989) and Strusz (1989). The third of these forms a chapter in the ISSS compilation by Holland & Bassett (1989). The fourth (Strusz 1989), of which the present chapter is an abridged and updated version, should be consulted for additional data and more detailed discussion than appears here. A Silurian Palaeogeographic Atlas for Australia was released in 1990 (Walley et al. 1990).

COLUMN 1: GEOCHRONOLOGICAL SCALE

The geochronological scale with which the standard subdivisions of the Silurian Period are calibrated can in no way be regarded as precise or settled. This work takes the date of the Ordovician–Silurian boundary as 434 Ma, and the date of the Silurian–Devonian boundary as 410 Ma.

COLUMNS 2–4: SERIES AND STAGES

The series and stages of the Silurian have been defined, after considerable international discussion, by the Subcommission on Silurian Stratigraphy. The definitions have been summarised by Holland (1985), and their history discussed by Holland (in Holland & Bassett 1989). These subdivisions comprise a global standard, and are used in Australia as elsewhere. The period is now divided into two epochs, the older Early Silurian now including the Wenlockian, which used to comprise the Middle Silurian Epoch in the previous tripartite scheme.

Llandovery Series

The base of the Silurian System, taken at the base of the Llandovery Series, coincides with the base of the Rhuddanian Stage. It is defined as the base of the *Parakidograptus acuminatus* Biozone in the Birkhill Shale at Dobb's Linn, southern Scotland. This is clearly above the widespread *Hirnantia* fauna (which was not the case with the previously used base of the *Glyptograptus persculptus* Biozone).

The initial subdivision of the Llandovery Series by Cocks et al. (1970) was into four stages, but only three are now recognised, of which the Rhuddanian is the first. The second is the Aeronian, defined as beginning at the base of the *Monograptus triangulatus* Biozone in the Trefawr Formation of the type Llandovery area of southern Wales. The last stage is the Telychian, which is defined as beginning at the base of the *Monograptus turriculatus* Biozone in the Wormwood Formation, Llandovery area.

Wenlock Series

The Wenlock Series was revised and its subdivisions defined by Bassett et al. (1975). The base of the series, coincident with the base of the Sheinwoodian Stage, is defined as the base of the Buildwas Shale in Hughley Brook, near Much Wenlock, Shropshire, England. This is correlated with the base of the *Cyrtograptus centrifugus* Biozone. The upper stage is the Homerian, whose base is defined as the base of the *Cyrtograptus lundgreni* Biozone within the Apedale Member of the Coalbrookdale Formation in Whitwell Coppice, near Much Wenlock. Bassett's chronozones within the Homerian Stage have not been recognised in Australia.

Ludlow Series

For the third series of the Silurian System, the Ludlow, international agreement had been reached by 1981 (Martinsson et al. 1981). The base coincides with the base of the Gorstian Stage, defined as the base of the Lower Elton Formation in a quarry in Pitch Coppice near Ludlow, Shropshire. It is correlated with the base of the *Neodiversograptus nilssoni* Biozone. The base of the Ludfordian Stage is defined as the base of the Lower Leintwardine Formation in Sunnyhill Quarry near Ludlow, and is correlated with the base of the *Saetograptus leintwardinensis* Biozone.

Přídolí Series

The fourth series is the most recently recognised, after realisation in 1958 of a long standing miscorrelation between the largely non-marine strata above the Ludlow Bone-bed of the United Kingdom and the marine rocks of central Europe. It is the Přídolí Series, the base defined as the base of the Požáry (formerly Přídolí) Formation and the base of the *Monograptus parultimus* Biozone in the Požáry Quarry near Řeporyje, Prague Basin (Holland 1985). There are no stages for the Přídolí Series and little prospect that any will be recognised, at least in the Prague Basin.

The top of the Silurian is defined by the base of the Devonian System and coincides with the base of the Lochkovian Stage, placed at the base of the zone of *Monograptus uniformis* at Klonk, near Beroun, in the Prague Basin. At that locality (and seemingly elsewhere in the region) the conodont zone of *Icriodus woschmidti* starts very slightly earlier, but for normal purposes of correlation it may also be taken to define the base of the Devonian.

COLUMNS 5–6: GRAPTOLITE ZONES

As noted by Sherwin (in Pickett 1982), most of the graptolite zones that have been defined and widely used in Europe have been recognised also in Australia (see Pickett 1982; Garratt 1983*a,b*; VandenBerg et al. 1984; Jenkins et al. 1986; and Jell & Talent 1989). The zonation shown is therefore essentially the standard European one, derived from Rickards (1976), Rickards et al. (1977), Jaeger (1977) and Holland (1985), modified after Pickett (1982), Garratt (1983*a,b*), and VandenBerg et al. (1984). In the 1989 chart (see Column 6), the relative duration of the Llandoverian zones was taken from Cocks et al. (1984), that of the Wenlockian zones from Bassett et al. (1975) and McKerrow et al. (1985), and of the Ludlovian zones from Pickett (1982). Because of lack of information, the Přídolían zones were shown as of equal duration. Subsequently a correlation chart of 'standard' graptolite and conodont biozones has been published by Cocks & Nowlan (1993), and the relative durations shown there have been taken into account in the drafting of the chart herein.

In Australia there are few zones in sequence, so it has often not been possible to confirm the relative ranges of individual taxa, or determine whether the zonal boundaries are coeval with those in Europe. Because of this uncertainty it is probably better to use the more broadly defined 'Standard' Graptolite Biozones suggested by members of the Subcommission on Silurian Stratigraphy (Koren & Cocks in Cocks & Nowlan 1993), adapted from those suggested by Koren (Koren & Karpinsky 1984; Koren & Modzalevskaya 1991), until a reliable regional zonation can be constructed. The 'standard' biozones are shown in Column 5.

Up to the Přídolí Series, the zones (but not subzones) defined in Rickards (1976) have been used in Australia, with the following exceptions:

- The *Coronograptus cyphus* Zone s.l. has been divided into a lower *Lagarograptus acinaces* Zone and a higher *C. cyphus* Zone s.s., as in Rickards et al. (1977) and Pickett (1982). This does not correspond to the 'standard' *C. cyphus* Biozone, which excludes the *L. acinaces* Zone.
- The zone of *Coronograptus gregarius*, comprising Rickards' zones of *Monograptus triangulatus, Diplograptus magnus* and *Monograptus argenteus*, cannot be subdivided in most Australian sections. It is retained as a 'standard' biozone by Koren & Cocks (in Cocks & Nowlan 1993).
- Following Garratt (1983*a*), a single zone of *Monograptus exiguus* is shown, corresponding to the less often distinguished zones of *M. crispus* and *Monoclimacis griestoniensis.* This overlaps the 'standard' biozones of *turriculatus-crispus* and *griestoniensis.*
- The zone of *Cyrtograptus lundgreni* is the same as Garratt's zone of *Testograptus testis, Cyrtograptus* rarely being found in Australia.
- The *Neodiversograptus nilssoni* Zone s.l. is subdivided into the zones of *N. nilssoni* s.s. and *Lobograptus progenitor.* The situation is discussed by Rickards (1976).
- The *Bohemograptus* Proliferation Zone of Rickards et al. (1977, Fig. 31), Teller (1969) and Holland & Palmer (1974) corresponds to the 'standard' biozone of *Bohemograptus bohemicus–B. kozlowskii.*

In Europe *Bohemograptus bohemicus* s.l. appears in the *N. nilssoni* Zone, and Rickards et al. (1977) give a range for the genus of lower *nilssoni* Zone to *fecundus* Zone (=*Bohemograptus* Zone), that is, essentially the whole of the Ludlow. As suggested by Garratt & Wright (1989), careful study of the Australian bohemograptids may be rewarding.

For the Přídolí Series, the Australian zonation is as in Rickards et al. (1977), taking into account the specific synonymies put forward by Jaeger (1977) and the information in Holland (1985). Several of these zones have been combined by Koren & Cocks (in Cocks & Nowlan 1993).

COLUMNS 7–8: CONODONT ZONES

The only international series of conodont zones for the Silurian is the set of 'Standard' Conodont Biozones recently proposed by Nowlan for the Subcommission on Silurian Stratigraphy (in Cocks & Nowlan 1993). Until the publication of those zones, there have been different systems for Europe and North America, and some evidence for facies-dependence, provinciality and disparate ranges. In this chapter, the 'Standard' Conodont Biozones of Nowlan are shown in Column 7, and a synthetic Australian zonation in Column 8. For the Australian Llandovery and Sheinwoodian, the zonation proposed by Bischoff (1986) is used, although there remain some difficulties (see below). For the Homerian onwards, the zonation is essentially based on the recognition of zones established elsewhere (see Link & Druce 1972; Ziegler et al. 1974; Cooper 1980; Schönlaub 1980; Pickett 1982; Aldridge in Higgins & Austin 1985). Occurrences in Australia have been compiled from Talent et al. (1975), Pickett (1982), Jenkins et al. (1986), Bischoff (1986) and Jell & Talent (1989). The two conodont zonations are aligned indirectly, by correlation with the graptolite zones.

The zones defined by Bischoff (1986) are now discussed.

Zone of **Distomodus combinatus** *plus* **D. tridens**

This zone is defined by the appearance of *D. combinatus*, and characterised by the occurrence of that species with *D. tridens* and *Ozarko-*

dina australiensis. Other important associates are *Oulodus angullongensis* and *Oulodus* sp. A. This fauna comes from the first conodont-bearing limestone above an unconformity.

Zone of **Distomodus pseudopesavis** *plus* **Ozarkodina masurenensis** This zone is defined by the entry and co-occurrence of the eponymous species; also significant are *O. panuarensis* and *Pterospathodus cadiaensis*, and in the lower part, *Distomodus calcar*. The top of the zone is truncated by a recognisable hiatus.

Zone of **Aulacognathus antiquus** *plus* **Distomodus staurognathoides** *morphotype* α The base of the zone is defined by the entry of *D. staurognathoides* α, the top by the appearance of *Astropentagnathus irregularis*. *A. antiquus* occurs only in the lowest part of the zone. Important components of the zonal fauna are *Aulacognathus angulatus*, *A. bifurcatus*, *Ozarkodina waugoolaensis* and *Oulodus planus planus*. There is a gap between this and the previous zone, preventing precise correlation of the lower zonal boundary.

Zone of **Astropentagnathus irregularis** *plus* **Pterospathodus pennatus** This zone begins with the entry of *A. irregularis* and is defined by the association of that species with *P. pennatus*. Entering at the same time is *Aulacognathus kuehni* while *Aulacognathus bifurcatus* and *Ozarkodina waugoolaensis* continue from the previous zone. Common in the zone also are *Distomodus staurognathoides*, *Oulodus australis* and *Oulodus planus*.

Zone of **Pterospathodus celloni** This zone, originally recognised at Cellon in the Carnic Alps, was redefined by Bischoff following reassessment of conodont ranges in the Cellon section, and observations in New South Wales. There, it is defined as the interval between the entries of *P. celloni* and of *P. latus*. Also present are the conodonts *Aulacognathus kuehni*, *A. liscombensis*, *Ozarkodina excavata eosilurica* and *O. bathurstensis*. *P. celloni* and *Aulacognathus liscombensis* are restricted to the zone. There is no apparent continuity with the preceding zone, and no graptolite control over its earliest occurrences. As redefined, the lower boundary is assumed to correlate with that of the zone of *Monoclimacis griestoniensis*. In New South Wales the upper boundary is not in sequence with the succeeding zone, and cannot readily be related to the graptolite zonation, so its correlation is uncertain.

Zone of **Pterospathodus amorphognathoides** *plus* **P. latus** This zone replaces part or all of Walliser's *amorphognathoides* Zone, in view of the recovery of that species from the early *celloni* Zone at Cellon, and absence in New South Wales of the eponymous species of the following Cellon zone (*Kockelella patula*) in sediments overlying those with *P. amorphognathoides*. Bischoff (1986) defines the zone as being the interval characterised by the co-occurrence of *P. amorphognathoides*, *P. latus*, *Apsidognathus tuberculatus* and *Carniodus carnulus*. The lower boundary is defined by the entry of *P. latus*, the upper by the entry of *Oulodus sinuosus* and the disappearance of *Aulacognathus borenorensis*, *Oulodus planus borenorensis*, *Pyrsognathus obliquus* and *P. latus*. Important within the zone are *Johnognathus huddlei*, *Ozarkodina cadiaensis*, *Pseudopygodus scaber* and *Pterospathodus procerus*. There is no continuity with the preceding zone, and no graptolite control over the lower boundary, but as a result of conodont-based correlation between the Quarry Creek Limestone and the sequence in Boree Creek, Bischoff suggested a level above the *Monoclimacis griestoniensis* Zone, that is, towards the end of the Llandovery. Unfortunately this was not in accord with the available graptolite evidence at Quarry Creek, which placed the limestone below the *M. griestoniensis* Zone. The problem was discussed in detail by Strusz (1989, pp.5–6). Recent re-examination of the graptolites in Quarry Creek (Rickards et al. 1994) has removed the discrepancy: the fauna closest above the limestone, previously placed below the *M. griestoniensis* Zone, is now known to be within the mid-Sheinwoodian zone of *Monograptus riccartonensis*. The Quarry Creek Limestone is thus of early Sheinwoodian, not early Telychian, age.

Zone of ***Kockelella ranuliformis*** This zone was defined by Barrick & Klapper (1976) to be the interval between the entries of *Pseudooneotodus bicornis* and *Kockelella amsdeni*. Bischoff has redefined the base, for New South Wales at least, as being the appearance of *Ozarkodina sinuosus*, replacing *Apsidognathus tuberculatus tuberculatus, A. tuberculatus lobatus, Aulacognathus borenorensis, Carniodus carnulus, Oulodus planus borenorensis, Pyrsognathus latus* and *P. obliquus*. Both boundaries are in a continuous limestone sequence, and lack close graptolite control.

Zone of ***Kockelella amsdeni*** This zone is as defined by Barrick & Klapper (1976), the base being defined by the entry of *K. amsdeni*, the top by the entry of *K. stauros*. The latter is not known from New South Wales; so the upper limit of the zone is indefinite. Again, it cannot be closely correlated with the graptolite zonation.

The remaining zonation draws more heavily on Europe, as the few extended Australian limestone sequences in the later Silurian either have very sparse conodont faunas or have not been studied in detail.

Zone of ***Ozarkodina sagitta*** This has been reported only by Owen & Wyborn (1979), from the lower Cooleman Limestone.

Zone of ***Ozarkodina crassa*** Link & Druce (1972) recognised two informal assemblages low in the Yass sequence: *Neoprioniodus excavatus* and *Spathognathodus* (now *Kockelella*) cf. *ranuliformis*, which have been correlated with the *crassa* Zone (see Cooper 1980).

Zone of ***Ancoradella ploeckensis*** This is the same as the *Kockelella variabilis–A. ploeckensis* Assemblage-Zone of Link & Druce (1972).

Zone of ***Polygnathoides siluricus*** This is the *Belodella triangularis–P. siluricus* Assemblage-Zone of Link & Druce (1972). The upper limit is indefinite in the Yass sequence.

Zone of ***Ozarkodina snajdri*** See Chlupáč et al. (1980). Link & Druce (1972) only tentatively correlated the fauna of the Yarwood Siltstone with this level, because the diagnostic species are absent.

Zone of ***Ozarkodina remscheidensis eosteinhornensis*** *and subzone of* ***Ozarkodina crispa*** Away from the Cellon Pass section, *O. remscheidensis eosteinhornensis* enters at or near the same level as *O. crispa* (Chlupáč et al. 1980). Thus it has been proposed that Walliser's *eosteinhornensis* Zone be extended down to the base of his *crispa* Zone, the latter becoming a basal subzone. This was not adopted by Nowlan (in Cocks & Nowlan 1993).

Zone of ***Icriodus woschmidti woschmidti*** This was proposed by Walliser (1964) as a basal Devonian zone. The subspecies is not known at Klonk, but at Cellon and elsewhere in central Europe it occurs slightly below the entry of *Monograptus uniformis*, which defines the base of the Devonian. However, for international purposes, in the absence of closely interbedded graptolites and conodonts the zone may be taken as marking the Silurian–Devonian boundary.

Prior to the publication of Bischoff's work, Cooper (1980) proposed the recognition of a series of 'datum planes' based on key conodont occurrences, in an attempt to establish some degree of synchroneity between existing European and North American zonal schemes. The relationship of those datum planes to the current zonation in Australia is discussed by Strusz (1989).

COLUMN 9: CORAL ASSEMBLAGES

As noted by Pickett (1982), many coral faunas have been recorded, but few have been described. McLean (1974*a,b,c*, 1975*a,b*, 1977, 1985) has described the Llandovery rugosans and some younger faunas from New South Wales. A study of the halysitids by J. Byrnes remains unpublished, but was summarised by Pickett (1982). The earlier Silurian corals of northern Queensland have been studied by T. Munson, but the results are not yet available. Tasmanian data are sparse, while there are very few Silurian corals known from Victoria, with small faunas being listed, not described, by Talent et al. (1975) and VandenBerg (1976).

Vandyke & Byrnes (1976) formally recognised a 'Dripstone Fauna' ('Assemblage' herein), based on the faunas of the Dripstone Group south of Wellington, NSW. The other assemblages on the chart have been compiled from published descriptions and from Pickett (1982). For the halysitids the distribution data in Byrnes' study have been used, but not his taxonomic scheme. These assemblages are imprecise in content and extent and cannot be regarded as a formal zonation. There is insufficient biostratigraphic control and too little information on facies relationships of the few described faunas, for a formal zonation to be attempted. To emphasise the informal nature of the assemblages, they are named after the stratigraphic units in which they occur.

Bridge Creek Assemblage This is based on the rugosan and tabulate faunas of the Bridge Creek Limestone and underlying Wire Gully Limestone Members of the Bagdad Formation, southwest of Orange (Etheridge 1904*a*; McLean 1974*b,c*, 1977, 1985; Byrnes in Pickett 1982), these being of Rhuddanian to possibly early Aeronian age.

Quarry Creek Assemblage This assemblage is more widespread than the previous one; it is based on the corals of the Quarry Creek Limestone west of Orange, and of the Waugoola Group (Cobbler's Creek Limestone, and Burly Jack Sandstone Member, Glendalough Formation) farther south (McLean 1974*a,b*, 1975*b*, 1977). It is of Telychian to early Sheinwoodian age. There is no sequence that clearly shows the transition from the Bridge Creek Assemblage, or into the Dripstone Assemblage, as there are widespread hiatuses both before and after.

Dripstone Assemblage Originally proposed by Vandyke & Byrnes (1976), this is based on the faunas of the Homerian to early Gorstian Dripstone Group (Wylinga and Catombal Park Formations) near Wellington, NSW. Those faunas have been described and revised by Strusz (1961) and McLean (1975*a*).

Hatton's Corner Assemblage This is based on the corals of the Yass sequence, now placed mainly in Link & Druce's (1972) Hatton's Corner Group (for the confused nomenclatural history of this sequence see Owen & Wyborn 1979; the current equivalent is the Silverdale Formation). These were described by Hill (1940) and have been recently reviewed by McLean (1976). The Hatton's Corner Assemblage is of Ludlow and possibly Přídolí age; the upper limit is indefinite.

BRACHIOPODS

The situation regarding described Silurian brachiopod faunas is even less satisfactory than that for corals. The late Silurian faunas of the Yass sequence were largely but not completely described many decades ago (e.g. Mitchell 1923; Brown 1949) and have seen little subsequent revision. The Llandovery fauna of the '*Illaenus* Band' (basal Wapentake Formation) at Heathcote in Victoria was described by Öpik (1953), and part of the Wenlock–Ludlow fauna of Canberra has been recently described by Strusz (1982, 1983, 1984, 1985). Gill has described elements of the Victorian faunas in a number of papers; their stratigraphic import has been summarised by VandenBerg et al. (1976). Finally, McKellar (1969) described a small fauna from the Craigilee Anticline near Rockhampton in Queensland, of late Přídolí age. The only succession covering a reasonable timespan and so offering the possibility of consistent zonation is the lithologically monotonous and structurally complex sequence of the Melbourne Trough. Garratt (1983*a*) has recognised only two late Silurian zones, which have been further discussed by Garratt & Wright (1989) and strongly criticised by Jell & Talent (1989); they are not shown on the current chart. Strusz (1989) gives a full discussion of the succession of brachiopod species of the Australian Silurian.

TRILOBITES

Despite a number of recent taxonomic revisions (see Pickett 1982), little attempt has been made to bring together the information necessary for a trilobite biostratigraphy for either New South Wales or Victoria. This is largely because nearly all the described taxa are known from very few localities or horizons. A detailed discussion of known trilobite ranges is given by Strusz (1989), and summarised below.

In general, the faunas of the Yass–Canberra region numerically dominate the list of described species (Mitchell 1887, 1888, 1919; Etheridge & Mitchell 1890, 1892, 1894, 1896, 1916, 1917; Chatterton 1971; Chatterton & Campbell 1980; Strusz 1980). Further trilobites are recorded from other parts of New South Wales (Gill 1940; Sherwin 1968, 1971; Holloway & Campbell 1974; Fletcher 1975), Victoria (Gill 1945, 1949; Öpik 1953; Holloway & Neil 1982), Queensland (Lane & Thomas 1978), and Tasmania (Gill 1948; Holloway & Sandford 1993). It is noticeable that there are few Silurian species in common between the sequences of the Melbourne Trough, the Canberra–Yass Shelf, the Molong High, and Queensland. This is probably caused by facies differences but also by insufficient collection and description.

OTHER TAXONOMIC GROUPS

No other taxonomic groups are known in sufficient detail to be of significant use, although in terms of international correlation, some, such as tentaculitids, microvertebrates and ostracods, may provide reliable indicators of age for isolated samples.

Perhaps the most interesting occurrence is of land plants in Victoria — the *Baragwanathia* flora. There has been much debate on the age of the first occurrence of this flora, but it seems now to be accepted as Late Silurian (Garratt 1978; Garratt & Rickards 1984; Tims & Chambers 1984; Richardson & Edwards 1989).

Also in Victoria, distinctive occurrences of Echinodermata have been used for regional correlation in the past. As these probably represent an unusual environment, and hence a community group, this correlation is not reliable without independent evidence.

COLUMN 10: SEA-LEVEL CURVE

Until fairly recently, little has been published on relative sea-levels through the Silurian. However, it is now possible to generate a curve showing changes on a global scale. The one used here is based on that of Johnson et al. (1991). It shows a general trend of deepening during the earlier part of the Silurian, then shallowing through the later Silurian. Superimposed on this is a sequence of high and low stands, which the authors have recognised globally.

ACKNOWLEDGMENTS

A number of co-workers on the Silurian of Australia have provided useful information and criticism. I thank John Jell (Univ. Qld), Chris Jenkins (Univ. Syd.), Günther Bischoff (Macquarie Univ.), John Pickett, Lawrence Sherwin and John Byrnes (NSW Geol. Surv.), Tony Wright (Univ. Wollongong), Mike Garratt (formerly Vic. Geol. Surv.), Fons VandenBerg (Vic. Geol. Surv.), Dave Holloway (Vic. Mus.) and Barry Cooper (SA Geol. Surv.). We have not always agreed, and the end result is my responsibility, but their help was most welcome.

2.4

Devonian (Chart 4)

G.C. Young

INTRODUCTION

Fossiliferous limestones subsequently included in the Devonian were first recorded in Australia by Cunningham (1825), Mitchell (1838), and Leichhardt (1847), at about the time the Devonian System was erected by Sedgwick & Murchison (1839). The lepidodendroid plant *Leptophloeum australe* was first collected in the 1830s by Thomas Mitchell, and its association with the brachiopod *Cyrtospirifer* from near Bathurst, NSW, led David & Pittman (1893) to conclude that they came from Upper Devonian rather than Lower Carboniferous strata. The discovery of placoderm fish remains by Hills (1929, 1931, 1959) established a Late Devonian age for volcanic and sedimentary sequences of eastern Victoria, and extensive non-marine sequences of the Georgina and Amadeus Basins in central Australia. Fossiliferous limestones in the Kimberley region of WA were initially assigned Devonian and Carboniferous ages by Foord (1890), and the reef limestones of Late Devonian age in the Canning Basin are now well known (e.g. Playford & Lowry 1966; Playford 1980). Today, Devonian sedimentary strata are known from all of the Palaeozoic sedimentary basins of Australia (Palfreyman 1984), as both surface outcrop and in the subsurface in both onshore (e.g. Adavale Basin, Fitzroy Trough) and offshore areas (Perth, Carnarvon, offshore Canning and Bonaparte, and Arafura Basins). Early summaries of the Devonian of Australia were by Benson (1922) and Hill (1968). A set of charts and explanatory notes for Devonian correlation in the Australasian region was produced by Pickett (1972), Roberts et al. (1972), and Strusz (1972). Recent contributions include updates of the Devonian geology of Western Australia (Cockbain & Playford 1989) and Queensland (Jell 1989), and summaries of Early Devonian biostratigraphy and Devonian stage and zonal boundaries (Garratt & Wright 1989; Mawson et al. 1989). Overviews of the Devonian geology of Victoria and Tasmania are given in Douglas & Ferguson (1988) and Burrett & Martin (1989) respectively. A synthesis of the palaeogeography of the Devonian of Australia is presented by Vearncombe & Young (in prep.).

This Devonian chart is extensively revised and updated from the preliminary edition of Young (1989*a*) and summarises currently available reliable isotopic ages for points on the Devonian timescale (with emphasis on Australian data) to give a numerical calibration for various zonal schemes or preliminary biostratigraphic range data for groups of Devonian fossils useful in biocorrelation of Australian sedimentary sequences. An attempt is made to correlate with various 'international' zonal schemes, of which the conodont (marine) and palynomorph (non-marine) zonations form the overall framework of age control for the Devonian System. Establishing correlations from the non-marine to the standard marine zonations is a major research goal.

COLUMN 1: GEOCHRONOLOGY

The calibrated scale on the left is based on currently accepted numerical best estimates for the base (410 Ma) and top (354 Ma) of the Devonian, which differ only slightly from the previous chart, and correspond closely with the IUGS global stratigraphic chart (Cowie & Bassett 1989) of 410–355 Ma for the Devonian. However, the scheme of Harland et al. (1990) has a considerably shorter Devonian Period duration of 46 Ma, with significant differences in some boundary ages (about 9 Ma older for the Devonian–Carboniferous boundary, 9–10 Ma younger for the Lochkovian–Pragian and Pragian–Emsian boundaries, 7 Ma older for the Givetian–Frasnian boundary, etc.). Few isotopic dates for the Devonian are currently well constrained biochronologically, and they therefore lack the precision of existing biological age control (some fifty zonal subdivisions based on conodonts or ammonoids, an average zonal duration of about 1 Ma). For the Palaeozoic generally, the greatest precision currently is about 1% on isotopically determined ages (J. Claoué-Long, pers. comm.), which for the Devonian gives a margin of error encompassing several conodont or ammonoid zones. The timescale calibration therefore depends also on quantitative and qualitative assessments of the duration of Devonian stages (e.g. Boucot 1975; Ziegler 1978), and conodont zones (e.g. Sandberg & Poole 1977; Sandberg et al. 1983, 1989*b*; Fordham 1992). Research currently in progress includes a Frasnian composite standard for the Late Devonian conodont zonation using graphic correlation of twenty-eight sections in the Montagne Noire, western Canada, mid-continent and eastern USA, Canning Basin, Western Australia, and the Russian Platform (Klapper & Foster 1993).

COLUMN 2: DEVONIAN STAGES

Names are those adopted by the Subcommission on Devonian Stratigraphy (SDS) and ratified by the IUGS International Commission on Stratigraphy (Bassett 1985; Cowie et al. 1989; Oliver & Chlupáč 1991). As a result of these decisions, the name 'Tournaisian' is restricted to the Carboniferous. However 'zones' Tn1a and the lower part of Tn1b of the classic Belgian succession across the D–C boundary remain within the late Famennian, as does the Strunian or Etroeungtian 'stage' (see Jones 1985; Oliver & Chlupáč 1991).

Stage and series boundaries adopted by the SDS have been defined in terms of standard conodont zones (abbreviated CZ; see Column 3), as summarised by Ziegler & Klapper (1982, 1985). The base of the Givetian is the only Devonian boundary awaiting definition in terms of conodont zones (Oliver & Chlupáč 1991). Agreed definitions and Global Stratotype Section and Point (GSSP) for each boundary are as follows.

Base of Lochkovian Stage This coincides with the base of the Devonian System (Siluro–Devonian boundary) as defined by the first appearance of the graptolite *Monograptus uniformis* (McLaren 1973). This approximates to (is slightly above) the base of the *Icriodus woschmidti* CZ. The GSSP was defined in 1972 within bed 20 in the section at Klonk, Czech Republic (see Martinsson 1977).

Base of the Pragian Stage This is defined at the lower boundary of the *Eognathus sulcatus* CZ (first appearance of the conodont *E. sulcatus sulcatus*), with the GSSP at the base of bed 12, Homolka Hill quarry, Velka Chuchle, Prague, Czech Republic. This approximates to the base of the Siegenian of previous usage.

Base of the Emsian Stage This is defined at the lower boundary of the *Polygnathus dehiscens* CZ, with proposed GSSP in bed 5 of interval 9 in the Zinzilban Gorge section, Zerafshan, Uzbekistan.

Base of the Eifelian Stage (and of the Middle Devonian Series) This is defined at the base of the *Polygnathus costatus partitus* CZ. GSSP is 1.9 m below the Heisdorf–Lauch formational boundary (the traditional Emsian–Eifelian boundary) in the Welleldorf Trench, Eifel Hills, Germany.

Base of the Givetian Stage This is provisionally placed within the *Polygnathus ensensis* CZ, at the first appearance of *P. hemiansatus*, which is shown in Column 3 as an additional zone above the *ensensis* CZ (*sensu stricto*). This is the level considered most likely to be adopted by the SDS. The previous chart (Young 1989*a*) placed this boundary lower, at the base of the *ensensis* Zone. The GSSP submitted by SDS for ratification by IUGS is at Mech Irdane, near Erfoud, Morocco.

Base of the Frasnian Stage (and of the Upper Devonian Series) Defined at the lower boundary of the Lower *Polygnathus asymmetricus* CZ, this is represented on this chart as the *norrisi* CZ of Klapper & Johnson (1990). The GSSP is bed 42a of the Col du Puech de la Suque section at Montagne Noire, southern France (Klapper et al. 1987).

Base of the Famennian Stage This is defined at the base of the *Palmatolepis triangularis* CZ, which is one subzone lower than the level recommended by Ziegler & Klapper (1985), and used on the previous chart. As now defined, the Frasnian–Famennian boundary corresponds to the main extinction event of the Late Devonian (the 'Kellwasser event'). The GSSP is the base of bed 32a in the Coumiac section, Montagne Noire, southern France (e.g. Becker et al. 1989).

COLUMN 3: CONODONT ZONATION (PELAGIC BIOFACIES)

The standard conodont zonation is based on the work of Ziegler & Sandberg (1984) and Sandberg et al. (1989*b*), with various recent contributions. Ziegler & Sandberg (1984) proposed the replacement name '*linguiformis* CZ' for the terminal zone of the Frasnian (previously 'uppermost *gigas* CZ'). Klapper & Johnson (1990) proposed the *norrisi* CZ for the uppermost Middle Devonian CZ (cf. *falsiovalis* CZ of Sandberg et al. 1989*b*). They also identified upper and lower subdivisions of the *disparilis* CZ, and suggested the simpler name for the '*hermanni-cristatus*' CZ. A more extensive zonal revision for the Late Devonian by Sandberg et al. (1989*b*) as modified by Ziegler & Sandberg (1990) is used for the Frasnian, with the original zones of Ziegler (1971) shown on the right (*asymmetricus, triangularis, gigas* CZs), but with the *falsiovalis* CZ restricted to the basal Frasnian, rather than straddling the Givetian–Frasnian boundary.

The standard zonation is not readily applied to Australian sequences where shallow-water carbonates may be characterised by much lower conodont yields than in overseas studies (see Mawson et al. 1985, 1989, 1992).

COLUMN 4: SUPPLEMENTARY CONODONT–GRAPTOLITE ZONATION

For the Famennian the previous international conodont zonation of Ziegler (1962, 1971) is given for comparison with the *Palmatolepis* zonation of Column 3. The Late Devonian conodont zonation based on shallow-water icriodid species is given down the right side (after Sandberg & Dreesen 1984).

For the Frasnian, the original zones of Ziegler (1962, 1971) are shown on the left in relation to the new zonation of Ziegler & Sandberg (1990) given in Column 3. Klapper's (1989) detailed subdivision of the Montagne Noire succession involving thirteen zones spanning the whole of the Frasnian stage (MN 1–13) is indicated in the centre of the column. However, there are currently only two clear tie-points to the standard succession (base of MN 5 is equivalent to the base of the *punctata* CZ, and the incoming of *P. linguiformis* near the top of MN 13; see Klapper & Johnson 1990, Fig. 52; Klapper & Foster 1993, Fig. 2).

The *Monograptus* zonation for the Early Devonian of Victoria follows Garratt & Wright (1989). The *M. kayseri* (late Lochkovian) and *M. fanicus* (Pragian) zones of Jaeger (1989) have not been recognised in Australia. Jenkins (1982) discussed the relationship between graptolite and condont zones near the Siluro–Devonian boundary in the Yass area.

COLUMN 5: EURAMERICAN PALYNOMORPH ZONATION

Two zonal schemes are compared: Richardson & McGregor's (1986) zonation for the Old Red Sandstone continent on the left, and on the right the scheme constructed by Streel et al. (1987) for the Ardenne–Rhenish region of Europe.

Integration of palynomorph zones with the standard conodont zonation for the marine Devonian remains provisional. The correlations given are from Young (1989*a*), which were based on Streel et al. (1987). The same scheme was adopted by Grey (1991) and Young (1993*a*).

COLUMN 6: AUSTRALIAN MIOSPORE ZONATION

An informal Australian scheme widely used in the petroleum industry is based on the work of de Jersey (1966), Price (1980) and Price et al. (1985) on the subsurface sequence in the Adavale Basin. Subsequent investigation of the palynology of the Gneudna Formation by Balme (1988) has supported de Jersey's (1966) conclusion that the youngest part of the Adavale Basin sequence does not extend into the Late Devonian. McGregor & Playford (1992) confirmed Balme's (1988) opinion that palynostratigraphic unit PD5 of Price (1980) corresponds to the late Givetian *optivus–triangulatus* Zone of Richardson & McGregor (1986). Beneath this, key elements of the *devonicus–naumovae* and *velata–langii* assemblage-zones have been recognised (McGregor & Playford 1992), indicating an Eifelian age (cf. Price et al. 1985, who assigned the PD2 and lower PD3 zones to the Early Devonian). Macrofossil evidence from the Log Creek Formation in the Adavale Basin also indicates an Eifelian age (Pickett 1972).

There remains some uncertainty about the actual ranges of some key miospore species in Australia in relation to the conodont zonation. *Retispora lepidophyta* characterises the two uppermost assemblage-zones, and corresponds to the uppermost *expansa* and *praesulcata* CZs of the latest Famennian. Streel & Loboziak (1993) recently placed the first appearance of *R. lepidophyta* somewhere within the middle *expansa CZ* or lower part of the upper *expansa* CZ. Previously, Paproth & Streel (1979) correlated the base of the *lepidophyta* assemblage-zone to just above the base of the lower *costatus* CZ, while Conil et al. (1986) placed it higher, within the upper *expansa* CZ. However in the Canning Basin (Playford 1976, p.8) *Retispora lepidophyta* is associated with the *Icriodus platys* conodont assemblage of Nicoll & Druce (1979). *I. platys* is a junior synonym of *I. raymondi*, which disappears at the top of the middle *expansa* CZ (Sandberg & Dreesen 1984). Playford (1982, p.155) considered that the first appearance of *R. lepidophyta* post-dated the middle *styriacus* CZ (equivalent to the base of the *expansa* CZ), suggesting that the '*I. platys*' association in the Canning must be near the first appearance of *R. lepidophyta*. However, in the opinion of Grey (1992, p.4), the lower boundary of the range of *R. lepidophyta* had not been identified. Other evidence from the Canning (cores 14, 15, in Babrongan No.1 well) suggest age equivalence with the next oldest *postera* CZ, based on the association of *Retispora lepidophyta* with *eocostata* Zone entomozoacean ostracods (P.J. Jones, pers. comm.). On this evidence a provisional approximation of the base of *Retispora lepidophyta* with the base of the *expansa* CZ is used for Column 6.

The spore *Geminospora lemurata* was reviewed by Playford (1983), who suggested a total range from the *varcus* CZ (middle Givetian) or younger, through the Frasnian and possibly into the early Famennian. In Europe, however, this species is recorded from the earlier *ensensis* CZ (Streel et al. 1987; Streel & Loboziak 1993). Balme (1988) suggested an Eifelian–middle Frasnian range, possibly extending into the basal Famennian, for *G. lemurata. Geminospora* sp. is recorded from the Frasnian Bellbird Creek Formation of the Merrimbula Group (Anan-Yorke 1975), and *G. lemurata* is abundant in the Brewer Conglomerate palynoflora of the Amadeus Basin (Playford et al. 1976). In the Canning Basin it is associated with conodonts said to be equivalent to the *varcus* or *ensensis* CZs (Grey 1991). The Brewer palynoflora also contains six species (*Grandispora clandestina, Hystricosporites porrectus* etc.) in common with the late Famennian Fairfield Formation palynoflora of the Canning Basin (Playford 1976), suggesting that *G. lemurata* may be reworked in this assemblage (Balme 1988). It is therefore shown with a more restricted upper range in Column 6.

COLUMN 7: OSTRACOD ZONATION

On the left side is a zonation for entomozacean ostracods based on the work of Groos-Uffenorde & Wang (1989) from the Devonian of southern China and Europe. The biostratigraphic utility of this pelagic ostracod group is well established (e.g. Gooday & Becker 1979). Entomozoacean ostracods were recorded from Australia by Jones (1968, 1974), but remain to be studied in detail, although the nominate species of the middle Famennian *eocostata* Zone was recently identified in the subsurface of the Canning Basin (Jones & Young 1992).

The local zones of Jones (1968, 1989), shown in the right column, are based on benthic ostracods from the Late Devonian of the Bonaparte Basin. The Famennian zones are now known (Playford 1982; Jones 1985) to cover only the latest Famennian. They correspond approximately to assemblage A of Jones (in Veevers & Wells 1961), recorded from the 'Fairfield Beds' (now Gumhole Formation) in the Canning Basin. The '*hanaicus*' Zone occurs in the Westwood Member (late Frasnian) of the Cockatoo Formation (Jones 1968).

COLUMN 8: MICROVERTEBRATE ZONATION

Many new microvertebrate taxa have been discovered in Australian sequences in recent years, most of which remain undescribed (for recent reviews see Turner 1991, 1993). Column 8 shows on the left a provisional scheme for turiniid thelodont 'assemblages' ranging in age from basal Devonian to Frasnian. This group is well represented in marine, marginal and non-marine deposits. Interpolated is the Taemas–Buchan microvertebrate 'fauna', a marine assemblage in which thelodonts are rare or absent, but other major vertebrate groups (placoderms, sharks, acanthodians, osteichthyans) are abundantly represented. Little taxonomic work has been done on these groups in Australian sequences generally, but boxed ranges on the right side of Column 8 summarise some recent work on microvertebrate taxonomy and biostratigraphy for the Early Devonian of central New South Wales and the Middle Devonian of the Broken River sequence.

For the Late Devonian, a large quantity of undescribed microvertebrate material from throughout Australia awaits detailed study (e.g. Turner 1993, pp.193–5). Sharks (chondrichthyans) become increasingly abundant (protacrodont and cladodont teeth and scales,

Thrinacodus, Holmesella, Stethacanthus, etc.), and various osteichthyans are common. Placoderm scales are much diminished in these assemblages (but the scale form '*Artenolepis*' occurs in the Frasnian (Gogo, Canning Basin)). Acanthodian scales of various types occur throughout the Late Devonian, but await detailed study. The only group for which a provisional zonation related to the standard conodont zones is available are the phoebodont sharks. The oldest record of this group from marine strata in Australia comes from a Givetian (*varcus* CZ) sample (Papilio Formation) in the Broken River sequence (Turner 1993), which is comparable to the Givetian form *Phoebodus sophiae* from North America (S. Turner, pers. comm.). This group diversified in the Late Devonian, but they have not yet been found in early Frasnian strata, and they are rare in the late Frasnian *linguiformis* CZ (Ginter & Ivanov 1992). Their Famennian radiation was coincident with the radiation of palmatolepid conodonts, and Ginter & Ivanov (1992) showed known ranges of various species for the eastern European Platform and elsewhere, which are summarised here in a provisional scheme of assemblages (local zones have not yet been analysed in detail). The genus *Thrinacodus* (*Harpagodens* on the previous chart; Young 1989*a*) is known from the late Famennian of Australia (Turner 1982*a*), and many other areas (South-East Asia, China, etc.; see Wang & Turner 1985, Derycke 1992, Long 1990). The species *Thrinacodus ferox* and *Phoebodus australiensis* are the only two in this group known to persist into the Tournaisian (Ginter & Ivanov 1992).

COLUMN 9: RADIOLARIAN–CHITINOZOAN ZONATION

Devonian Radiolaria have long been known from eastern Australia, but it is only in the past decade that their biostratigraphic utility has been developed. An Upper Devonian zonation was proposed by Holdsworth & Jones (1980), and Braun (1990) has summarised recent German studies. Late Devonian (Frasnian) Radiolaria from the Gogo Formation of the Canning Basin were described by Nazarov et al. (1982), Nazarov & Ormiston (1983), and Aitchison (1993*a*,*b*), and in eastern Australia Radiolaria have helped unravel the stratigraphy of siliceous rocks in the New England Fold Belt (Ishiga et al. 1987, 1988; Aitchison 1988*a*,*b*,*c*,*d*; Flood & Aitchison 1992). Ishiga et al. (1988, Fig. 3) proposed a preliminary zonation for the Late Devonian of this region, which is summarised in Column 9, with the Holdsworth & Jones (1980) scheme shown on the right side.

Chitinozoans are poorly known for the Australian Devonian, but recent work (Winchester-Seeto & Paris 1989; Winchester-Seeto 1993*a*–*c*, 1994) has provided a preliminary zonation for the Early Devonian of eastern Australia. Ranges in Column 9 are based on Lochkovian–Pragian assemblages from the Garra Limestone (Winchester-Seeto 1993*a*), Shield Creek Formation, and Coopers Creek Limestone (Winchester-Seeto 1993*b*), and Emsian chitinozoans from the Taravale Limestone (Winchester-Seeto, in prep.). A high degree of endemism is evident in the Emsian assemblages from Victoria (Winchester-Seeto 1994), but Pragian assemblages correlate well with existing biozonations from overseas (e.g. Paris 1981).

COLUMN 10: DACRYOCONARID ZONATION

Lutke's (1979) scheme of twelve named dacryoconarid zones tied to the standard conodont zonation was expanded to eighteen named zones by Alberti (1984). This scheme has been updated with taxonomic documentation by Alberti (1993), but precise alignment of some of these with CZs is not available. The numbered zones in Column 10 come from Alberti (1993, Table 1).

Dacryoconarids in Australia were first recorded from Victoria and are widely distributed in the Lower Devonian of the Melbourne

Trough (Garratt 1983*a*). Alberti (1993) records index taxa of his zones 2 (*Homoctenowakia bohemica*) to 6 (*Guerichina infundibulum*) in the Melbourne Trough, and taxa of zones 7 (*Nowakia zlichovensis*) to 11 (*Nowakia richteri*; associated with *Polygnathus serotinus*) in the Buchan sequence (Taravale Formation, as previously noted by Mawson et al. 1985). In New South Wales, Sherrard (1967) recorded *Nowakia* aff. *acuaria* from the Garra Formation (*pesavis–dehiscens* CZs; Strusz 1972, Wilson 1989), and lower Taemas Formation (*dehiscens–serotinus* CZs; Mawson et al. 1985), but these determinations need to be updated. In Queensland, the Broken River sequence has yielded zones 3–5 of Alberti (1993) in the Martins Well Limestone, and zones 10–13 in the '*Bracteata* Mudstone'–Jessey Springs Limestone. In Tasmania, abundant dacryoconarids indicative of an early–late Pragian age, including *Viriatellina* sp., *Metastyliolina* sp. and *Nowakia matlockiensis*, are reported from the Mathinna beds (Banks & Baillie 1989, p.236).

COLUMN 11: AMMONOID ZONATION

Correlation of the standard ammonoid and conodont zonations was discussed by Klapper & Ziegler (1979, Fig. 8), House et al. (1985), Ziegler & Klapper (1985), Klapper et al. (1987), and Becker & House (1994). Recent work by Becker et al. (1991, 1993) builds on the earlier studies of Glenister (1958) and Petersen (1975) in the Devonian of the Canning Basin, where equivalents of the classic German 'Stufen' for the Upper Devonian have been recognised. Famennian ammonoids are also known from the New England Fold Belt in eastern Australia, with *Cheiloceras* from the Baldwin Formation (Jenkins 1966), *Platyclymenia annulata* and other forms from the Mandowa Mudstone (Jenkins 1968), and younger *Wocklumeria* Zone faunas reported by Pickett (1960). Column 11 shows the scheme comprising thirty-six genozones outlined by Becker (1993), which spans a period of about 15 Ma, giving average durations of zones of about 400 Ka, one of the finest time-resolutions known for the Palaeozoic. The detail of resolution is high near boundaries that have been the focus of research effort, for example the Emsian–Eifelian boundary, where three zones (A–C of MDI) are recognised within the *partitus* CZ, which on this chart is assigned a very short duration.

COLUMN 12: BRACHIOPOD ZONATION

The Early Devonian zonation gives the Victorian *Boucotia* zones of Garratt (1983*a*) as revised by Garratt & Wright (1989). For the shallow-water assemblages of western New South Wales, Sherwin (in Glen et al. 1985, Figs 10, 11) outlined an informal scheme of five assemblage zones for the Cobar Supergroup, which has been revised for central New South Wales by Sherwin (1992). Previous assemblages now dated on associated conodonts give ages about half a stage older than before (cf. Young 1989*a*). The Early Devonian zonation on the right side of the chart combines information from Garratt & Wright (1989) and Sherwin (1992).

For the Middle–Late Devonian, the zones of Veevers (1959) for the Canning Basin are shown as interpreted by Roberts et al. (1972). The top of the *Stringocephalus* Zone is placed within the *hermanni* CZ, although most stringocephalids disappeared within the *varcus* CZ (e.g. Talent et al. 1993). The upper two of Veevers' zones (*scopimus, proteus*) are boxed on the left side of the column. The centre shows the five Famennian productid zones of McKellar (1970) from the Star Basin in Queensland. Pickett (1981) showed that the boundary between the *profunda* and *minuta* Zones lay within the Lower *marginifera* CZ, suggesting an older age for all five brachiopod zones than interpreted by McKellar (1970). Indicated on the right is the approximate position of the *Cyrtospirifer, Sulcatospirifer* and *Tenticospirifer* zones erected by Maxwell (1954) for the sequence at Mt Morgan in Queensland. Their precise relationship to the productid zones is unclear, as discussed by McKellar (1970, p.9). Maxwell's *Tenticospirifer* Zone is older than his Tournaisian assessment (beneath the *tenuistriata* Zone; Roberts 1975), and his *Cyrtospirifer* Zone younger (probably early Famennian; McKellar 1970; see also Dear 1968). Roberts et al. (1993*a*) discuss the basal Carboniferous *Tulcumbella tenuistriata* brachiopod zone, which in the Mt Morgan sequence straddles the D–C boundary, and is thus aligned with the upper *praesulcata* CZ of the latest Famennian.

COLUMN 13: CORAL ZONATION

The coral–conodont faunal scheme of Philip & Pedder (1967) was used by Strusz (1972) and Pickett (1972) to provide eleven zones (A–K) for the Early–Middle Devonian. The '*Spongophyllum*' Zone was changed to the *Carlinastrea halysitoides* Zone by Pedder (1985), and this zone in eastern Australia at least extends up to include the *pesavis* CZ (see Yu & Jell 1990; cf. Garratt & Wright 1989, who equate it with their *janaea* brachiopod zone). Garratt & Wright (1989) have revised the Early Devonian part of the Philip & Pedder scheme, as shown in Column 13. Faunas G (*touti*) and H (*callosum*) of the previous scheme are now considered to overlap, with a late Emsian rather than Eifelian age. A mid–late Frasnian (*gigas–linguiformis* CZ) coral fauna was described from the Mostyn Vale Formation of eastern Australia by Wright et al. (1990). Middle–Late Devonian corals from the Canning Basin were described by Hill & Jell (1970), and Brownlaw & Jell (1994) provide a preliminary biostratigraphic framework of three coral assemblages: *Argutastrea hullensis*, *Donia brevilamellata*, and *Aulopora–Disphyllum–Temnophyllum* assemblages. These have not been included since they cannot yet be aligned with the conodont zonation.

COLUMN 14: MACROVERTEBRATE ZONATION

The scheme presented is modified from that of Young (1993*a*), which showed fifteen macrovertebrate assemblages ranging in age from Emsian to Famennian. They are indicated as either marine (M) or non-marine (NM), and numbered as in Young (1993*a*). Older macrovertebrate remains are known from south-eastern Australia as isolated occurrences not yet useful biostratigraphically. The main changes from Young (1993*a*) are in the relationship between marine and non-marine faunas. Thus the Taemas–Buchan fauna [1] is at least as old as Pragian, and extends well into the Emsian (*dehiscens* to *serotinus* CZs; Mawson et al. 1985, 1989), and is now assumed to be largely contemporaneous with the non-marine *Wuttagoonaspis* fauna [2] from the Mulga Downs Group in western New South Wales. This fauna has also been recognised in the Georgina (Turner et al. 1981; Young 1984) and Amadeus Basins (Young 1985, 1988, Fig. 13; Young et al. 1987). Young (1993*a*,*b*) previously assessed a maximum age for the Mulga Downs Group based on conodont faunas from the underlying marine Cobar Supergroup, which lack *Polygnathus*, and thus indicate a pre-Emsian age (see Pickett 1980; Pickett & McClatchie 1991). Garratt & Wright (1989) considered it more likely that the *jaqueti* brachiopod assemblage underlying the Mulga Downs Group was Lochkovian, rather than Pragian as earlier suggested by Sherwin (1980), and Sherwin (1992) revised the age assessment of his faunas downwards by at least half a stage. This evidence is consistent with a *pesavis–sulcatus* CZ age concluded above for the associated *Turinia australiensis* microvertebrate assemblage (see Column 8). An assumed younger assemblage from the Cravens Peak Beds, including antiarchs (Young 1984) and *Mcmurdodus* (Turner & Young 1987), is now provisionally grouped with the non-marine Hatchery Creek fauna [3], previously assigned an approximate late Eifelian age by Young & Gorter (1981). A considerable overlap is assumed with the next assemblage, the marine *Wurungulepis* fauna [4] from the Broken River sequence, for which an age range of early Eifelian (*partitus* CZ) to early Givetian was proposed by Young (1993*a*). The Hatchery Creek, Cravens Peak, and *Wurungulepis* faunas are all characterised by the presence of pterichthyodid antiarchs (Young & Gorter 1981; Young 1984, 1990). The Givetian *Nawagiaspis* fauna [5] from the Papilio Mudstone of the Broken River Formation is considered to overlap with the non-marine Aztec fauna [6] from southern Victoria Land, Antarctica, now interpreted as mainly or entirely of Middle Devonian age (Young 1989*b*, 1993*a*,*b*).

The remaining nine zones of Column 14 are all based primarily on non-marine assemblages from eastern Australia. The fine subdivision of the Givetian–Frasnian part of the zonation is a direct result of systematic palaeontology in Victoria and south-eastern New South Wales (e.g. Young 1979, 1982, 1983, 1989*c*; Long 1983*a*,*b*, 1984*b*, 1985, 1986, 1987, 1988*a*, 1989, 1992; Ritchie 1984; Long & Werdelin 1986). The Tatong fauna [7] may be the oldest, combining phyllolepid placoderms with elements from the Hatchery Creek fauna (?sherbonaspid antiarchs), but lacking thelodonts. Succeeding faunas 8–12 are known to be older than the late Frasnian Ettrema–Westwood trangression (see Column 16), and are arbitrarily assigned approximately equal length for the late Givetian–late Frasnian interval. The Taggerty–Howitt fauna [8] is significant in being constrained above and below by isotopically dated rhyolites, which give the same (minimum) age of 367 ± 2 Ma (Williams et al. 1982). This aligns with a Givetian–Frasnian boundary position on the chart.

The three Famennian macrovertebrate zones are relatively long ranging, due to lack of taxonomic description of Famennian fish faunas from Australia. They are also arbitrarily assigned approximately equal durations, in the absence of detailed systematic accounts of these faunas. Palynomorphs of the *Retispora lepidophyta* Zone, from Pondie Range No. 1 well in western New South Wales (Evans 1968), are associated with fragmentary fish remains apparently belonging to the *Bothriolepis–Remigolepis* association, indicating a younger age limit for the Worange Point fauna [14] near the top of the Famennian. The youngest macrovertebrate fauna (the Grenfell fauna [15]) is assigned a latest Famennian age, and may exend into the earliest Carboniferous (Young 1993*a*).

COLUMN 15: MACROPLANT ZONATION

The oldest Devonian macroplant assemblage from Australia is the *Baragwanathia* flora of Victoria, characterised by the early vascular plant *Baragwanathia longifolia* (Gould 1975; Hueber 1983; Tims & Chambers 1984), which is noted for its association in the Melbourne Trough with Late Silurian and Early Devonian graptolites (e.g. Garratt & Rickards 1984, 1987). Jaeger (1966) first described the Early Devonian graptolite *Monograptus thomasi* in association with *Baragwanathia* from the Wilson Creek Shale, and facies relationships with limestones containing the conodont *Eognathodus sulcatus* demonstrate that the range of *M. thomasi* must overlap the *sulcatus* CZ, at least in part (Garratt & Wright 1989). Elements of the *Baragwanathia* flora are therefore shown in Column 15 ranging up to about the middle of the early Pragian *Boucotia loyolensis–Nadiostrophia* Assemblage-Zone, a correlation suggested by Garratt (1983*a*) and Garratt & Wright (1989).

Plant remains from the Emsian–Eifelian interval in Australia are poorly documented. A flora of probable late Middle Devonian age from south-eastern Australia includes the lycopods *Protolepidodendron lineare*, *P. yalwalense*, and *?Lepidodendron clarkei* (Yalwal, Bunga Beds, also identified recently from central New South Wales). *Protolepidodendron* was assigned an Emsian–Givetian range by Chaloner & Sheerin (1979). *P. scharianum* is recorded with *Astralo-*

caulis in the Frasnian Dotswood Formation of Queensland (Gould 1975), the latter given a Givetian–Frasnian range by Chaloner & Sheerin (1979). Other Frasnian records from Australia are ?*Cordaites* and ?*Archaeosigillaria* listed from the Twofold Bay Formation by Fergusson et al. (1979), the latter having a Givetian–Tournaisian range according to Chaloner & Sheerin (1979). In south-eastern Australia there is no evidence for this floral assemblage above the late Frasnian *gigas* Zone transgression (J. Pickett, pers. comm.), and this is assumed on the chart.

A more widespread and presumably younger Late Devonian lycopod assemblage in Australia is represented by *Leptophloeum australe*. The range of *L. australe* has been considered to encompass the whole of the Late Devonian, and possibly extending into the Early Carboniferous, as suggested for the Canning Basin (Veevers et al. 1967), and eastern Australia (Roberts et al., in press *a*). However, nowhere is this upper limit actually demonstrated (Jones et al. 1973; Roberts et al. 1991*a*,*b*). Mory (1981) recorded the upper limit of *L. australe* in the Tamworth Belt to be marked by the Borah (Kiah) Limestone, which contains Upper *praesulcata* Zone conodonts. In this region *L. australe* is common below this level, but has never been seen above it (J. Pickett, pers. comm.), and this evidence is used to fix a provisional upper limit on the chart.

A lower limit for *L. australe* is more difficult to determine, but in sequences manifesting the late Frasnian *gigas* Zone transgression it is not recorded from beneath this level (J. Pickett, pers. comm.). More diverse Frasnian and Famennian floras from eastern Victoria are summarised by Marsden (1988).

COLUMNS 16–17: EVENT STRATIGRAPHY/TRANSGRESSION–REGRESSION PATTERN

Column 17 summarises selected Devonian 'events' of possible global extent, and therefore potentially useful for correlation purposes. Included are various bioevents (e.g. extinctions, radiations; House 1989; Kauffman & Walliser 1990), isotopic markers (e.g. Andrew et al. 1994), anoxic events, rapid sea-level changes attributed to eustasy, etc., which may be discerned in or predicted for Australian Devonian sequences. The eleven events are numbered as in the following summary, which is based largely on the recent work of Becker (1993), Talent et al. (1993) and Andrew et al. (1994).

1 *Silurian–Devonian boundary event* Andrew et al. (1994) record a shift in carbon and oxygen isotope values about 10 m beneath the incoming of the conodont *Icriodus woschmidti hesperius* in the Jack Limestone, Broken River, Queensland. Possibly this corresponds to an anoxic event supposedly connected with the S–D boundary in Europe (Schönlaub 1986).

2 *End-**pesavis** event* A conspicuous reduction in conodont diversity in the latest Lochkovian *pesavis* CZ (Ziegler & Lane 1987) is also evident in eastern Australia, with a significant regional regression just before the end of the zone.

3 ***pireneae-dehiscens** T–R interlude* A major regressive event is followed by transgression before the first appearance of the conodont *Polygnathus dehiscens*. This is the eponymous form for the basal Emsian *dehiscens* CZ. Emsian limestones of *dehiscens–serotinus* CZ age (e.g. Buchan, Taemas/Wee Jasper) are well developed in the south-east of the continent, although whether this is a local tectonic phenomenon or has a eustatic component is not yet clear.

4 *Daleje event* The Daleje event was first documented in Bohemia (House 1985; Chlupáč & Kukal 1986), and appears to correspond to a mid-Emsian global transgression (Talent et al. 1993). In Australia this may be correlated

with the mid-Emsian upper Taravale transgression, initiated just below the *perbonus–inversus* boundary (Talent 1989).

5 *Kacak event* Again first identified in Bohemia (House 1985), this event is manifested by the high generic turnover of brachiopods across the Eifelian–Givetian boundary. Depletion in $\delta^{13}C$ at about this level (*kockelianus* CZ) in several sections from the Broken River area (Andrew et al. 1994) may be related to this event, and an association has also been suggested with the *otomari* event (Walliser 1984; Boucot 1990), representing the appearance of the dacryoconarid *Nowakia otomari* (but this species has not yet been found in Australia; Talent et al. 1993).

6 *Taghanic event* This is an extinction event marking the disappearance of the widespread brachiopod *Stringocephalus*, which occurred during the middle Givetian *varcus* CZ, and may be associated with significant decreases in coral and conodont diversity during the Givetian, and major rearrangements in global biogeography (Talent et al. 1993). Based on Australian sequences (Mawson & Talent 1989), a marked drop in brachiopod diversity occurred at about the end of the middle *varcus* CZ (Talent et al. 1993), even though a few *Stringocephalus* apparently survived into the succeeding *hermanni* CZ in North America (see above). In Australia this event approximates to the main transgression onto the Kimberley Block (Becker et al. 1993), and in the Broken River area an abrupt and short-lived depletion in $\delta^{13}C$ within the upper *varcus* CZ has been identified by Andrew et al. (1994). In Column 16 this event is shown as coincident with the major trangsression marking the base of cycle IIa of the Euramerican sea-level curve, placed at about the middle of the *varcus* CZ (cf. Talent et al. 1993, Fig. 2, who align their 'Stringocephalid event' above the transgression, at the base of the upper *varcus* CZ).

7 *Kellwasser events* These closely spaced events are associated with the celebrated Frasnian– Famennian extinction, one of the major extinctions of marine invertebrates during the Palaeozoic. Two distinctive anoxic horizons and associated geochemical signatures have been documented in late Frasnian sequences from Europe and Morocco east to central Asia and China, but have not so far been identified in Australia (Nicoll & Playford 1993). The lower Kellwasser dark shale horizon is dated as within the youngest part of the lower *gigas* CZ, and the upper horizon defines the end of the *linguiformis* CZ, and the end of the Frasnian Stage (both horizons probably lie within Zone MH13 of Klapper 1989). These horizons, with the intervening interval (the 'Kellwasser Crisis' of Schindler 1993), saw major extinctions in several groups, including brachiopods, goniatites, trilobites, corals, stromatoporoids, palmatolepid conodonts, etc. The disappearance of coral/stromatoporoid biohermal reefs occurred mainly in late *rhenana* CZ time, but in some regions during the latest Frasnian *linguiformis* CZ. Sea-level fluctuations are documented during the Kellwasser interval, with a major end-Frasnian regression to the sea-level low of the earliest Famennian (Becker et al. 1993, Fig. 10).

8 *Condroz event* Following the sharp terminal-Frasnian regression was a period of fluctuating but increasing global sea-level of the early Famennian, with the distribution of hypoxic goniatite shales reaching its acme in the upper *crepida* CZ. This was preceded by a short regression causing regional red bed deposition in some areas and hardgrounds and stromatolitic microreefs in the Canning Basin (Becker 1993). The major regression that brought this early Famennian phase of high sea-level to an end is termed the Condroz event, which according to Becker (1993) is characterised by red bed development in the Rhenish Slate Mountains (the Nehden

Sandstone) and in Virginia, with the same event recognised in Poland, and in China ('Oujiachong regression'), and marked by basinward progradation of algal reefs in the Canning Basin.

9 *Enkeberg event* Becker (1993) assigns this name to the global transgression at the base of the *Maeneceras* Genozone (*marginifera* CZ), which he correlates with event 12 on the sea-level curve of Johnson & Sandberg (1989). As noted above this seems not to correspond to the Westwood transgression in the Bonaparte Basin, which is older, but perhaps the *marginifera* CZ fauna identified by Pickett (1981) from Myrtlevale in Queensland may be aligned with this event.

10 ***Annulata*** *event* Anoxic intervals in the lower *Prionoceras* Genozone of the mid-Famennian were named the *Annulata* event by House (1985). Hypoxic equivalents in other regions include the haematitic ammonoid faunas of the Piker Hills Formation in the Canning Basin. Becker (1993) equates this event with the rapid short-term eustatic rise 14 of Johnson & Sandberg 1989 (upper *velifer*/upper *trachytera* CZ), and with *Platyclymenia* faunas in otherwise thick successions lacking this group in California and New South Wales (Jenkins 1968). Note that the upper *expansa* to middle *praesulcata* CZ fauna reported from the relevant formation (Mandowa Mudstone) by Wright et al. (1990, p.223) is from a different locality and horizon (J. Pickett, pers. comm.)

11 *Hangenberg event* The extermination of most ammonoid groups near the end of the Famennian in the middle *praesulcata* CZ is termed the Hangenberg event (e.g. House 1985, 1988), although brachiopods were seemingly less affected. A sharp regression at this time in the Euramerican sea-level curve is identified as the Yellow Drum regression in the Canning Basin by Talent et al. (1993). The Devonian–Carboniferous boundary is placed about 7.5 m above the base of the Yellow Drum Formation in a trench section in the Canning Basin (Talent et al., in prep.), and 2–4 m beneath the boundary a marked depletion in $\delta^{13}C$ has been recognised by Andrew et al. (1994, Fig. 8). A similar change in carbon isotope values occurs in about the same position (bed E in upper *praesulcata* CZ) in the Nanbiancun section in China (Jones, in press; see Wei & Ji 1989, Fig. 9, Table 11).

Devonian transgression–regression pattern

Column 16 is a qualitative representation of the transgression–regression pattern for the Devonian as inferred from Australian sedimentary sequences, derived from various sources. The 'inundation curve' for the Australian Devonian developed by Struckmeyer & Brown (1990) showed the following broad features: late Lochkovian regression, mid-Emsian transgression and late Emsian regression, a Givetian–Frasnian transgressive phase, the Frasnian highstand, and a Famennian regressive phase. The identification of eustatic as opposed to local causes was based on comparison with the fourteen transgressive–regressive (T–R) cycles of the Devonian sea-level curve for Euramerica identified by Johnson et al. (1985). Talent & Yolkin (1987) compared this curve as modified by Dennison (1985) with sea-level changes interpreted from sediments in Australia and West Siberia. This work was updated by Talent (1989) and Talent et al. (1993), taking into account the more detailed Euramerican curve of Johnson & Sandberg (1989). A eustatic curve for the Late Devonian of the Canning Basin was outlined in Becker et al. (1993, Fig. 10). The Late Devonian eustatic curve, labelled CB in Column 16, is derived from subsurface data from the Lennard Shelf, Canning Basin (see Young 1995). There is approximate alignment of these eustatic cycles (G–F, F1–4) with those of Becker et al. (1993, Fig. 10). Major T–R events identified so far for the Australian Devonian are now discussed.

Early Devonian In south-eastern Australia, a Derringullen/Bowning regression in the basal *hesperius* CZ (which may be related to local tectonics; Talent 1989) was followed by two transgressive events manifested in the Garra Limestone (Garra 1 and 2 of Talent 1989). There was some uncertainty about the age of the younger of these (*sulcatus* CZ, or 'late Pragian'; Talent 1989) but Talent et al. (1993) now align it with the latest Lochkovian *pesavis* CZ. In the Broken River sequence of Queensland the Garra 2 transgression is apparently reflected in the change from clastic to carbonate deposition of the Martins Well and Arch Creek Limestone members (Talent 1989). Within the Pragian the regression manifested by platform exposure and development of sub-marine fans in the Red Hill Limestone (Mawson et al. 1992) is aligned with the base of the *pireneae* CZ (Talent et al. 1993). This was followed by the Buchan Caves Limestone transgression of Talent (1989), originally placed at the base of the *dehiscens* CZ, and correlated with the base of cycle Ib of Johnson et al. (1985). However, Mawson et al. (1992) suggested that it may have occurred slightly earlier, in the early part of the latest Pragian *pireneae* CZ, which makes it too old for cycle Ib, but also too young for the transgression initiated near the base of the *kindlei* CZ (mid-Pragian), within cycle Ia of the Euramerican curve. The differences in the Pragian sections of the Euramerican and Australian T–R patterns (e.g. Ta!ent et al. 1993, Fig. 2) may indicate local tectonic effects. In the Emsian two transgressions have been suggested at Buchan (e.g. Talent 1989) and Bindi (Mawson 1987; Webb 1992) in Victoria: at the change from the Buchan Caves Limestone to mudstones of the Taravale Formation, interpreted to indicate a deepening event; and higher in the Murrindal Limestone within the Taravale Formation. The lower event (Basal Taravale transgression) is placed high in the *dehiscens* CZ (Mawson et al. 1992).

In the Broken River sequence of Queensland a major transgression beginning late in the *inversus* CZ (Mawson et al. 1985) seems to align with the upper part of cycle Ib of the Euramerican curve (e.g. Johnson & Sandberg 1989, Fig. 1), rather than cycle Ic as suggested by Talent (1989). It must be assumed that the major transgression in the latest Emsian *patulus* CZ, which initiated cycle Ic of the Euramerican curve, is masked in south-eastern Australia by local tectonism with onset of the Tabberabberan orogenic phase (e.g. the 'Hatchery Creek regression' of Talent 1989).

Middle Devonian The Euramerican curve shows an overall rise in sea-level through the Middle Devonian (T–R cycles Ic–f), with a marked middle *varcus* CZ transgression initiating cycle IIa (Taghanic onlap and event; see above). Talent & Yolkin (1987) could not recognise this mid-Givetian transgression, which they suggested may have been masked by orogenic activity, and this may apply to the Middle Devonian generally for eastern Australia. However, in Western Australia Becker et al. (1993, Fig. 10) equate this event with the initial major transgression onto the Kimberley Block. The overall trend of rising sea-level through the Middle Devonian is also suggested in the marine flooding data of Struckmeyer & Brown (1990, Fig. 5) for basins in the west, which were not affected by the Tabberabberan Orogeny. The lower part of the Australian T–R pattern for the Middle Devonian has been generalised after the Euramerican curve, with T–R cycles labelled as in Johnson & Sandberg (1989), on the assumption that its broad features reflect global eustatic changes. A marked mid-Givetian (early *varcus* CZ) regression, which terminated cycle If, may be represented in eastern Australia by the Papilio–Mytton regression of the Broken River sequence (Talent 1989), although this may be better placed higher in the Givetian (Talent et al. 1993, Fig. 2). The late Givetian part of the T–R pattern in Column 16 follows Becker et al. (1993, Fig. 10).

Late Devonian In the Canning Basin a series of five 'anoxic pulses' (labelled 1–5) related to a fluctuating increase in sea-level during the early Frasnian are identified, with further rapid fluctuations before the major regression of the terminal Frasnian Kellwasser event (Becker et al. 1993). A major flooding event on the northern margin of Gondwana and elsewhere, at the base of the *Maeneceras* Genozone, is correlated by Becker (1993) with event 12 on the curve of Johnson & Sandberg (1989), and approximated to the Westwood transgression in the Bonaparte Basin by Talent et al. (1993). However, conodont dating of the relevant limestones in the Cockatoo Formation of the Bonaparte indicate a late Frasnian age (P.J. Jones, pers. comm.; Roberts 1972; Mory & Beere 1988), and this transgression more likely relates to the *gigas–linguiformis* CZ T–R cycle IId of Johnson et al. (1985), also recognised in eastern Australia (see Wright et al. 1990, p.223; Young 1993*a*, p.216). Other T–R cycles in the early–middle Famennian that may be manifested in Australian sequences were summarised above using the event terminology of Becker (1993) and Talent et al. (1993): the *Cheiloceras* Genozone age Condroz event T–R (*crepida* CZ), the Enkeberg and *Annulata* transgressions (*marginifera–trachytera* CZs), and the Hangenberg event (*praesulcata* CZ). The beginning of the Dasberg event transgression (event 16 of Johnson & Sandberg 1989) is not yet clearly identified, but the Teddy Mountain transgression of the late Famennian in the Canning Basin is equated with maximum onlap of the upper part of cycle IIf (Talent 1989) before the marked Yellow Drum regression, which evidently corresponds to the Hangenberg event at the end of the Famennian.

OTHER GROUPS

Various other macrofossil and microfossil groups have proved useful for age control overseas, but have not yet been studied in sufficient detail to be applied for Australian Devonian sequences. These include some well known groups of great biostratigraphic utility in other periods of the timescale, such as forams (e.g. Conkin & Conkin 1968), nautiloids (Teichert et al. 1979), and trilobites (Alberti 1979; Chatterton et al. 1979; Holloway & Neil 1982; Chatterton & Wright 1986). Among other arthropods only eurypterids, conchostracans and phyllocarids are sufficiently abundant to yield biostratigraphic data (Rolfe & Edwards 1979). Briggs & Rolfe (1983) described phyllocarids from Western Australia. The biostratigraphic utility of these and other groups in Australian sequences will depend on detailed systematics being carried out to provide a basis for determining the ranges of described taxa.

ACKNOWLEDGMENTS

Many experts on the Australian Devonian contributed to the compilation of this and the previous chart. Dr P.J. Jones (AGSO) gave constant advice on all aspects of the chart, and helped greatly with the conodont and ostracod zonations. Drs R. Mawson and J. Talent (MUCEP) and J. Pickett (NSWGS) advised on conodont zonations and zone durations, T. Winchester-Seeto, C. Burrow and A. De Pomeroy provided unpublished data, and Dr S. Turner provided much advice on microvertebrates. For helpful comments on earlier versions of the chart I thank Dr B.E. Balme (University of WA), Dr J. Douglas (Victorian Dept Mines), Dr J. Long (WA Museum), Dr L. Sherwin (New South Wales Dept of Mines), Dr G. Playford and Dr J. Jell (University of Queensland), Dr B. Webby (University of Sydney), and Dr A. Wright (University of Wollongong). Drs J. Claoué-Long and I. Williams discussed isotopic data, and Dr E.M. Truswell, Dr D. Strusz (AGSO), and Dr N. Morris (University of Newcastle) provided information on macroplant and brachiopod occurrences.

2.5

Carboniferous (Chart 5)

P.J. Jones

INTRODUCTION

The first Carboniferous fossils from Australia were collected in late 1842 by Strzelecki from the Booral district, in the Hunter Valley of the Colony of New South Wales, and were described by Morris (in Strzelecki 1845). In present-day terms, this fauna probably represents the *Levipustula levis* Zone of the Booral Formation (Campbell 1955, 1961). Both M'Coy (1847) and De Koninck (1877) described part of the Rev. W.B. Clarke's extensive fossil collections from the Hunter Valley, and compared them with Lower Carboniferous species from Europe. Systematic descriptions of the Carboniferous invertebrate faunas of eastern Australia have continued with such authors as Jack & Etheridge (1892), Dun (1902), Dun & Benson (1920), Carey (1937), Carey & Browne (1938), Delepine (1941), Maxwell (1954, 1961*a*,*b*, 1964), Campbell (1955, 1956, 1957, 1961, 1962), Campbell & Engel (1963), Roberts (1963, 1965, 1975, 1976), McKellar (1967), Dear (1968), Jenkins (1974), Engel (1975, 1980).

In Western Australia Carboniferous rocks were first recorded from the Burt Range of the Bonaparte Basin by Matheson & Teichert (1948), and then discovered in the Carnarvon Basin (Teichert 1950). In the Canning Basin erroneous records of Carboniferous fossils date back to Hardman (1884, 1885) and Blatchford (1927); confirmation that strata of this age occurred in the Canning came from both subsurface and outcrop in the 1950s (Thomas 1957, 1959).

Carboniferous strata are now known from all of the Palaeozoic sedimentary basins in Australia (Palfreyman 1984). Two distinct regions of mainly marine sedimentation are present in the intracratonic basins of the west, and within the Tasman Mobile Belt on the eastern margin (see Jones et al. 1973, Fig. 1). In central Australia (e.g. Amadeus and Ngalia Basins) is a third region of dominantly terrestrial sedimentation. Detailed correlation between the two regions of marine sedimentation is limited by differences in the composition of the faunas. Roberts (1985*a*,*b*) provided correlation charts in a comprehensive review of Australian Carboniferous palaeontology and stratigraphy. Most of the major fossil groups known from the Late Palaeozoic fossil record are represented in the Carboniferous rocks of Australia, and those most widely applied in biochronological analysis of the Australian Carboniferous (e.g. brachiopods, ammonoids, conodonts, palynofloras) are represented by various zonal schemes in the text-figures and on the chart. More recent studies have demonstrated the biostratigraphic value of other groups (e.g. trilobites, radiolarians, ostracods, and fish). Some groups may be rare (e.g. insects; Riek 1973, 1976), or of limited use in age dating and correlation, although with further study they may prove of some biostratigraphic significance (e.g. Bryozoa, Engel 1989; plant macrofossils, Gould 1975; Retallack 1980; Rigby 1985).

COLUMN 1: GEOCHRONOLOGY

The numerical limits used for the Carboniferous Chart (354–298 Ma) give a duration of 56 Ma, considerably less than the 73 Ma duration used by Harland et al. (1990), who place both the D–C boundary older (at 362.5 Ma), and the C–P boundary younger (290 Ma). The Tournaisian–Viséan boundary is about 343 Ma, and the Viséan–Namurian (Dinantian–Silesian) boundary is about 325 Ma. A more detailed summary of the ages used in the construction of the Carboniferous scale is provided in Jones (1995).

COLUMN 2: WESTERN EUROPE

Historically, three standard stratigraphic scales have been developed for the Carboniferous of the Northern Hemisphere—western Europe, the former USSR, and USA—all within the palaeoequatorial belt. Although the use of a two-fold subdivision of the Carboniferous was well established in western Europe and North America, a three-fold division (as series) continued to be used in Russia. The two-fold subdivision was complicated by the fact that the boundary between the Mississippian and Pennsylvanian subsystems of North America does not coincide with that between the Dinantian and Silesian of western Europe. Moreover, the lowermost Silesian (Namurian A) is equivalent to the Serpukhovian, the upper stage of the Lower Carboniferous of Russia. In this respect, the timespan of the Russian Lower Carboniferous is closer to that of the Mississippian rather than the Dinantian.

Attempts to integrate all three schemes (e.g. Bouroz et al. 1978; Rotai 1979), by placing the Lower, Middle, and Upper series of the Russian Carboniferous within the two-fold Mississippian and Pennsylvanian Subsystems of North America, have led to the general acceptance of a mid-Carboniferous boundary (Lane et al. 1985*b*; Wagner et al. 1985). This boundary, now defined as the first appearance of the conodont species *Declinognathodus noduliferus*, has been the focus of much research (Lane et al. 1985*b*). Palaeoequatorial correlations about the mid-Carboniferous boundary are summarised in Figure 24.

The mid-Carboniferous or Mississippian–Pennsylvanian boundary represents a major event in Phanerozoic history, when major climatic and sea-level changes occurred, as a result of the presumed Late Carboniferous Ice Age. These changes had a profound effect on the fauna and flora of Gondwana and, as a result, pose the problem of correlating Gondwanan sequences with the standard chronostratigraphic units of western Europe.

There are no formally designated local stages for the Carboniferous of Australia (see discussion in Clarke & Farmer 1976; Jones & Roberts 1976; Jones 1991); so the standard stratigraphic scales within the palaeoequatorial belt of the Northern Hemisphere must be used for comparisons with the Australian Carboniferous succession.

The western European scale is the most appropriate standard for the Carboniferous of Australia, where cosmopolitan shelly faunas of the Early Carboniferous (Dinantian) are replaced by endemic (Gondwanan), poorly represented faunas of Late Carboniferous (Silesian) age. The scarcity of conodonts and absence of fusulinids in the latter inhibit correlation with North America and Russia, where the main biozonations are based on these groups. The western European scale is also favoured by Australian palynostratigraphers (e.g. Helby & Playford in Kemp et al. 1977; Powis 1984).

The western European scale given here is a compilation based on a composite for the Early

1	2					3			4	5				6			
Ma	WESTERN EUROPE					Former USSR				USA							
	BELGIUM, BRITISH ISLES, FRANCE, GERMANY COMPOSITE					EAST EUROPEAN PLATFORM			URAL MOUNTAINS	ILLINOIS BASIN				ARKANSAS, APPALACHIAN BASIN			
311	LATE CARBONIFEROUS (SILESIAN)	WESTPHALIAN (part)	B		DUCKMANTIAN	MIDDLE CARBONIFEROUS	MOSCOVIAN (part)	VEREISKY	KIROVSKY (part)	PENNSYLVANIAN (part)	ATOKAN (part)	ABBOTT (part)		ATOKA FM	TRACE CREEK MBR	KANAWHA	POTTSVILLIAN
312			A	G2	LANGSETTIAN		BASHKIRIAN	MELEKESSKY	ASATAUSKY		MORROWAN	CASEYVILLE FM	POUND SST	BLOYD FM	KESSLER	NEW RIVER FM	
															DYE		
								CHEREMSHANSKY	TASHASTINSKY				DRURY SHALE		WOOLSEY		
															BRENTWOOD		
		NAMURIAN (part)	C	G1	YEODONIAN			PRIKAMSKY	ASKYNBASHSKY				BATTERY ROCK SANDSTONE	HALE FM	PRAIRIE GROVE		
			B	R2	MARSDENIAN			SEVEROKELTMENSKY	AKAVASSKY								
313				R1	KINDERSCOUTIAN			KRASNOPOLYANSKY	SIURANSKY				WAYSIDE SST & LUSK SHALE		CANE H	POCAHONTAS FM	
				H2	ALPORTIAN	EARLY CARBONIFEROUS	SERPUKHOVIAN (part)	VOZNESENSKY	BOGDANOVSKY								
314				H1	CHOKIERIAN			ZAPALTYUBINSKY									
315			A (part)	E2c (part)	ARNSBERGIAN (top)					?	?						

Figure 24: Details of the mid-Carboniferous correlation.

Carboniferous (Dinantian) of Belgian and British stages (George et al. 1976; Conil et al. 1977; Paproth et al. 1983), and for the Late Carboniferous (Silesian) of British, German, French and Spanish stages (Ramsbottom et al. 1978; Owens et al. 1985; Riley et al. 1985; Wagner & Winkler Prins 1985). The Devonian–Carboniferous boundary in Australia is taken at the internationally accepted definition, viz. at the first appearance of the conodont *Siphonodella sulcata.* This level is slightly below the lowermost record of *Gattendorfia subinvoluta* in the Hönnetal section, the earlier definition based on goniatites (Paproth 1980). The Global Boundary Stratotype is at La Serre in the Montagne Noire (Paproth et al. 1991); auxiliary boundary stratotypes are at Hasselbachtal (western Germany; Becker et al. 1984; Becker 1988) and Nanbiancun (southern China; Yu 1988). In terms of western European stages, the D–C boundary coincides with the Wocklumerian–Balvian boundary of Germany, which is close to the Strunian–Hastarian boundary of Belgium.

The biostratigraphic criteria for the recognition and correlation of the Tournaisian–Viséan boundary in Belgium has been discussed by Conil et al. (1989, 1991) and Jones (1991). Correlation with the boundary stratotype sec-

7	8	9	10		11	12	13	14
NORTHERN HEMISPHERE BIOZONES								
AMMONOIDEA	FORAMINIFERIDA				CONODONTS		MICROFLORA	
WESTERN EUROPE, USA, FORMER USSR COMPOSITE	FORMER USSR	USA				BRITISH ISLES	WESTERN EUROPE	ILLINOIS BASIN
Diaboloceras - Winslowoceras	*Aljutovella aljutovica - Schubertella pauciseptata*	22 (part)	*Profusulinella*	ATOKAN	*Neognathodus atokaensis*		NJ	NG
					De. marginodosus			
					Diplognathodus spp.			
Diaboloceras - Axinolobus					*Id. ouachitensis*		RA	SR
	Verella spicata - A. tikhonovichi				*Id. convexus*	*Idiognathoides sulcatus parvus*		
					I. klapperi			
Branneroceras - Gastrioceras	*Ozawainella pararhomboidalis - Profusulinella primitiva*				*Idiognathodus sinuosis*	*Idiognathoides sinuatus - I. primulus*	SS	LP
					Neognathodus bassleri			
Bilinguites - Cancelloceras	*Pseudostaffella praegorskyi Pr. staffellaeformis*	21	*Millerella*		*N. symmetricus*		FR	
	Pseudostaffella antiqua				*Idiognathoides sinuatus- Rhachistognathus minutus*		KV	
Reticuloceras - Bashkortoceras	*Eostaffella pseudostruvei - E. postmosquensis*	20		MORROWAN	U *Declinognathodus noduliferus- Rhachistognathus primus*	*Id. corrugatus - Id. sulcatus*		
Homoceras - Hudsonoceras	*Plectostaffella bogdanovkensis*				L	*D. noduliferus*	SO (part)	
Eumorphoceras (top) E2c4, E2c3 — *Nuculoceras nuculum*	*Eosigmoilina explicata- Monotaxinoides subplana* (part)	19 (part)	*Eosigmoilina robertsoni- Brenckleina rugosa* (part)	?	U *Rhachistognathus muricatus* (part)	*Rhach. minutus* — *Gnathodus bilineatus bollandensis* (upper part)		

20/O/31

tion at Dinant shows that the base of the Belgian Stage (Moliniacian) formerly taken to be the base of the Viséan in fact lies within the Tournaisian (Conil et al. 1989). Riley (1990*b*, 1991) has discussed this question from the point of view of the British Chadian Stage, which he regarded as mainly Tournaisian. The base of the Viséan corresponds *approximately* with the base of the German goniatite zone CuIIγ, and within the British Chadian stage. Recognition of this horizon in Australia is discussed below.

The base of the Late Carboniferous (Silesian) of the western European scale coincides with the base of the Namurian, which is taken as the base of the range zone of *Cravenoceras leion*. Horn (1960) has shown that this zone overlaps the upper part of the German goniatite zone CuIIIγ, viz. CuIIIγ2, which had previously been considered as the latest unit in the Viséan.

The Carboniferous–Permian boundary on the western European scale, on palynological evidence, may be taken at the Stephanian–Autunian boundary (*sensu* Bouroz & Doubinger 1977). Davydov et al. (1992) have claimed that the Stephanian–Autunian boundary (of the Donetz scale) coincides with the base of the *Daixina bosbytauensis–D. robusta* Zone of the fusulinid scale, and therefore is slightly older.

COLUMNS 3–4: FORMER USSR

The tripartite Carboniferous scale of the former Soviet Union is shown for two major regions: the East European Platform (Column 3) and the Urals (Column 4). Correlation between them is based mainly on smaller foraminifers in the Early Carboniferous, and fusulinids in the Middle and Late Carboniferous (e.g. Vissarionova 1975; Yabolkov 1975; Rotai 1979; Vdovenko et al. 1987). Inclusion of the Tsninsky suite between the Kashirsky and the Vereisky follows Solovieva (1985*a*; Solovieva et al. 1985*a*,*b*). The base of the Permian System as used here corresponds to the Russian boundary stratotype (base of the Asselian Stage), at the boundary between the zones of *Daixina bosbytauensis–D. robusta* and *Sphaeroschwagerina vulgaris–S. fusiformis* (Davydov et al. 1992).

COLUMNS 5–6: USA

Column 5 shows the four provincial series of the Mississippian established for the Upper Mississippi Valley, with a number of stratigraphic breaks that are often difficult to correlate within the USA and elsewhere. The redefined Osagean–Meramecian boundary for the type Meramecian follows Kammer et al. (1991). The position of the base of the Chesterian Series is defined at the base of the Ste Genevieve (Maples & Waters 1987). Midcontinent (Arkansas) and Appalachian representatives of the Mississippian are depicted in Column 6. The Pennsylvanian (Morrowan–Virgilian) succession of the Illinois Basin (Column 5) is correlated with the Silesian of western Europe on the basis of miospores (Peppers 1984); the Midcontinent (Arkansas) type section of the Morrowan (Column 6) is correlated with European sections on miospore (Loboziak et al. 1984; Owens et al. 1984), ammonoid (Manger & Saunders 1980; Saunders & Ramsbottom 1993) and conodont evidence (Lane et al. 1985*a*); and the Appalachian subdivisions of the Pennsylvanian stratotype (Column 6) are correlated with the Silesian of western Europe on the basis of plants (Englund et al. 1985).

The term 'Atokan' is used in the restricted sense, after Shaver (1984). Two distinctly differing opinions concerning the international correlation of the Atokan–Morrowan boundary have been summarised by Manger & Sutherland (1991), and are shown both in the chart and Figure 1. The correlation based on ammonoid and conodont evidence differs significantly from that based on foraminifers. Ammonoid, palynomorph and conodont evidence suggests that the Atokan–Morrowan boundary is within the Westphalian A Stage (Ramsbottom et al. 1978; Loboziak et al. 1984; Lane et al. 1985*a*,*b*) in Arkansas and in the Cordilleran Sections. There the base of the Atokan is drawn at the base of Zone 21 of Mamet's foraminiferal scheme, which is thought to coincide with the first appearance of the ammonoid *Winslowceras*, and the conodont *Diplognathodus*. The base of the Atokan (*sensu* Groves 1988), that is, the base of Mamet Zone 21 (see Fig. 1), is at an older level, at the base of the *Pseudostaffella antiqua* Zone, near the base of the Akavassky of the Urals. As the base of the *P. antiqua* Zone is now thought to approximate to the base of the Marsdenian of Britain (Wagner & Winkler Prins 1991; Winkler Prins 1991, p.303), it follows that this is an alternative position for the Atokan–Morrowan boundary.

Post-Virgilian units are shown for Kansas (Column 5), Texas and the Appalachian Basin (Column 6). Correlation of these follows Wilde (1975, 1984, 1990) and Baars et al. (1992), who redefined the Pennsylvanian–Permian boundary in Kansas, by extending the Virgilian up to the first appearance of Early Permian conodonts and inflated schwagerinids of the *Pseudoschwagerina* Biozone at the base of the Neva Limestone (within the Council Grove Group). In this chart the base of the Permian System in North America is taken at the base of the *Pseudoschwagerina uddeni* Zone in the Neal Ranch Formation of Texas, at the base of the Wolfcampian Series (*sensu* Ross 1963).

COLUMN 7: AMMONOIDEA

The major ammonoid genozones are shown, after Ramsbottom & Saunders (1984), with modifications to accommodate the *Goniocyclus–Protocanites* Zone of Kullmann et al. (1991) for the middle Tournaisian (Tn2), and the Viséan zones of Riley (1990*a*,*b*, 1991, 1993).

COLUMN 8: FORAMINIFERA (FORMER USSR)

Foraminiferal zonations of the former Soviet Union are based on Lipina & Reitlinger (1970), Lipina & Tschigova (1979), Rotai (1979), Vdovenko et al. (1987), Davydov (1988) and Davydov et al. (1992).

COLUMN 9: FORAMINIFERA (NORTH AMERICA AND FORMER USSR)

Mamet's global foraminiferal Zones 3–23 (Mamet 1974) are shown in Column 9. These are 'inferred Mamet zones' as recalibrated against the Mississippi Valley formations by various authors (e.g. Brenckle et al. 1974, 1982; Baxter & Brenckle 1982). Above these is shown Wilde's (1984) fusulinid zonation for the latest Carboniferous–earliest Permian sequences of North America and the former Soviet Union.

COLUMN 10: FORAMINIFERA (USA)

The Mississippian zonations are mainly based on Mamet (1974), Baxter & Brenckle (1982) and Brenckle & Groves (1986). The first appearance of tuberculate foraminifers (Brenckle 1991) is an unsuitable marker for the Kinderhookian–Osagean boundary, because their occurrence is sporadic and probably diachronous in relation to the siphonodellid zonations between the Midcontinent and the Cordillera (Brenckle & Groves 1986; Webster et al. 1993).

Groves (1988) established an approximate correlation between the base of the *Pseudostaffella antiqua* Zone and the base of Mamet's Zone 21 (base of Atokan in Texas). The Desmoinesian fusulinid zonation is after Douglass (1987), which, in toto, probably represents Mamet's undefined Zone 23. Above this the Missourian, Virgilian, post-Virgilian and Wolfcampian fusulinid zonations are taken from Wilde (1975, 1984).

COLUMN 11: CONODONTS (USA)

This column gives a composite conodont zonation. The Kinderhookian and Osagean are from Sandberg et al. (1978) and Lane et al. (1980); the Meramecian and Chesterian from Collinson et al. (1971); and the Pennsylvanian from Lane et al. (1971), Lane & Straka (1974) and Merrill (1975).

COLUMN 12: CONODONTS (WESTERN EUROPE, FORMER USSR)

The Hastarian part of the column is from Paproth et al. (1983), the Ivorian to Warnantian part from Conil et al. (1991). More detail on how the Dinantian conodont zones correlate with the Australian Carboniferous conodont zonation is given in Figure 25. The Namurian zonation follows that of Higgins (1975, 1981, 1985). The remaining zonation from Moscovian to the earliest Permian (Asselian) is after Movshovich et al. (1979), Chernykh & Reshetkova (1987), Wang (1991) and Nemirovskaya (in Winkler Prins 1991).

COLUMN 13: RADIOLARIA (GERMANY, USA, RUSSIA)

The Early Carboniferous part of the radiolarian biozonation follows Braun (1991), Braun & Gusky (1991) and Braun & Schmidt-Effing (1993) for the Rheinisches Schiefergebirge. The right-hand side of the Early Carboniferous part of the column depicts the *Albaillella* zonation of Cheng (1986, Fig. 3), based on successions in Arkansas and Oklahoma. Both zonations have been calibrated against selected conodont zones and events, which have been tied into North American and European stages. The *A. paradoxa* group first appears in the Upper *duplicata* Zone (i.e. Tn1b) in North America, which is somewhat earlier than in Germany (Lower *crenulata* Zone; i.e. Tn2a), as reported by Braun & Schmidt-Effing (1993). In the present paper, the older event is taken as the base of the *A. paradoxa* Zone (= Ab2A Zone of Cheng 1986), and is thus depicted in the chart (Column 13).

The base of the last *Albaillella* zone (*A. nazarovi* Zone) is defined by the entry of the eponymous species, and *A. pennata* in the uppermost Brigantian (P2) stage. The top of the zone remains to be established; however, it is at least as high as the Kinderscoutian (R1), and the zone appears to be the equivalent of the *Albaillella* - 3 assemblage of Holdsworth & Jones (1980).

The Late Carboniferous–earliest Permian (Moscovian–Asselian) part of the zonation is after the broad Palaeozoic radiolarian associations of Nazarov & Ormiston (1985, Fig. 7) from Russia.

COLUMN 14: MEGAFLORA

The megafloral zonation for Europe follows Wagner (1984) and Wagner & Winkler Prins (1985), who correlated the base of the *Odontopteris cantabrica* Zone with the base of the Cantabrian Stage, at the base of the Villanueva Marine Formation in the boundary stratotype section at Velilla de Tarilonte, Palencia, Spain.

COLUMN 15: MICROFLORA

The microfloral zonation for the Northern Hemisphere is based on Owens (1984) and Peppers (1984).

COLUMN 16: CONODONTS (EASTERN AUSTRALIA)

Conodonts are a widely used key group for international correlation. In Australia, Carboniferous conodonts were first discovered in the Bonaparte Basin (McWhae et al. 1958, Glenister 1960); their biostratigraphic significance was discussed by Jones & Druce (1966), and Roberts et al. (1967), and they were later described by Druce (1969). Subsequent studies on Lower Carboniferous conodont faunas in Western Australia concentrated on the Canning Basin (Nicoll & Druce 1979). Other faunas were also described from eastern Australia (Druce 1970*a,b*; Jenkins 1974; Webb 1977; Pickett 1981, 1993; Mory & Crane 1982). Earlier summaries and review compilations have been provided by Jones et al. (1973), Druce (1974), Jones & Roberts (1976), Nicoll & Jenkins (1985) and Jones (1991).

Apart from the descriptions of small faunas from Queensland (Druce 1970*a,b*; Webb 1977; Pickett 1981), the first major work on Carboniferous conodont faunas from eastern Australia was the conodont zonation proposed by Jenkins (1974) for New South Wales. Jenkins (1974) proposed six informal biostratigraphic zones above the lowest fauna, which was characterised by *Siphonodella*. Subsequently, Mory & Crane (1982) developed a siphonodellid zonation that could be tied into international siphonodellid schemes (e.g. Sandberg et al. 1978). Siphonodellids have also been recognised in Western Australia (Druce 1969; Nicoll & Druce 1979) but they are poorly represented.

In this column (also Fig. 25), the base of the *Gnathodus punctatus* Zone of Jenkins (1974) is correlated with the base of the *Polygnathus communis carina* Zone at the base of the Ivorian

Stage (Tn3a) of Belgium as advocated by Mory & Crane (1982). In terms of standard conodont zonations, this horizon corresponds to the base of the Lower *typicus* Zone of Lane et al. (1980), and to the base of the *cuneiformis* Zone of Belka & Groessens (1986). This differs from the earlier suggestion (Jones 1991) that the *Gnathodus punctatus* Zone of Jenkins (1974) correlates with the uppermost siphonodellid zone (*Gnathodus punctatus* [Cc 1γ] Zone), at the top of the Hastarian Stage (Tn2c) of the scheme of Paproth et al. (1983).

Jenkins et al. (1993) described the conodont sequence of the Viséan rocks above the *Scaliognathus anchoralis* Zone in New South Wales and Queensland. Four Viséan conodont biozones were established above the *Scaliognathus anchoralis* Zone (previously zones 1–4 of Jones 1991 and Roberts et al. 1993*b,c*). The three lower zones are based on the ranges of species in the *Patrognathus–Montognathus* lineage, and the fourth on *Gnathodus texanus* and *G. bilineatus*.

The lowest biozone (*Patrognathus conjunctus* Zone) corresponds to the lower part of the informal *Patrognathus? capricornis* Zone of Jenkins (1974). It is marked by the entry of the eponymous species which, in northern New South Wales, follows closely on the *Scaliognathus anchoralis* Zone. The informal *Pseudopolygnathus* cf. *nodomarginatus* Zone of Jenkins (1974) corresponds to the upper part of the *anchoralis* Zone, and is now regarded as a local teilzone (Jenkins et al. 1993). Taxa that first appear in the top of the *Scaliognathus anchoralis* Zone include key species such as *Polygnathus bischoffi* and *Gnathodus subbilineatus* that indicate an early Viséan age.

The base of the *Montognathus semicarinatus* Zone (Zone 2) is taken at the lowest abundant appearance of the eponymous species, disregarding its rare occurrence in the *conjunctus* Zone. Important taxa include species of *Adetognathus*. Difficulties in correlating at this level because of the endemic nature of the Australian conodont faunas are discussed in detail by Jenkins et al. (1993). The extinction of *Polygnathus bischoffi* in the lower part of the zone indicates that the base of the zone lies within the lower (V1b) part of the Arundian.

The base of the *Montognathus carinatus* Zone is considered to approximate to the base of the Holkerian (Roberts et al. 1993*c*). Important taxa in this zone include the eponymous species, *Gnathodus girtyi*, *Lochriea commutata*, *Adetognathus subunicornis* and *Mestognathus convexus*.

The base of the *Gnathodus texanus–G. bilineatus* Zone is characterised by the eponymous species, as revised by Jenkins et al. (1993). The first entry of *G. bilineatus* in western Europe establishes an early Asbian age for this level. The top of this zone contains *Rhachistognathus prolixus* Baesemann & Lane, a species reported by Higgins et al. (1991) from the *Gnathodus girtyi collinsoni* and *G.* cf. *texanus* zones of western Canada, indicating a Brigantian age. Jenkins et al. (1993) demonstrated that *R. prolixus* first appears in Australia in rocks equivalent to a topmost V3c or E1a age. On this basis, the *Gnathodus texanus–G. bilineatus* Zone is topmost Viséan and may extend into the early Namurian.

The youngest known conodonts in the Carboniferous of Australia were described by Palmieri (1969) from the Murgon district, south-eastern Queensland (see Jones et al. 1973; Druce 1974; Lane & Straka 1974; Jones & Roberts 1976; Nicoll & Jenkins 1985). These samples contained poorly preserved faunas, possibly of different ages, ranging from early Namurian to early Westphalian.

COLUMNS 17–18: BRACHIOPODS

Brachiopods are well represented in the Carboniferous marine shelf sediments of Australia, and have been utilised for biostratigraphic studies and correlation in eastern and Western Australia (Roberts 1975, 1985*a*). Column 17 gives the zonal scheme for the New England and Yarrol orogens and the Broken River Embayment of eastern Australia (Roberts 1975); Column 18 is the scheme for the Bonaparte Basin of Western Australia (Roberts 1971; Thomas 1971).

Some of the zones established for the Bonaparte Basin can be identified in other basins in Western Australia (Canning and Carnarvon basins). Both brachiopod schemes have been used, together with conodont zones in Western Australia, in the correlation of the Carboniferous System of Australia (Jones et al. 1973). Compositional differences between the brachiopod faunas in eastern and Western Australia have been explained palaeobiogeographically (Roberts 1971), or palaeoenvironmentally (Runnegar & Campbell 1976). A recent study of benthic marine communities in the *Rhipidomella fortimuscula* Zone (Lavering 1993) shows scope for further work on the Carboniferous brachiopod succession in terms of community associations and ecostratigraphy.

Roberts et al. (1993*c*) in a recent review of the international significance of the brachiopod, ammonoid, conodont and foraminiferal evidence from the Early Carboniferous of eastern Australia, resolved most of the apparent anomalies between the conodont and brachiopod zones (Jones 1991). Recent conodont evidence (Jenkins et al. 1993) supports the correlation of the boundary between the zones of *Marginirugus barringtonensis* and *Levipustula levis* within the early Namurian Pendleian Stage of Britain (Roberts et al. 1976). However, there still remains the age problem of the top of the *Levipustula levis* Zone. The surprisingly young (Permian) isotopic ages reported for this level in the Southern New England Orogen (Roberts et al. 1993*b*) dated thin sills, rather than erupted units; so the younger ages do not constrain the biostratigraphy (Roberts et al. 1995). Although this level cannot be dated precisely, recent work suggests it is unlikely that it extends above the Namurian (Roberts et al. 1995).

In Queensland, the *Levipustula levis* Zone is followed by the younger *Auriculispina levis* Zone, within the Neerkol Formation. The *A. levis* Zone was originally introduced by Engel (1975) as the *Cancrinella levis* Zone for the association of the eponymous brachiopod species and the bryozoans *Septatopora flemingi* and *S. pustulosa*. In the type locality, in central Queensland, Fleming (1969) recognised two faunal units within the Neerkol Formation: a lower one characterised by *Levipustula levis*, and an upper one characterised by *Cancrinella levis*. The Neerkol Formation is separated from the overlying Dinner Creek Conglomerate (with a *Glossopteris* flora) by a strong, erosional unconformity.

A few remnant species from the *L. levis* Zone persist through most of the sparsely fossiliferous Rands Formation, and are followed in the upper part of that unit (*sensu* Dear et al. 1971) by the *Auriculispina levis* Zone. The basal conglomerate of the overlying Burnett Formation (*sensu* Dear et al. 1971) with the *Eurydesma* fauna probably marks an hiatus, comparable with the erosional unconformity in the Rockhampton area (Fleming 1969) that separates the upper Neerkol Formation from the overlying Dinner Creek Conglomerate. Although the hiatus may not necessarily be of the same magnitude, this tectostratigraphic comparison, and the fact that the *Auriculispina levis* Zone contains Carboniferous rather than Permian bryozoans (Engel 1989), indicate that the zone is more likely to be at the older (Namurian–Westphalian) limit of its total possible age range.

COLUMN 19: OSTRACODA

Ostracoda have been used since the late 1950s, to provide correlations of Carboniferous and Late Devonian rocks encountered in petroleum exploration wells in the Canning and Bonaparte Basins, Western Australia (Jones 1958, 1959, 1961*a*, 1962*a,b,c*). So far the analysis of these extensive assemblages is limited to some eridostracans (Jones 1962*c*; 1968), and Carboniferous ostracods from the Bonaparte Basin (Jones 1989).

The ostracod scale given in Column 19 is internally controlled by conodont and foraminiferal zonations within the Bonaparte Basin, and is also calibrated against the Dinantian timescale by these means, and by cognate and conspecific ostracod species (Jones 1989).

To date, the older three assemblages, of early and middle Tournaisian age, have been recognised in the Laurel Formation of the Canning Basin. The alphabetic notation refers to a provisional ostracod scale set up for the Bonaparte and Canning basins (Jones 1974).

COLUMN 20: RADIOLARIA

Despite the recognition of radiolarians in Palaeozoic rocks from the New England Fold Belt last century (David & Pittman 1899) and the detailed description of a well preserved and varied Middle Devonian fauna (Hinde 1899), studies of Carboniferous Radiolaria did not start in this region until the late 1980s (e.g. Ishiga et al. 1988; Aitchison 1988*b,c,d*). Apart from those reported from the New England Fold Belt, including the Neranleigh–Fernvale beds in the Brisbane area (Aitchison 1988*a*), no Carboniferous radiolarians have been recorded from elsewhere in Australia. Thus, the study of the biostratigraphy of Carboniferous Radiolaria in Australia is still at an early stage and, until recently, the New England assemblages were calibrated against the preliminary Late Devonian to Permian radiolarian zonation of Holdsworth & Jones (1980).

Over the past five years radiolarian age data have contributed towards the interpretation of the complex geological history of the New England Fold Belt (Aitchison 1989, 1990, 1993*b*; Aitchison & Flood 1990). Aitchison (1993*b*) has recently introduced a radiolarian biostratigraphy for the Late Devonian–Early Carboniferous, consisting of ten assemblages, seven of which are within the Early Carboniferous. He noted that the presence of the *Albaillella paradoxa* group in his *Protoalbaillella anaiwanensis* assemblage indicated that this fauna can be assigned to the Ab2A assemblage-zone of Cheng (1986). This suggests that the *P. anaiwanensis* assemblage is equivalent to an early part of the *A. paradoxa* Zone in North America. It also supports the correlation of Cheng (1986), which has the entry of the *A. paradoxa* group earlier in North America than in Germany (see notes for Column 13).

COLUMN 21: PLANT MACROFOSSILS

Most accounts of the Carboniferous macrofloral succession in Australia (e.g. Gould 1975; Morris 1975, 1985; Retallack 1980; Rigby 1985; White 1986), including the first report by Feistmantel (1890), have concentrated on those of eastern Australia, mainly from the Southern New England Orogen (SNEO). Here, the *Leptophloeum australe* flora of the Late Devonian is succeeded by the early Tournaisian *Lepidodendropsis* flora. The so-called *Lepidodendron* flora of Morris (1975: *Lepidodendropsis* or *Sublepidodendron*, according to Rigby 1973) ranges from the *Spirofer sol* Zone to the earliest part of the lower *Inflatia elegans* subzone of the *Delepinea aspinosa* brachiopod zone (Tn2a–V1b).

The *Pitus* flora of Morris (1975), which consists of silicified logs of *P. sussmilchii*, appears to range throughout the remainder of the *aspinosa* Zone in the SNEO (Roberts et al., in press *a*); however, according to Gould (1975) the *Pitus* flora does not have the biostratigraphic significance implied by Morris, and he quotes a possible younger record from the Clifden Formation (Walkom 1928*b*) in the northern part of the SNEO.

The *Nothorhacopteris argentinica* flora has been previously interpreted as entirely Late Carboniferous in age. However, its earliest definite appearance (as the 'enriched *Nothorhacopteris* flora' of Morris 1985) in the upper Mount Johnstone Formation near Paterson, has now been shown, on the basis of zircon-dating (Roberts et al., in press *a,b*), to be as old as late Viséan (V3b). This enriched flora, also known as the *Sphenopteridium* flora (Gould 1975; Retallack 1980) and the *Fedekurtzia intermedia* flora (Rigby 1985), appears to be present throughout most of the range of *Nothorhacopteris argentinica*.

Retallack (1980) recognised a *Botrychiopsis* megafossil florule in rocks containing the

Potonieisporites or Stage 1 assemblage (Kemp et al. 1977), a palynological assemblage that is now correlated with Biozone D of the Galilee Basin sequence (Jones & Truswell 1992). According to Rigby (1985), this plant assemblage, which he named the *Botrychiopsis ovata* assemblage, succeeds that of *Nothorhacopteris argentinica*. It is present in the Seaham Formation at Lochinvar in the SNEO. Stratigraphically, it succeeds the *Sphenopteridium* florule, and is followed by the Permian *Gangamopteris* florule (of Retallack 1980). The *Botrychiopsis ovata* and *Gangomopteris* floras are separated in the SNEO by a strong erosional unconformity, which represents an hiatus of about 15 Ma (Roberts et al., in press *b*).

COLUMN 22: PLANT MICROFOSSILS

Since Carboniferous plant microfossils were first reported from Australia (Balme 1960), they have proven to be widely distributed and of considerable stratigraphic value. Key references include Playford & Helby (1968), Playford (1971, 1972, 1976), Kemp et al. (1977), Powis (1979) and Truswell (1980). Balme (1980*a*) discussed palynological evidence bearing on the Carboniferous–Permian boundary problem in Gondwana sequences, and Playford (1985, 1991) reviewed Early Carboniferous palynomorphs from Australia. The summary zonation (Column 22) is a composite attempting to integrate western and eastern Australian sequences. More confidence is placed in the lower part of the column, which is founded on detailed taxonomic work (Playford 1971, 1972, 1976; Playford & Satterthwait 1985, 1986, 1988) from the Bonaparte and Canning Basins in Western Australia. These broad zones are also identified in the Early Carboniferous of eastern Australia, for example in the Drummond Basin (Playford 1977, 1978). Foster (1989) has recently introduced a new zone, the *Grandispora* cf. *G. praecipua* Zone, situated between the *Grandispora spiculifera* and the *Anapiculatisporites largus* Zones based on material from the offshore Bonaparte Basin.

The post-*maculosa* part of the zonation follows the recent work by Jones & Truswell (1992) in the Galilee Basin, where four Late Carboniferous palynofloral zones (A–D) were recognised, being approximately equivalent to the *S. ybertii* and *Potonieisporites* assemblages of Kemp et al. (1977). As the first appearance of monosaccate pollen in the earliest Namurian (Clayton et al. 1991) also defines the base of the *Spelaeotriletes ybertii* assemblage, it follows that the base of the oldest biozone (Zone A) of Jones & Truswell (1992) is also earliest Namurian. Jones & Truswell (1992) estimated that the youngest zone (Zone D) lay within the range Westphalian D to Late Autunian. This biozone may be as old as Westphalian C, given the similarities between its palynofloras and those in the Seaham Formation of the Hunter Valley, which from zircon data is unlikely to be younger than about 309 Ma (Roberts et al., in press *b*). Also, in view of the erosional unconformity between the Carboniferous and Permian sequences in the Yarrol Basin, Queensland, Zone D may be no younger than Westphalian C.

Playford (1986) has published the only account of megaspores from the Australian Carboniferous. He reported that they are rarely encountered and fragmentary in the Carboniferous of Australia, mostly in marine sequences examined for miospores and pollen. At present, megaspores are not important for biostratigraphic studies in the Australian Carboniferous, but increased knowledge of these larger palynomorphs could be of palaeobotanical, if not stratigraphical, significance.

COLUMNS 23–26: CORRELATION WITH ARGENTINA

The marine invertebrate, megafloral and microfloral zonations from Argentina are after Gonzalez (1985, 1989, 1993), Taboada (1989), Archangelsky et al. (1980), and Azcuy et al. (1990). The hiatus shown for the late Famennian and most of the Tournaisian represents the

Devonian–Carboniferous unconformity related to the Eo-Hercynian Orogeny in southern South America (Lopez-Gamundi & Rossello 1993). In the Calingasta–Uspallata Basin, floodplain deposits in the El Raton Formation contain the Viséan flora of the *Archeosigillaria–Lepidodendropsis* Zone (= AL Zone of Sessarego & Césari 1989). An unnamed microfloral association within the AL Zone is thought to share species in common with the *Grandispora maculosa* and *Anapiculatisporites largus* assemblages of Australia (Sessarego & Césari 1989). The AL Zone is also present in the Maliman Formation of the Rio Blanco Basin, which contains marine interbeds with representatives of the '*Protocanites*'–*Rosirhynchus* Zone of Gonzalez (1993). The '*Protocanites*' ammonoid fauna, known from only one locality, is considered to be late Tournaisian (Gonzalez 1985); however, according to Kullmann (1993) the specimen figured by Antelo (1969) as *Protocanites* may belong to the early Viséan genus *Michiganites*.

The major glacial episode in the Carboniferous of Argentina is represented by the San Eduardo Group (El Paso and Hoyada Verde formations) in the Calingasta–Uspallata Basin. Gonzales (1990) recognised two glaciations in the El Paso Formation (EPI, EPII) which are separated by an interglacial period (IG), and a third within the Hoyada Verde Formation (HV). An alternative interpretation of the glacial episode in the San Eduardo Group has been provided by Lopez-Gamundi & Espejo (1993). Taboada's (1989) *Rugosochonetes–Bulahdelia* Zone, known from a single locality within the EPII glacial episode, contains species in common with, or closely comparable to, species in the *Rhipidomella fortimuscula* and *Marginirugus barringtonensis* zones of New South Wales. The *Levipustula levis* Zone is present in the HV glacial event in the Hoyada Verde Formation, which is approximately synchronous with the onset of cold climate conditions in eastern Australia. Sessarego & Césari (1989) regarded the extremely sparse megaflora at this time as a 'sterile interzone', which separates the AL Zone from the NBG Zone (see below).

The *L. levis* Zone is succeeded by a warm water fauna (the 'Intermediate fauna', which includes the *Buxtonia–Heteralosia* fauna) regarded as late Westphalian–Stephanian in age, as it is overlain by Early Permian (Asselian) glacigene deposits and the *Cancrinella* aff. *farleyensis* Zone (Gonzalez 1990, 1993). During this Late Carboniferous postglacial period, an important megaflora—the NBG Zone (*Nothorhacopteris, Botrychiopis, Ginkgophyllum* Zone)—was present in the coal seams and carbonaceous shales of the Tupe and Lagares formations. Associated with the NBG zone are the AN (*Ancistrospora*) and PO (*Potonieisporites*) palynozones of Azcuy & Jelin (1980), which are correlated with the *Spelaeotriletes* assemblage and the *Potonieisporites* assemblage of Australia respectively (Azcuy & Jelin 1980; Truswell 1980; Jones & Truswell 1992). Thus the AN Palynozone may be as old as earliest Namurian, and age equivalent to the marine *Levipustula levis* fauna.

The base of the Permian is taken at the upper limit of the NBG Zone, marked by the incoming of the *Nothorhacopteris chubutiana–Gangamopteris* flora, and the base of the *Cristatisporites* Zone (Palynozone III of Azcuy & Jelin 1980). The latter zone, in Australian terms, is equivalent to the *Granulatisporites confluens* Zone of Foster (Foster & Waterhouse 1988).

COLUMNS 27–29: CORRELATION WITH SOUTH CHINA

The standard stratigraphic subdivision for the Carboniferous of China has been obtained from the southern China region (Hunan, Guizhou and Guangxi provinces), after Zhang (1987) and Wang (1987), with the mid-Aikuanian Event after Ji (1987). The Lower Carboniferous (Fengningian) is commonly divided into two series: the Aikuanian (Tournaisian) and Tatangian (Viséan, Namurian) series.

Two competing levels for the top of the Tatangian are shown: (i) defined by the base of the Dewan Stage at the Viséan–Namurian boundary (Wu et al. 1987), and (ii) defined by the Luosuan Stage at the mid-Carboniferous boundary (Rui & Zhang 1987). The Upper Carboniferous (Hutian) is commonly divided into two series: the Weiningian and Mapingian series. The Weiningian consists of the Huashiban and Dalan Stages (Yang et al. 1980). The base of the Huashiban Stage is defined by the entry of *Pseudostaffella* and *Reticuloceras*, indicative of the Marsdenian; and the base of the Dalan Stage is defined by the incoming of *Idiognathoides sulcatus parvus*, indicative of late Westphalian A (Rui & Zhang 1991). The Xiaodushanian Stage, the base of which is marked by the incoming of *Protriticites*, was established by Zhou et al. (1987) for the lower part (*Montiparus* and *Triticites* zones) of the original Mapingian Stage (Yang et al. 1979). Zhou et al. 1987 restricted the Mapingian Stage to the upper part (*Pseudoschwagerina–Zellia* Zone) of the original Mapingian, which correlates with the Asselian stage of Russia.

OTHER GROUPS

Ammonoidea (Fig. 26) Australian Carboniferous ammonoids are mainly from the eastern states, where they occur in at least sixteen different stratigraphic horizons (Campbell et al. 1983; and Campbell in Roberts 1985*a*). Glenister (1960) recorded a single species of *Imitoceras* from the Canning Basin of Western Australia. The stratigraphic distribution of all named Australian species from the Early Carboniferous (Dinantian) are plotted up in Figure 26 against the eastern Australian brachiopod scale as calibrated by the latest conodont zonation (Roberts et al. 1993*c*). European goniatite and foram zones are given for reference purposes.

Trilobites Only one trilobite superfamily (the Proetacea) survived into the Carboniferous, but in parts of the Carboniferous trilobites can be biostratigraphically useful. This has been borne out by recent trilobite studies in Europe (e.g. Osmolska 1970; Thomas et al. 1984; Owens 1986; Hahn & Hahn 1988), and in eastern Australia (Engel & Morris 1975, 1980, 1983, 1984, 1985, 1989, 1991*a*,*b*).

For the Australian Carboniferous, species ranges for various genera have been documented (e.g. Engel & Morris 1985, 1991*a*), although no trilobite zonation has been constructed.

Corals These have a more limited geographic and stratigraphic distribution than brachiopods (Campbell & McKellar 1969), and no coral biozonation has been established. Pickett & Wu (1990) have recently reviewed the coral faunas of eastern Australia, and have plotted a succession of five faunas within the framework of the brachiopod zonation. Webb (1989, 1990) studied the systematics, biostratigraphy, and palaeoecology of Lower Carboniferous coral faunas in the northern Yarrol Basin.

Foraminifera and algae Carboniferous foraminifera from Australia were first reported from the Bonaparte Basin by Jones (1958), and have been described by Belford (1968, 1970) and Mamet & Belford (1968). Mamet & Playford (1968) reported foraminifera from the Canning Basin. Work on Carboniferous algae from the west is limited to Veevers (1970) and Mamet & Roux (1983) on material from the Bonaparte Basin.

No Australian zonations based on foraminifera and algae have been proposed, but their biostratigraphic potential is shown by a few taxonomic studies that have permitted assignment to the global zonations established by Mamet (1974). The study of Australian Carboniferous algae by Mamet & Roux (1983) was mainly taxonomic, and the results of this work have not yet been put into a biostratigraphic context.

In the volcanogenic provinces of eastern Australia there are few recorded occurrences of foraminifera and algae, but where present they have proved biostratigraphically useful. For example, foraminifera and algae identified by B.L. Mamet (in Roberts 1975) from near the top

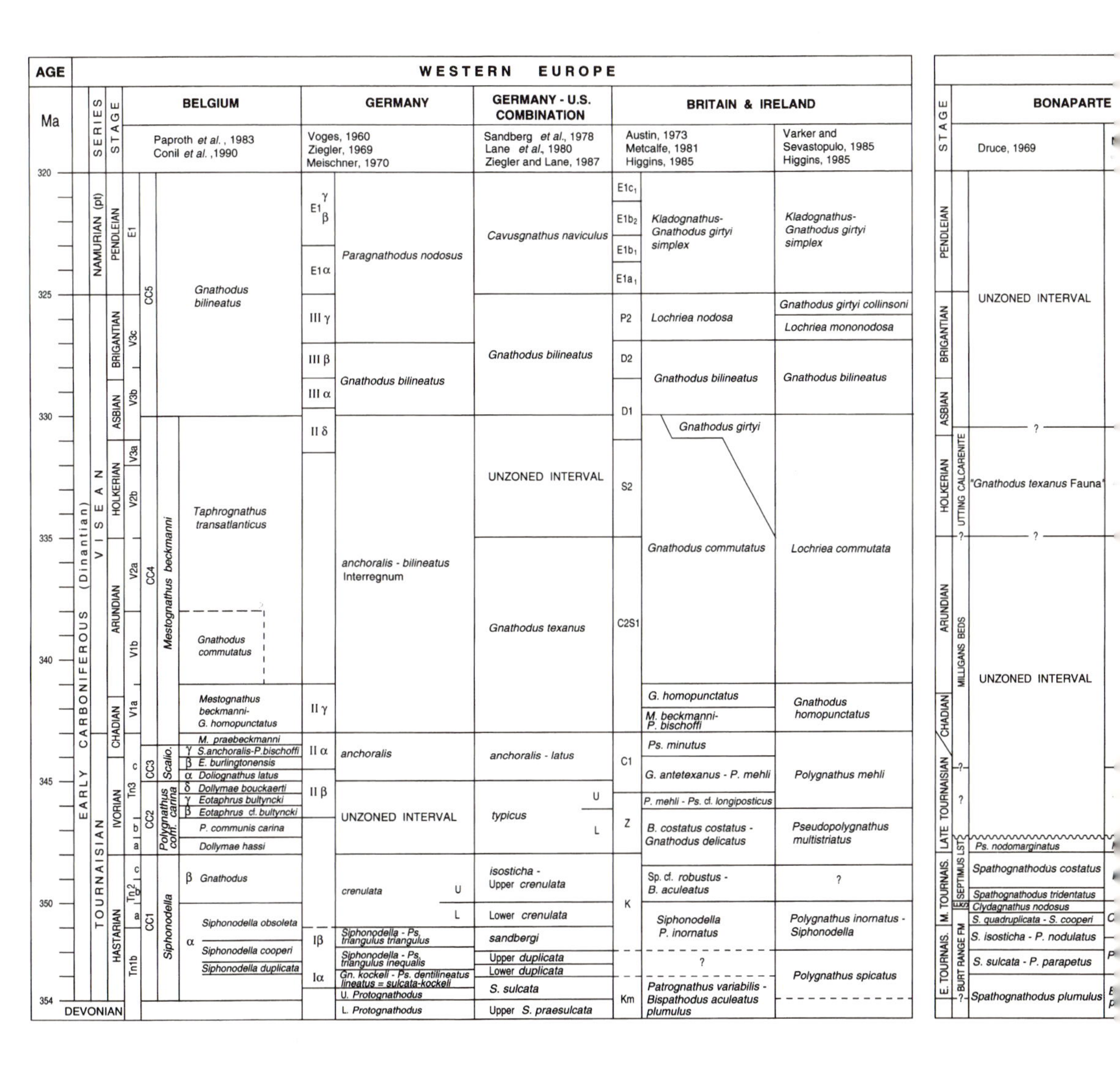

Figure 25: **Conodont zonations:** Proposed correlation of Early Carboniferous conodont zonations from **Western Europe,** after Voges (1960), Ziegler (1969), Meischner (1970), Sandberg et al. (1978), Lane et al. (1980), Metcalfe (1981), Paproth et al. (1983), and Varker & Sevastopulo (1985); **Australia,** after Druce (1969), Jenkins (1974), Nicoll & Druce (1979), Nicoll & Jones (1981), Mory & Crane (1982), Jones (1989), and Jenkins et al. (1993); **North America,** after Collinson et al. (1962, 1971), Thompson (1967), Thompson & Fellows (1970), and Baxter & von Bitter (1984).

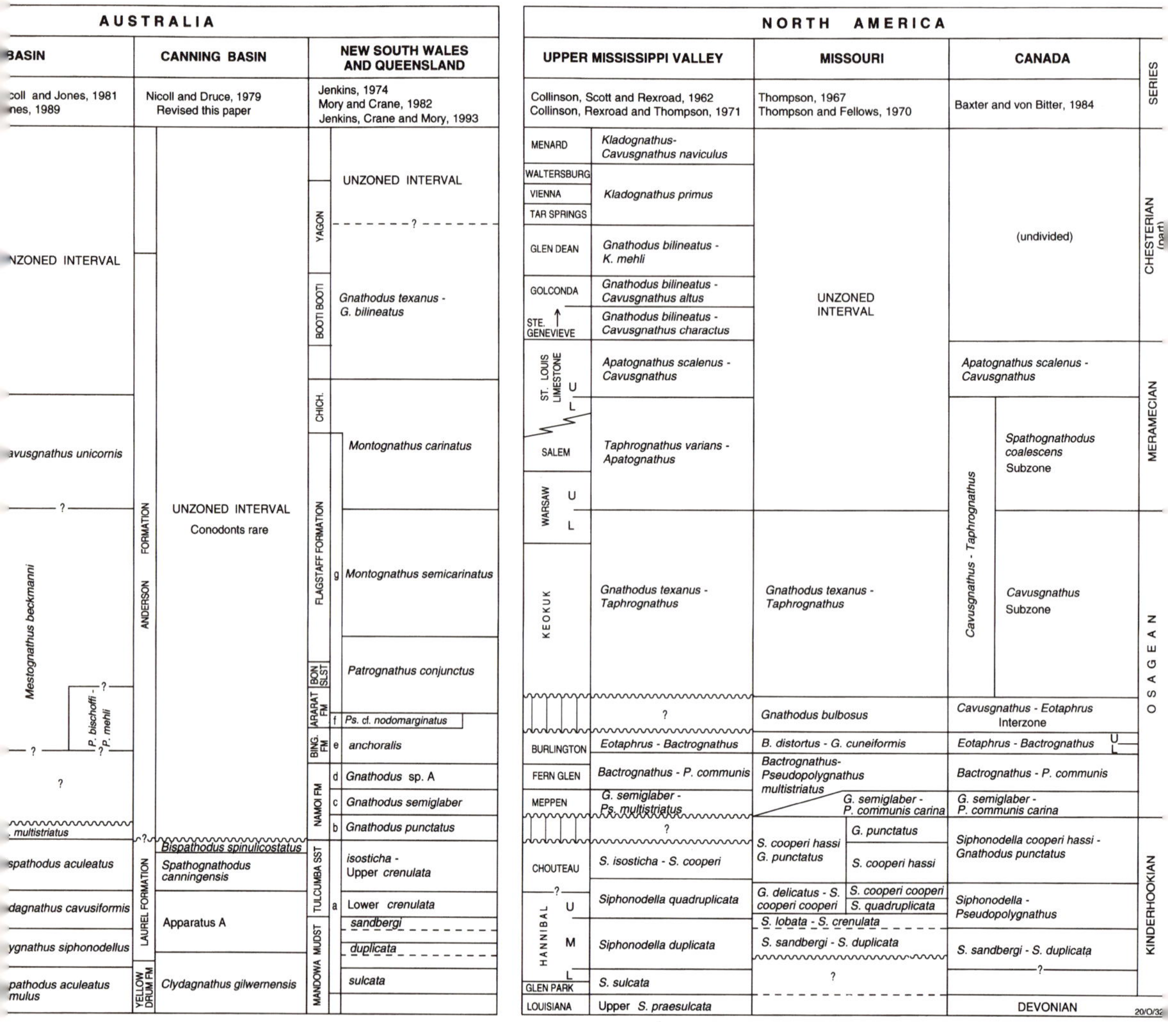

AUSTRALIA

[?]BASIN	CANNING BASIN		NEW SOUTH WALES AND QUEENSLAND		
[Nic]coll and Jones, 1981; [Jo]nes, 1989	Nicoll and Druce, 1979; Revised this paper		Jenkins, 1974; Mory and Crane, 1982; Jenkins, Crane and Mory, 1993		
[U]NZONED INTERVAL	ANDERSON FORMATION	UNZONED INTERVAL Conodonts rare	YAGON		UNZONED INTERVAL
					?
			BOOTI BOOTI		Gnathodus texanus - G. bilineatus
[C]avusgnathus unicornis			CHICH.		Montognathus carinatus
?					
Mestognathus beckmanni			FLAGSTAFF FORMATION	g	Montognathus semicarinatus
			BON SLST		Patrognathus conjunctus
P. bischoffi - P. mehli ?			ARARAT FM	f	Ps. cf. nodomarginatus
?			BING. FM	e	anchoralis
?			NAMOI FM	d	Gnathodus sp. A
				c	Gnathodus semiglaber
[..] multistriatus	?			b	Gnathodus punctatus
	LAUREL FORMATION	Bispathodus spinulicostatus			
[Bi]spathodus aculeatus		Spathognathodus canningensis	TULCUMBA SST		isosticha - Upper crenulata
[Cly]dagnathus cavusiformis		Apparatus A		a	Lower crenulata
			MANDOWA MUDST		sandbergi
[Pol]ygnathus siphonodellus					duplicata
[Bis]pathodus aculeatus [..]mulus	YELLOW DRUM FM	Clydagnathus gilwernensis			sulcata

NORTH AMERICA

UPPER MISSISSIPPI VALLEY		MISSOURI		CANADA	SERIES
Collinson, Scott and Rexroad, 1962; Collinson, Rexroad and Thompson, 1971		Thompson, 1967; Thompson and Fellows, 1970		Baxter and von Bitter, 1984	
MENARD	Kladognathus- Cavusgnathus naviculus	UNZONED INTERVAL		(undivided)	CHESTERIAN (part)
WALTERSBURG / VIENNA / TAR SPRINGS	Kladognathus primus				
GLEN DEAN	Gnathodus bilineatus - K. mehli				
GOLCONDA	Gnathodus bilineatus - Cavusgnathus altus				
STE. GENEVIEVE ↑	Gnathodus bilineatus - Cavusgnathus charactus				
ST. LOUIS LIMESTONE U	Apatognathus scalenus - Cavusgnathus			Apatognathus scalenus - Cavusgnathus	MERAMECIAN
ST. LOUIS LIMESTONE L / SALEM	Taphrognathus varians - Apatognathus			Cavusgnathus - Taphrognathus: Spathognathodus coalescens Subzone	
WARSAW U					
WARSAW L / KEOKUK	Gnathodus texanus - Taphrognathus	Gnathodus texanus - Taphrognathus		Cavusgnathus - Taphrognathus: Cavusgnathus Subzone	OSAGEAN
	?	Gnathodus bulbosus		Cavusgnathus - Eotaphrus Interzone	
BURLINGTON	Eotaphrus - Bactrognathus	B. distortus - G. cuneiformis		Eotaphrus - Bactrognathus (U / L)	
FERN GLEN	Bactrognathus - P. communis	Bactrognathus- Pseudopolygnathus multistriatus		Bactrognathus - P. communis	
MEPPEN	G. semiglaber - Ps. multistriatus		G. semiglaber - P. communis carina	G. semiglaber - P. communis carina	
	?	S. cooperi hassi G. punctatus	G. punctatus	Siphonodella cooperi hassi - Gnathodus punctatus	KINDERHOOKIAN
CHOUTEAU	S. isosticha - S. cooperi		S. cooperi hassi		
? HANNIBAL U	Siphonodella quadruplicata	G. delicatus - S. cooperi cooperi	S. cooperi cooperi / S. quadruplicata	Siphonodella - Pseudopolygnathus	
		S. lobata - S. crenulata			
HANNIBAL M	Siphonodella duplicata	S. sandbergi - S. duplicata		S. sandbergi - S. duplicata	
HANNIBAL L / GLEN PARK	S. sulcata	?		?	
LOUISIANA	Upper S. praesulcata			DEVONIAN	

20/O/32

One is present in the subsurface Anderson Formation (Viséan) in the Canning Basin, Western Australia (Tasch & Jones 1979*a*), and the other is present in the Raymond Formation (?late Tournaisian–early Viséan) of the Drummond Basin, Queensland (Tasch 1979). The dominant taxa of both faunas are the leaiids (*Hemicycloleaia* and *Rostroleaia*), which are currently being revised by Jones & Chen (in prep.).

Vertebrates Fish faunas are not well known in the Carboniferous of Australia, and consequently they have not received the same biostratigraphic attention. Important recent studies include the well known Lower Carboniferous (Tournaisian) Mansfield fauna of Victoria (Smith Woodward 1906; Hills 1958; Long 1988*a*). In Queensland, an extremely rich fish fauna has been recently found in the Drummond Basin (Turner & Long 1987; Turner 1993). Early Carboniferous shark remains have been reported by Turner (1990) from the Rockhampton district and by Thomas (1959) from the Lower Carboniferous (Tournaisian) Laurel Formation in the Canning Basin of Western Australia.

EUSTASY

Ramsbottom (1973, 1977, 1979) analysed the major transgressions and regressions within the Carboniferous of north-western Europe, and established a nomenclature of eustatic cycles. The largest units were regarded as synthems (Chang 1975), these being composed of several mesothems, each comprising several cyclothems. Ramsbottom (1979) regarded these eustatic units as time-significant, existing in parallel with chronostratigraphic series and stages, because their boundaries, which are at unconformities on shelf areas, are actually defined at the bases of chronozones in basinal areas. Thus his classification was a pioneer analysis of the Carboniferous in terms of modern sequence stratigraphy (Riley 1993).

Ross & Ross (1985) demonstrated the synchronous distribution of these transgressive–regressive depositional sequences in north-western Europe, Russia and North America, and later (Ross & Ross 1987) used such sequence stratigraphy methods to construct a eustatic curve from North American transgressive–regressive depositional sequences. They identified more than fifty third-order sequences in the Carboniferous. Most of these occur in the Late Carboniferous (Pennsylvanian), for which they suggested a glacio-eustatic origin, with other (tectonically induced) changes in sea-level being superimposed on the smaller cycles. However, some trangressions are still difficult to fit into a biostratigraphy and cannot be identified easily within a continuous marine succession, especially where the correlation of major chronostratigraphic schemes are based on the stratigraphic ranges of single species (Wagner & Winkler Prins 1991).

Roberts (1985*b*) interpreted Carboniferous sea-level changes from depositional patterns in various areas of Australia, and Veevers & Powell (1987) proposed a model relating Late Palaeozoic glacial episodes of Gondwana with the transgressive–regressive depositional sequences in Euramerica. The timing of such eustatic events, as interpreted in these papers, now requires recalibration against a revised timescale, as evidenced from later biostratigraphic and geochronologic studies from eastern Australia (Roberts et al. 1993*a,b*, 1995; Roberts et al., in press). Although the resolution of the Australia-wide curve does not permit comparison with the level of the third-order cycles, some possible correlations to the second-order curve of Ross & Ross (1987) are indicated.

A sharp global regression in the latest Devonian, which is recognised in the Canning Basin as the Yellow Drum regression, is followed by the Laurel transgression, the first major transgression of the Early Carboniferous (Talent et al. 1993). Roberts (1985*b*) identified a regression about the Tournaisian–Viséan boundary, which he interpreted as a major eustatic fall in sea-level throughout an hiatus spanning most of the early and middle Viséan. New biostratigraphic studies (Jenkins et al. 1993; Roberts et al. 1993*c*) now show that this hiatus is very short, and is probably due to local uplift associated with the onset of widespread magmatic activity, rather than the result of falling sea-level (Roberts et al., in press).

A major regression in the late Viséan, marked by a sharp decrease in diversity of the brachiopod faunas at the end of the *Rhipidomella fortimuscula* Zone (Roberts 1981), was associated with the culmination of regressive volcanogenic deposition within the SNEO, and arching along the western margin of the continent. This event was possibly enhanced by glacio-eustatic fluctuations, as the main Gondwanan glaciation started in Argentina in the late Viséan (Gonzalez 1990). In Euramerica, the inception of Gondwanan glaciation is marked by the start of cyclothemic deposition near the base of the Chester Series of the Upper Mississippi Valley, and the base of the Brigantian (Veevers & Powell 1987). Carboniferous glacial deposits in Australia are not older than Namurian, and occur at the time of the *Levipustula levis* Zone (Roberts et al. 1995).

ACKNOWLEDGMENTS

Many colleagues have contributed information and discussion, the results of which have been included in the charts and notes. I especially thank Professor John Roberts (University of NSW) for many discussions on the Carboniferous biostratigraphy of eastern Australia, and Dr Huw Jenkins (University of Sydney) for his advice on Carboniferous conodont biostratigraphy. I am indebted to my AGSO colleagues Dr J.M. Dickins, Dr Clinton Foster, Dr R.S. Nicoll and Dr Gavin C. Young for many helpful comments on earlier versions of the chart and figures.

2.6

Permian (Chart 6)

N.W. Archbold and J.M. Dickins

INTRODUCTION

The first fossil species described from the Australian continent was the Tasmanian Permian brachiopod *Trigonotreta stokesii* documented by Charles Koenig in 1825 (see Archbold 1985). Despite major contributions to Australian Permian palaeontology by such nineteenth century workers as Frederick M'Coy, John Morris, Alcide d'Orbigny, Laurent Guillaume de Koninck, James Dwight Dana and subsequently Robert Etheridge Junior, reliable recognition of the Permian System in Australia was not achieved until studies on Western Australian Permian ammonoids were undertaken (see Archbold 1985 for a detailed review).

Use of the term 'Permian', substantially in its present scope, followed the detailed consideration of both eastern and western Australian marine faunas by Raggatt & Fletcher (1937; see Murray 1983). Wade (1937) and Hill (1937; see Clarke 1937) had independently reached similar conclusions with respect to the Permian sequences of Western Australia. Correlation with overseas, especially of the Western Australian sequences, was placed on a firm basis by the work of Teichert (e.g. 1941), although the recognition of Upper Permian marine deposits was delayed until established first by G.A. Thomas (Thomas & Dickins 1954).

The Permian System is extensively developed in Australia. Large coal reserves and important oil and gas reserves have resulted in the system receiving attention since the initial discoveries of coal in the 1790s. Occurrences of the distinctive Permian marine bivalve *Eurydesma*, the coal plant *Glossopteris*, and strata of glacial origin within the Permian sequences, have caused great international interest, and were used as evidence for the theory of continental drift during the first half of the twentieth century. However, the interpretation of the distribution of faunas and floras remains controversial (Dickins & Shah 1981, 1987; Archbold 1983*a*; Runnegar 1984; Chatterjee & Hotton 1986).

A comprehensive correlation chart based largely on the marine faunas was produced along with charts for several other systems for the 1st Gondwana Symposium in Buenos Aires in 1967, subsequently brought up to date and published by the Bureau of Mineral Resources (Dickins 1976). Correlation by spores and pollen has been subsequently undertaken (Kemp et al. 1977). The revised Permian Biostratigraphic Chart presented here indicates the degree to which progress has been made. Activity has varied but critical studies have been heavily reliant on the marine groups including the Ammonoidea, Bivalvia, Gastropoda and Brachiopoda with important studies on the Bryozoa and Foraminifera. Terrestrial sequences have been subdivided by studies on spores and, to a lesser degree, mega-plant remains.

Permian correlation and biostratigraphy is a challenge to further development. Although much has been done, radiometric dating (geochronology) is still at a preliminary level, and palaeomagnetic stratigraphy is poorly resolved. Pertinent comments are summarised in the following notes. The framework for the chart was designed by N.W. Archbold, the western Australian sections are the joint responsibility of the two authors, and J.M. Dickins is mainly responsible for the eastern Australian sections.

COLUMN 1: GEOCHRONOLOGY

The 251 Ma age for the top of the Permian is similar to that of Forster & Warrington (1985) and is supported by the zircon age of 251 ± 3.6 Ma of Claoué-Long et al. (1991) from the clay at the top of the Changhsingian at Meishan, southern China. The Carboniferous–Permian boundary is taken as 298 Ma based on interpolation from the data given by Lippolt & Hess (1985). The boundary between the Early and Late Permian is tentatively taken as 270 Ma.

COLUMN 2: PERMIAN STAGES

The Lower Permian uses the four stages, and their subdivisions, recognised in the Ural–Russian Platform area (Kotlyar & Stepanov 1984). The detailed basis for the importance of the recognition of the Kungurian Stage has been given by Dickins et al. (1989).

The base of the Asselian, the lowest of the four stages, regarded here as making up the Lower Permian, is taken at the base of the *vulgaris–fusiformis* Zone, in conformity with Archbold (1982), Kotlyar et al. (1984) and Davydov (1988).

The Upper Permian utilises a combination of the Ural–Russian Platform subdivision, and the stages from the Armenian (Transcaucasian) sequence. This has the value of combining the scheme for the traditional type area, and the utility of using the more complete marine sequence in Armenia, which has the additional advantage that it is in the Tethyan Region.

The top of the Permian is taken as the top of the Changhsingian at Meishan in southern China (see Zhao Jinke et al. 1981).

The Midian has been recently defined (Kotlyar et al. 1989). This usage and that of Kotlyar & Stepanov (1984), and Kotlyar et al. (1987), have been followed here.

The Changhsingian and the Dorashamian are taken as representing the same interval, although this remains under discussion (see Zhao Jinke et al. 1981).

Almost certainly the Tatarian does not represent all of the uppermost Permian (Gomankov 1988) and the equivalence of its lower boundary above the Kazanian with the lower boundary of the Midian is tentative and subject to review. However, it appears to be approximately equivalent.

COLUMN 3: 'FAUNAL STAGES', WESTERN AUSTRALIA

The biostratigraphical stages of A to F are those of Dickins (1963). The validity of Stage C has been questioned by Runnegar (1969*a*), Waterhouse (1970) and Cockbain (1980), but recent work indicates the presence of distinctive faunal elements in the Perth Basin (Archbold 1988*a*), and the Carnarvon Basin. This provides substantial support for retaining the stage.

Two faunas are now discriminated on the basis of brachiopods within Stage F, and are shown as subdivisions F_1 and F_2 (Archbold 1988*b*).

COLUMN 4: BRACHIOPOD ZONES, WESTERN AUSTRALIA

Brachiopod zones are after Archbold (1993).

COLUMN 5: PALYNOMORPHS, WESTERN AUSTRALIA

The palynostratigraphical units for Western Australia are based on those documented by Kemp et al. (1977).

COLUMN 6: AMMONOIDS, WESTERN AUSTRALIA

Ammonoids are indicated as 'spot points' rather than ranges, because many species are represented by only a few specimens from a single locality within formations.

Occurrences are based on the documentation of species provided by Glenister & Furnish (1961), Glenister et al. (1973), Cockbain (1980) and Glenister et al. (1990 *a,b*).

The occurrence of *Paragastrioceras wandageense* is important for understanding the age of the upper limit of Substage D1, as this species is closely related to *P. kungurense* from the Kungurian of the Central Urals. Condon (1954) and Cockbain (1980), without giving any field evidence, conclude that *P. wandageense* came from the Nalbia Sandstone, placed at the top of D1. At present there seems no reason not to accept Teichert's view that the species came from low in the Coolkilya Sandstone (i.e. Baker Formation in current terms).

COLUMN 7: BIVALVES, WESTERN AUSTRALIA

Selected named species are included according to stratigraphical and other interest. The bivalves from the Hardman Formation and the equivalent sequence in the Bonaparte Basin have not been described, and only a minor part of the fauna from the Wooramel and Byro groups and the Noonkanbah Formation has been documented. This is reflected in the ranges shown.

COLUMN 8: BRACHIOPOD RANGES, WESTERN AUSTRALIA

Important brachiopod species (with ranges as determined by examination of collections of the AGSO (previously BMR) Geological Survey of Western Australia etc.) are provided in Column 8, and are based on summaries and descriptions by Archbold et al. (1993*a*), a summary of data up to the end of 1986, and Archbold (1988*b*), Archbold & Skwarko (1988) and Archbold & Thomas (1987).

Not included in the column are reports of species by Waterhouse (in Foster & Waterhouse 1988) many of which are doubtful or appear to be misidentified. In an attempt to force correlation between the Grant Formation of the Canning Basin and the Lower Permian of Tasmania, species from the two regions were compared with each other, while closely related Western Australian species were dismissed. Of the few illustrated species described by Waterhouse, we would reassign them as follows after examination of the specimens. The *Neochonetes* is re-identified as *Neochonetes (Sommeriella)* sp. nov.; the *Strophalosia* cf. *subcircularis* is re-identified as *Heteralosia (Etherilosia)* sp. nov., not juvenile *Strophalosia* cf. *irwinensis* (see Archbold 1986) as previously judged from the illustration provided by Waterhouse, and *Terrakea capillata* Waterhouse appears to be a representative of *Costatumulus* rather than *Lyonia*. Spines are not obviously present on the available specimens. In view of the isolated nature of the sequence that yielded the fauna, and the lack of complete faunal description, it is not feasible to assess the biostratigraphical importance of the assemblage. A pre-Sterlitamakian (pre-Late Sakmarian) age may be possible and is being reviewed.

COLUMN 9: PERTH BASIN STRATIGRAPHY

The Permian stratigraphy of the Perth Basin is based on the detailed account provided by Playford et al. (1976) and summarised in Skwarko (1993).

Although Playford et al. (1976) and other workers show the Fossil Cliff as a member presumably interbedded with the top of the Holmwood Shale, the relationships of the two units are not clear. The absence of the Fossil Cliff in some places could alternatively be due to the erosional unconformity at the base of the overlying High Cliff Sandstone or lack of outcrop. No observation of interbedding has been recorded. The usage of Fossil Cliff Formation by earlier workers has considerable merit.

COLUMN 10: CARNARVON BASIN STRATIGRAPHY

The Permian stratigraphy of the Carnarvon Basin has been fully described by Condon (1967) with some revision by Hocking et al. (1987). The stratigraphy provided in this column is simplified from those accounts.

COLUMN 11: CANNING BASIN STRATIGRAPHY

The stratigraphy of the Canning Basin is summarised in Towner & Gibson (1983). Morante (1993) has examined the carbon isotope composition of the Permian–Triassic sequence from boreholes (Paradise Station) and finds a distinct carbon 13 offset in what he regards as the lower third of the Hardman Formation, which he equates with the Permian–Triassic boundary on palynological information. The bore information including the palynology is unpublished and, therefore, difficult to assess. From his figure, however, it seems possible that the top of the Hardman Formation is wrongly identified and the anomaly may correspond to the boundary between the Permian Hardman Formation and the overlying Triassic Blina Shale. The youngest stage of the Permian (Changhsingian) and the oldest part of the Triassic are missing according to the palaeontological evidence.

COLUMN 12: MACROFAUNAS, EASTERN AUSTRALIA

The faunal stages in this column are based on those proposed by Runnegar (1967), Dickins & Malone (1973) and Dickins (1984). Although criticised by Waterhouse & Jell (1983) and Briggs (1987, 1989), the scheme shown here has proven reasonably useful, and is generally consistent. Reasonably well based developments or alternatives to these schemes remain to be proposed.

Foraminiferal assemblage zones have been defined for the Springsure area by Palmieri (1983). Some difficulty has been experienced in relating these zones to the present chart because it has not been possible to distinguish what part of the definitions is based on strict superpositional relationships, and what part is based on interpretative correlation.

Comments pertinent to the scheme are as follows:

- *Tomiopsis brevis* and *T. branxtonensis* are not known in the Bowen Basin, nor *T. plana* in the Sydney Basin so that the extent of the *T. brevis* range zone can only be rather arbitrarily determined in the Bowen Basin. In the Bowen Basin *T. plana* overlaps the range of *T. ovata* so that *T. plica* may be a better nominal species for this zone as suggested by Waterhouse & Jell (1983). Probably the range of *T. ovata* in the Bowen Basin overlaps that recorded for *T. ovata* and *T. branxtonensis* in the Sydney Basin.
- Ranges of the same species are known to differ considerably in the different basins, for example see Dickins (1968). The meaning of these differences are probably most satisfactorily examined in terms of faunizones such as those used in Columns 12 and 14. Indeed, practice

in Permian correlations in Australia is showing that faunizones, tied carefully into the stratigraphy, are much more reliable than zones based on a single or a few species.

The *Neospirifer campbelli* Zone of Runnegar & McClung (1975) is replaced by *Trigonotreta* sp. nov. *N. campbelli* Maxwell (1964) from the Namurian of the Yarrol Basin, Queensland, is a rather different species from that found in the basal Permian sequence in the Cranky Corner Basin containing the undescribed species referred here to *Trigonotreta* sp. nov. The Allandale fauna of Runnegar & McClung (1975) includes the *N. campbelli* (= *Trigonotreta* sp. nov. as used here), *Tomiopsis elongata* and *T. konincki* Zones. The *N. campbelli* Zone is regarded as being older than Faunizone 1 of Tasmania. The *T. elongata* Zone is regarded as being equivalent to Faunizone 1, and basal Faunizone 2 (Clarke 1990), and the *T. konincki* Zone is regarded as equivalent to most of Faunizone 2 and Faunizone 3 of the Tasmanian succession.

COLUMN 13: BRACHIOPOD ZONES, SYDNEY AND BOWEN BASINS

This is based on Runnegar & McClung (1975), who predominantly used spiriferid brachiopods of the family Ingelarellidae. A detailed scheme based primarily on productids has been outlined by Briggs (1987, 1989), but awaits publication of the supporting systematics.

Briggs (1993*a*,*b*) has also published some Australian Permian correlation charts. These schemes are difficult to assess because supporting taxonomic and stratigraphical data are lacking.

COLUMN 14: TASMANIAN FAUNIZONES

Detailed faunizones for Tasmania were set out by Clarke & Banks (1975) and Clarke & Farmer (1976). Modifications to details of ranges of some faunal elements have been documented in Calver et al. (1984) and Farmer (1985). Brachiopods from Faunizone 10 have been described by Clarke (1987), while those of Faunizones 1–3 are described in Clarke (1990). Invertebrates from the Hellyerian and Tamarian Stages are described by Clarke (1992).

COLUMN 15: AMMONOIDS, EASTERN AUSTRALIA

The occurrences of ammonoids in the eastern Australian Permian successions have been documented by Glenister & Furnish (1961) and Armstrong et al. (1967). Ammonoids are generally extremely rare and provide 'key points' for correlation only. The species *Uraloceras lobulatum* (Armstrong et al. 1967) is transferred to *Gobioceras* following Bogoslovskaya & Pavlova (1988).

The species *Neocrimites meridionalis* has been recently assigned to *Aricoceras* by Leonova & Bogoslovskaya (1990).

COLUMNS 16–17: BIVALVES AND BRACHIOPODS, EASTERN AUSTRALIA

The ranges shown are based on the conclusions about the correlations shown in Runnegar (1967) and Dickins (1976) and modified in Dickins (1983, 1989). These conclusions are also used in Columns 20, 21 and 22, supplemented by information from Armstrong et al. (1967), McClung (1978) and Archbold (1982). The conclusions are closely in line with those of Clarke & Banks (1975) and Clarke (1987, 1990). For alternative conclusions see McClung (1978, 1981), Waterhouse (1983, 1987) and Briggs (1989).

COLUMN 18: PALYNOSTRATIGRAPHY, EASTERN AUSTRALIA

Great difficulty has been experienced in relating the spore zones of the Bowen Basin, and in particular those of the Springsure area, which has been the main reference area for Permian palynostratigraphy in eastern Australia, to the overall stratigraphical sequence and the macrofossils. On the left side of the column the authors have shown a generalised relationship, and on the right are shown the zones after Price (1983), together with the first occurrence of the key species on which the zones are based from Kemp et al. (1977) and Price (1983). Certain species of *Dulhuntyispora* are rare and very sporadic, to the extent that great reservation is necessary in interpreting the presence or absence of single species or combinations of species. These difficulties are referred to by Price (1983) in his account of the history of the palynological zonal scheme, and by McMinn (1987). They are also referred to below. Presumably these spores were derived from a flora that required a special environment that was not necessarily widespread or common in eastern Australia.

The problems inherent in the Price (1983) zonal scheme for the upper part of the Permian sequence in the Springsure area are also considered by Fielding & McLoughlin (1992). It would seem that for the part of the sequence in this area from the upper Aldebaran Sandstone and above, the scheme has practical difficulties.

Fielding & McLoughlin (1992, Fig. 2), however, without explanation, place the boundary between the Lower and Upper Permian within the Aldebaran Sandstone instead of higher as in the present chart. Their correlation is also suspect, since, without supporting evidence or reference, they identify the interval 228–367 in GSQ Springsure 18 as Catherine Sandstone rather than Peawaddy Formation (see also note on Columns 20–21).

Because of the problems in correlating the surface and subsurface sequences in the southern Galilee Basin, the recent publication of Jones & Truswell (1992) seems to have done little to help in understanding the Carboniferous–Permian boundary beds.

COLUMN 19: MEGAFLORA, EASTERN AUSTRALIA

Data on megafloral distribution are modified from Retallack (1980).

COLUMNS 20–21: BOWEN AND SYDNEY BASIN STRATIGRAPHY

The stratigraphy in Column 20 is based on that in Mollan et al. (1969), Dickins & Malone (1973), and Dickins (1983, 1989). The subdivision of the Cattle Creek Formation is that of Balfe (1982). The stratigraphy in Column 21 follows that of Dickins (1976).

The correlations shown for the Bowen Basin have been supported by Dickins & Malone (1968, 1973), Mollan et al. (1969), Brakel (1983) and Dickins (1983, 1989), but alternative correlations have been offered by Dear (1972), McClung (1978), Draper (1983) and Briggs (1989).

Parfrey (1988) has published information based on McClung's field work at the southern end of Reid's Dome in Dry Creek, confirming that the lower Peawaddy Formation rests on the Ingelara Formation. This implies that the Catherine Sandstone and the upper part of the Ingelara Formation found to the north are missing, and that the lower sandstone of the Ingelara beds of Campbell (1953) together with its fauna belong in the Ingelara, and the upper mudstone with its fauna belong in the Peawaddy Formation. This information supports the correlations shown in Column 20.

However, the conclusions of Parfrey (1988) on the faunal relationships of the Ingelara and the lower Peawaddy Formation are puzzling, as apparently she regards the faunas from the two as so similar that they should belong in a single zone. At the same time she shows differences in the faunas that would not seem to confirm her conclusion, and that on the other hand would suggest that the Ingelara fauna can be placed in Fauna III and the lower Peawaddy fauna in Fauna IV.

Balfe (1982) shows different correlations of the Springsure area from those shown here. Although he claims to have solved this problem, examination of the details of the well logs in his report shows that apparently the Catherine Sandstone is miscorrelated with the upper part of the Peawaddy Formation between GSQ Springsure 17 and 18. This miscorrelation is also shown in AFO Arcturus 1 and is the same miscorrelation that caused such confusion earlier in the Dry Creek area, where the Catherine was correlated over an area of no outcrop to the top part of the Peawaddy Formation (see Mollan et al. 1969).

Inconsistencies appear to remain between the distributions in the sequence of the spore–pollen and the macrofossils, but this may be an artefact if, as seems almost certain, the occurrence of some of the spores, such as species of *Dulhuntyispora*, is very sporadic. This seems to be the position with the identification of Lower Stage C in relation to a marine fauna reported by Briggs (1989) in the Gloucester Trough, suggesting that the occurrence of *D. parvithola* may be older here than recorded in the Sydney Basin by McMinn (1985). Some further indication of such sporadic occurrence is tabulated by Foster (1982*a*) from the south-eastern part of the Bowen Basin, where *Dulhuntyispora dulhuntyi* is absent from most of the samples examined from GSQ Mundubbera 5, and by Wood (1984) from the south-western part of the basin, where it is entirely absent from above the Ingelara Formation in GSQ Springsure 19, the part of the sequence where it is recorded by Foster (1982*a*) from GSQ Mundubbera 5.

The claims by Waterhouse (e.g. 1987) that the lower part of the marine sequence in the south-western part of the Bowen Basin is Carboniferous, or even Asselian and Sakmarian, seem without a satisfactory basis whereas, on the other hand, there is very substantial existing evidence that Fauna II is entirely younger than the Allandale and Wasp Head Formations of the Sydney Basin (Runnegar 1967; Dickins 1968; Dickins et al. 1969; see also Draper et al. 1990). This has been borne out by subsequent work (Dear 1972; Clarke & Banks 1975; Runnegar & McClung 1975; Clarke & Baillie 1984). The Allandale and Wasp Head Formations have been most recently regarded as of Tastubian (Lower Sakmarian) age by Archbold et al. (1993*b*) and can hardly be younger than Sakmarian.

Also in the south-west of the Bowen Basin it seems in part that the Brae and Pindari Formations (see Waterhouse 1987) at least represent fault repetition of the Oxtrack and possibly Barfield Formations. This was the conclusion on these outcrops (later named Brae and Pindari) in earlier field work, and has been confirmed by re-examination of the aerial photographs (Mollan et al. 1971) and further field examination. The fauna from the Pindari Formation seems close to or the same as that of the Oxtrack—virtually no fauna has been found in the Brae. Certainly, in the syncline slightly to the north-west of Cracow Homestead, beds of unmistakable Oxtrack fauna and lithology rest with an hiatus directly on beds with Fauna II.

A modification compared with earlier correlations of the Sirius Shale is shown. This unit is now regarded as younger than Fauna II and equivalent to Fauna IIIA of the north-eastern part of the basin. This correlation has been difficult because the fauna of the Sirius Shale is largely made up of brachiopods whereas in the north-east, in the basal part of the Gebbie Subgroup (Fauna IIIA), bivalves and gastropods make up a more important part of the fauna. Dear (1972) suggested that the Sirius might be correlated with IIIA, and the description of the fauna in the north-east from below the Wall Sandstone by Waterhouse (1983) has provided new information. Many of the brachiopods of Fauna II appear to range into IIIA and apparently a more distinctive change occurs in the bivalves at this level. Among the brachiopods, the Sirius Shale (in Reids Dome the Sirius makes up the upper part of the Cattle Creek Formation) has only *Tomiopsis plica* to distinguish it from faunas lower in the sequence. Fauna IIIA from below the Wall Sandstone in addition has *Terrakea dickinsi* and *Notospirifer extensus tweedalei*, which apparently distinguish it from the underlying beds.

Both the Gebbie Subgroup, as originally proposed (Malone et al. 1966), and the Sirius Shale (see Heywood 1978, where in Eddystone No. 1 Sirius rests with hiatus on pre-Cattle Creek) are transgressively unconformable on the underlying sequences. The reasons given by Waterhouse (1983) and Waterhouse & Jell (1983) for removing the beds underlying the Wall Sandstone from the Gebbie Subgroup, and regarding them as Tiverton, are considered as unsatisfactory.

Field mapping and lithological considerations indicate that the interval between the Tiverton Subgroup, as originally defined, and the Gebbie Subgroup, marks an important change in the structure and development of the basin, and is supported by our faunal analysis. The basis for the conclusion is fully discussed by Dickins & Malone (1968, 1973). They placed the beds below the Wall Sandstone, lying above the transgressive unconformity, within the overlying Gebbie Subgroup, believing this solution led to the best understanding of the geology. In their proposal for placing these beds in the Tiverton, Waterhouse & Jell (1983, p.232) doubted the validity of the suggested faunal affinities; however, the conclusions of Dickins & Malone (1968, 1973) are accepted here.

Further evidence that the Sirius Shale belongs with the Gebbie Subgroup is indicated in GSQ Springsure 16 and 17 in the Springsure–Arcturus Downs area, where the Aldebaran rests directly on the Staircase Sandstone without the Sirius Shale (Balfe 1982), suggesting overlap in this area at the base of the Gebbie Subgroup as in the Collinsville area in the north of the Bowen Basin.

For reasons similar to those for the Sirius Shale, it seems likely that the lower part of the Branxton Formation (Elderslie Formation) of the Sydney Basin has a IIIA fauna rather than a Fauna II, as was previously considered in Dickins (1969).

The terminology for the Cattle Creek Formation shown for the south-western Bowen Basin follows that of Balfe (1982). The Stanleigh Formation is, however, shown in addition. This is a convenient grouping for the members underlying the Sirius Shale in the Springsure Anticline, and might also be useful in Reid's Dome to the south where these members have not so far been recognised for the part of the Cattle Creek Formation below the Sirius Shale.

From new information (Dickins 1989) it is now thought that the upper part of the Blenheim Subgroup is younger than the Mulbring Formation and its equivalents in the Sydney Basin.

The basal part of the Rewan Formation and the Narrabeen Group are taken as uppermost Permian following Helby (1973) and Foster (1982*b*), although the occurrence of *Dicroidium* at the base of the Munmorah Conglomerate may suggest an alternative Triassic age (Dickins & Campbell 1992).

COLUMN 22: TASMANIAN STRATIGRAPHY

Based on Clarke & Farmer (1976), Banks & Clarke (1987) and Clarke (1989), rocks of the Hellyerian Stage rest with marked hiatus and unconformity on Upper Devonian granites and older folded rocks. The Hellyerian Stage has been regarded as Carboniferous (see Banks & Clarke 1987). The overlying Tamarian, however, according to its fauna, appears likely to be equivalent to the Allandale Formation. Certainly it appears that the basal Tamarian at Maria Island, with Faunizone 1 resting unconformably on pre-Permian, is not older than Allandale and, therefore, that the Lochinvar equivalent of the Sydney Basin is missing. In this case Faunizone 1 may not be older than lower Sakmarian or slightly older and upper Asselian, if the occurrence of

Megadesmus pristinus and *Tomiopsis elongata* in equivalent beds elsewhere in Tasmania is of significance. Thus the underlying Hellyerian could well be Asselian and therefore Permian, since no marked break in sedimentation has been found.

It is not easy to understand the meaning of the Stage 1 palynoflora of Kemp et al. (1977), which has been taken to indicate that the Hellyerian is Carboniferous. The range of the macroplants in Australia in the uppermost Carboniferous and lowest Permian is not clear; for comment on the age of glacial beds in Queensland see Rigby (1973) and Dickins (1985). Stage 1 is a low diversity microflora apparently containing some long ranging taxa, and we are sceptical of its value in age determination. Foster, in Foster & Waterhouse (1988), has concluded that this microflora is not useful biostratigraphically (see also Foster 1992).

COLUMN 23: CENTRAL TETHYAN STAGES

The Murgabian is shown as equivalent to the Kazanian, the Kubergandian to the Ufimian, and the Bolorian to the Kungurian on a best-fit basis. However, some discrepancy between the boundaries may be present. The Yakhtashian is shown as spanning the whole interval between the Sakmarian and the Bolorian, although the limited complexity in the development of the fusulines might suggest this is not the case.

COLUMN 24: NORTH AMERICAN STAGES

Only the 'Upper Wolfcampian' is included within the Permian, following the correlation by Wilde (1984) of this interval in the Texas area with the Asselian and the underlying 'Lower Wolfcampian' with the Gzhelian of the sequence in Russia. Baars (1990) also concludes that the Wolfcampian contains post-Virgilian (i.e. post-Pennsylvanian), which predates the Permian of the Russian type section. This is discussed further by Jones (1991, this volume) on the Carboniferous Chart.

COLUMN 25: MAGNETOSTRATIGRAPHY

The polarity shown is from Menning (1986): black represents reversed polarity and shading normal. There is some difficulty in correlating to the base of the Asselian, and the precise stratigraphic position of the 'Illawarra Reversal' within the Illawarra Coal Measures remains elusive (see Klootwijk, this volume).

ACKNOWLEDGMENTS

Many colleagues have contributed work and discussion, the results of which have been included in the chart and the notes. We thank them for their interest and support. We especially thank, however, A.T. Brakel of the Australian Geological Survey Organisation, Canberra; M.J. Clarke of the Geological Survey of Tasmania; B.A. Engel and N. Morris of the Department of Geology of the University of Newcastle, New South Wales; C.B. Foster previously of Western Mining Corporation of Perth, Western Australia and now the Australian Geological Survey Organisation, Canberra; M.J. Jones, c/- Peppercorn Cottage, Windana Drive, Highton, Geelong; S.K. Skwarko formerly of the Geological Survey of Western Australia; and G.A. Thomas of the Department of Geology, University of Melbourne, Victoria, who have provided detailed information used in the preparation of the chart and the notes.

The contribution of G.V. Kotlyar of the A.P. Karpinsky Geological Research Institute, St Petersburg, Russia; K. Nakazawa of the Department of Geology, University of Kyoto, Japan; and T. Ozawa, Department of Earth Sciences, Nagoya University, Japan, to the International Permian Time-Scale, and thus the international correlation of the Australian Permian, is especially acknowledged.

2.7

Triassic (Chart 7)

B.E. Balme and C.B. Foster

INTRODUCTION

The Triassic System has an important association with the history of the development of Australian stratigraphy. European settlement of Australia, in 1788, was established at Sydney Cove on Triassic Hawkesbury Sandstone and, before the end of the eighteenth century, quarrying of sandstone and clay and the discovery of coal at Newcastle in 1791, had led to a crude understanding of the geology of the south-eastern Sydney Basin.

During the period 1839–44, W.B. Clarke collected a small number of poorly preserved plants from the Wianamatta Group near Sydney. These were described by M'Coy (1847) and included *Pecopteris odontopteroides* Morris (now *Dicroidium odontopteroides*), first recorded from an unknown locality in Tasmania (Morris in Strzelecki 1845). Although M'Coy believed the fossils to be Jurassic, his contribution was the first dealing with Australian Triassic fossils from a documented locality. Stephens (1886) appears to have been the first to suggest with any conviction that the 'Hawkesbury Series' was Triassic, when he described a labyrinthodont from the Narrabeen Group, and compared it to forms from the European Keuper. This dating was accepted with some reservation by David (1889), firmly by Feistmantel (1890) and subsequently by most workers.

Subsequent biostratigraphic refinements in the Sydney Basin succession have been based principally on palynology (Evans 1966*a*; Helby 1973; Helby et al. 1987), which has also been the principal biostratigraphic technique in the correlation of the other, mainly continental, successions of eastern Australia. Triassic strata are more extensive in Queensland, but there are important successions in South Australia and Tasmania and minor occurrences in Victoria (see Fig. 27).

The most complete Australian Triassic sequences are now known to be present in the marginal basins of the western craton, particularly on the North West Shelf. Lycopsids were collected from the Lower Triassic of the onshore Canning Basin last century (Foord 1890) but were misidentified and their stratigraphic significance not recognised. Antevs (1913) also recorded *Dicroidium* from the same basin, but made no stratigraphic inferences.

Brunnschweiler (1954) was the first to recognise that vertebrate and invertebrate faunas from the Blina Shale, in the Canning Basin, indicated a Triassic age and shortly afterwards important Early Triassic ammonite faunas were discovered in the Perth Basin. These, together with their associated amphibians, conodonts and palynomorphs, provided the first clear evidence enabling Australian successions to be correlated directly with the standard marine Triassic.

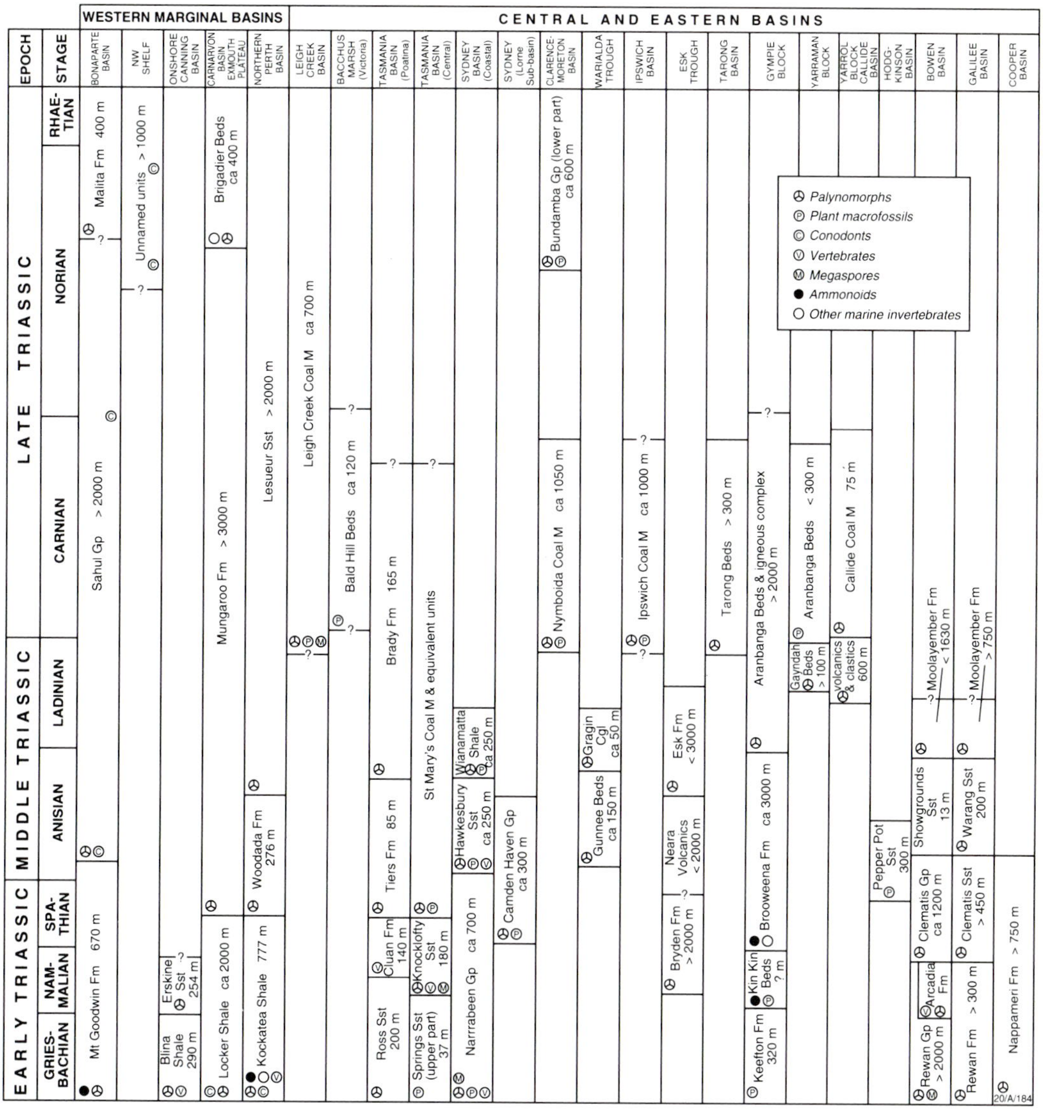

Figure 27: Australian Triassic lithostratigraphic units and their contained fossil types.

COLUMN 1: GEOCHRONOLOGY

This shows the estimates of numeric time for each of the biochronologic stages. The numbers are in millions of years; those in Column 1A are preferred by AGSO (see Chapter 1.6), Column 1B are from Harland et al. (1982), and Column 1C are from Forster & Warrington (1985). Some of the differences are marked; for example, our estimate of the age span of the Carnian is 10 Ma, whereas Harland et al. estimate 6 Ma, and the maximum range from Forster & Warrington is 20 Ma. These data of course affect estimates of sedimentation rates, and calculations of burial history. It should be emphasised that the relative positions of the zones themselves do not change with changing numeric scales, and that for correlation purposes the actual zones should be used rather than the numeric ages.

COLUMNS 2–5: TRIASSIC BIOCHRONOLOGY

Triassic biochronology was developed from early attempts to understand the stratigraphic succession of marine ammonite faunas in the Northern Calcareous Alps and Dolomites of Europe (Gümbel 1861; Mojsisovics 1869; Mojsisovics et al. 1895; Bittner 1892) and the Himalayas (Mojsisovics et al. 1895; see reviews by Zittel 1901 and Tozer 1984). Subsequently, it was shown that the sections in the Cordillera of North America were more complete and more easily decipherable (e.g. Smith 1904), and it is these that have provided the basis for the biochronological scheme proposed initially by Tozer (1967) and expanded by Silberling & Tozer (1968). The ammonoid biochronology shown here in Column 3 is taken from Tozer (1984) and is widely accepted today.

Since the mid-1950s, Triassic conodonts have been used to provide a biochronological scheme parallel to that of the ammonites. As a guide to the relationships between ammonite and conodont zonal schemes, the scale proposed by Sweet et al. (1971) and Orchard (1983, 1991*a*,*b*) has been incorporated in Column 6.

Other fossil groups that have been widely used as Triassic indices are pectinid bivalves, spores, pollen and, in the Upper Triassic, dinoflagellate cysts. A formal biochronology using plant microfossils is in use in Australia (Helby et al. 1987) and a basis of palynological biozonations has been established in the Alps (Visscher & Brugman 1981), North Sea (Geiger & Hopping 1968), United Kingdom (Fisher 1972), Canadian Arctic (Fisher 1979), Russia (Yaroshenko 1978; Tuzhikova 1985) and China (Qu Lifan et al. 1983).

There is general agreement that the Triassic–Jurassic boundary be defined by the base of the *Psiloceras planorbis* Zone, which occupies the lower part of the Hettangian Stage. However, in the absence of ammonite faunas it is not as easy to select a satisfactory boundary between the two systems.

There is, on the other hand, still spirited debate concerning the most appropriate datum for the base of the Triassic. Tradition and priority place the lower boundary at the base of the *Otoceras woodwardi* Zone in the central Himalayas. Tozer (1986, 1988) has strongly advocated this view and it has been adopted herein, although not with total conviction. On the other hand, Waterhouse (1978), Kozur et al. (1978) and Sweet (1979) have advocated redefining the Permian–Triassic boundary, usually with a view to including all or part of the Griesbachian in the Permian.

For most practical purposes, in marine Permian–Triassic sequences of low thermal maturity, palynology provides a satisfactory method of defining the base of the Triassic. The most striking feature of palynomorph assemblages recovered from basal Triassic marine clastic sediments throughout the world is their content of small acanthomorph acritarchs, lycopsid spores (including *Densoisporites* spp. and, commonly, the appearance of members of *Aratrisporites* spp.), and common occurrences of coniferalean pollen assigned to *Lunatisporites* spp. In the Salt Range and in Western Australia (Balme 1970; Dolby & Balme 1976) assemblages of this type range from the base of the Triassic section (i.e. probably lower Griesbachian) to the lower part of the Smithian. In Greenland (Balme 1979) the acanthomorph acritarch 'spike' probably occurs within the Griesbachian, and data from Alaska (Balme 1980*b*) and Arctic Canada (Utting 1989) suggest the same. In their global distribution and compositional similarity, these late Griesbachian–Nammalian plant microfossil assemblages are unlike any elsewhere in the geological column.

COLUMNS 6–10: AUSTRALIAN BIOCHRONOLOGY

Across Australia, many fossil groups, from vertebrates to insects, to plant spores and pollen, have been used for biochronology and correlation. Some, such as the ammonoids, provide direct links with the standard Triassic, but because of their limited geographic and stratigraphic distribution they are not used for local zonation. Figure 27 shows the stratigraphic distribution of all groups, and their contributions to dating are discussed below. Columns 6–10 show the zone fossils most widely used for dating: conodonts (marine), dinocysts (marine), plant macrofossils and spores and pollen (terrestrial & marine).

Banks (1978) prepared a correlation chart for the Australian Triassic and subsequent palaeontological studies have not greatly modified Banks' proposed correlations. The present summary of Australian Triassic biochronology has therefore drawn heavily on his account. Most of the papers dealing with Triassic palaeontology published since 1978 have dealt with vertebrates or palynology. Important new vertebrate finds have been reported from the Lower Triassic of the Tasmanian, Bowen and Sydney Basins and many new palynological data have become available, principally from the sedimentary basins of Queensland. From the biochronological viewpoint the most important contribution is that of Helby et al. (1987), which proposed a comprehensive zonal scheme based on spores, pollen and dinoflagellate cysts that has been widely adopted.

Molluscs

Ammonoids Triassic ammonoids are known from Western Australia and eastern Queensland. They are plentiful only near the base of the Kockatea Shale in the northern Perth Basin. Here the earliest record is of an ophiceratid recovered from a water bore at Geraldton (Glenister & Furnish 1961) in a core dated as Early Triassic on the basis of palynological comparison with the Blina Shale in the Canning Basin. Shortly afterwards Dickins & McTavish (1963) identified three ammonoid species from the Kockatea Shale in BMR 10 (Beagle Ridge) Bore. They referred the interval to the Lower Scythian.

Edgell (1964) was the first to recognise the Triassic age of the Mt Minchin fauna (Perth Basin). His tentative identifications were subsequently revised by Skwarko & Kummel (1974) who described and illustrated forms from Mt Minchin and nearby localities and tentatively identified several genera from the Kockatea Shale in Dongara No. 4 Well. They correlated this fauna with that from the Mittiwali Member of the Mianwali Formation (= Lower Ceratite Limestone) in the Salt Range and referred it to the Early Triassic. Skwarko & Kummel also recorded the only Western Australian ammonoids from outside the Perth Basin, these being from the Sahul Group, in Sahul Shoals No. 1 Well, offshore Bonaparte Basin. They were tentatively assigned to the Anisian.

Following Skwarko & Kummel's account, McTavish & Dickins (1974) revised their biochronological interpretation of the marine Lower Triassic sequence in BMR 10 (Beagle Ridge) Bore to extend from the early Griesbachian to the Smithian, taking into account conodont and bivalve evidence and employing the stage and substage nomenclature for the Lower Triassic proposed by Tozer (1965). Plant microfossils have been extracted from all the Western Australian subsurface samples containing ammonoids and palynostratigraphy for the Lower Triassic is therefore adequately controlled, at least from the Griesbachian to the Smithian. The basis for recognition of the Lower to Middle Anisian is less satisfactory and there are no younger Triassic ammonoid faunas known. This position is unlikely to change as upper Middle to Upper Triassic marine strata occur only in offshore areas of the North West Shelf. Future correlations of Australian post-Anisian sequences with standard international Triassic stages will probably depend on evidence from conodonts and dinoflagellate cysts (Nicoll & Foster 1994).

The only other Triassic ammonoid fauna known from Australia occurs in deformed strata in the vicinity of the Gympie Block, Queensland. Runnegar (1969*b*) described and illustrated representatives of several genera from sediments now referred to the Kin Kin Beds (Day et al. 1983). The association was correlated by Runnegar with the early Smithian *Euflemingites romunderi* Zone of the Canadian Arctic and is thus more or less coeval with faunas from the upper Kockatea Shale.

Bivalves and gastropods Triassic marine bivalves are known from the Perth Basin, the North West Shelf and the Gympie Block. The most significant faunas for biochronological purposes occur in cores from BMR 10 (Beagle Ridge) Bore in the Perth Basin. Here McTavish & Dickins (1974) recorded, among others, aviculopectinids belonging to *Claraia*, a reliable Early Triassic index especially characteristic of the early Griesbachian. Its occurrence about 20 m above the base of the Kockatea Shale in Beagle Ridge Bore therefore suggests that, in the central Perth Basin, the lower boundary of the Kockatea Shale corresponds closely to the base of the *Otoceras woodwardi* Zone. Skwarko & Kummel (1974) also recorded unidentified pteriacean bivalves from the Kockatea Shale and undetermined halobids in probable early Anisian strata in the offshore Bonaparte Basin. Diverse bivalves are present in uppermost Triassic marine carbonate sequences intersected in exploration wells and Ocean Drilling Program sites on the North West Shelf (Williamson et al. 1989; Shipboard Scientific Party 1990).

Bivalves of Smithian age have also been found in strata corresponding to part of the Brooweena Formation (Denmead 1964; Fleming 1966*a*) and associated with a bellerophontid from the Kin Kin Beds (Runnegar & Ferguson 1969) in the Gympie Block, Queensland.

Freshwater bivalves are surprisingly rare, considering the amount of continental sediments in basins of the Eastern Craton. They have been reported from the Leigh Creek and Springfield Coal Measures in South Australia (Ludbrook 1961), the Hawkesbury Sandstone (David 1950) and Wianamatta Shale (Etheridge Jr 1888) in the Sydney Basin, the Ipswich Coal Measures in the Clarence–Moreton Basin (Hill et al. 1965) and the Moolayember Formation, Bowen Basin (Alcock 1970). Gastropods are even rarer, having been reported from the Hawkesbury Sandstone, Sydney Basin (Standard 1962) and from the Brigadier Beds on the North West Shelf (Williamson et al. 1989).

Arthropods

Small arthropods, especially branchiopods and insects, are common in the Australian Triassic. They have clear palaeoecological significance but have not yet demonstrated biostratigraphic potential, although the application of modern systematics to conchostracans may enable them to be used with more confidence for this purpose. Their abundance at some localities was cited by Stephens (1886) as evidence for correlating the Narrabeen Group with the European Middle Triassic and coquinas of the same group, occurring near the base of the Blina Shale, led Brunnschweiler (1954) to assign that unit to the Keuper.

Conchostracans These were first reported from the Sydney Basin by Cox (1880) who identified *Estheria* in a borehole sunk in the Sydney Metropolitan Area. Subsequently they were frequently recorded and their abundance at some horizons led David (1889) to recognise a unit that he named the '*Estheria* Shales', which is now incorporated in the Collaroy Claystone. Mitchell (1927) described several species from the Wianamatta Group (Sydney Basin) and the Ipswich Coal Measures.

Recent discoveries have been made by Webb (1978) from the Hawkesbury Sandstone (Sydney Basin), Tasch (1975) from the Knocklofty Formation (Tasmania Basin) and Cockbain (1974) from the Kockatea Shale (Perth Basin). More detailed accounts of faunas from the Canning (Blina Shale), Bonaparte (Mount Goodwin Formation) and Bowen Basins (Sagittarius Sandstone) are given by Tasch & Jones (1979*a*,*b*) and Tasch (1979) respectively. Tasch (1979) assigned the Sagittarius Sandstone to the Early Triassic on the conchostracan evidence, although palynological data suggest that the unit is partly Permian.

Ostracods Lovering (1953) reported glauconitic casts of unidentifiable ostracods from the Wianamatta Group but these are now thought to be inorganic (Byrnes & Scheibnerová 1980) and the only undoubtedly marine species known from the Australian Triassic are from the Kockatea Shale, in the Perth Basin (Jones 1970) and from Sahul Shoals No. 1 Well, Bonaparte Basin (Kristan-Tollmann 1986). The latter assigned this association to the Rhaetian but palynological data indicate that it is no younger than Carnian. Ostracods were also reported from the uppermost Triassic Brigadier Beds at ODP Site 761 on the Wombat Plateau (Shipboard Scientific Party 1990) but no identifications have been published.

Insects and other arthropods Australian Triassic insect faunas are among the most diverse found anywhere in the world. The oldest is from the Anisian Hawkesbury Sandstone in Beacon Hill Quarry, near Brookvale, in the Sydney Basin (Tillyard 1925; McKeown 1937; Riek 1950, 1954; Evans 1963). A slightly younger, but probably still Anisian, assemblage occurs in the Ashfield Shale, in the lower Wianamatta Group (Tillyard 1916).

The most varied and best known Australian Triassic insect remains are those from the Ipswich Coal Measures. They occur at two stratigraphic horizons, both considered to be of Carnian age (de Jersey 1970, 1971, 1975): the lower in the Mt Crosby Formation in the Kholo Subgroup (Tillyard 1937; Dodds 1949; Riek 1955; Evans 1956, 1961, 1971; Fleming 1966*b*); and the other in the Blackstone Formation, the uppermost unit of the Brassall Subgroup (Tillyard 1916, 1919, 1922, 1923; Tillyard & Dunstan 1924), the youngest in the Ipswich Coal Measures.

Other fossil insect records are from the uppermost Triassic of the Clarence–Moreton Basin (Rozefelds 1985) and from the Upper Triassic of the Tasmania Basin (Riek 1962, 1967). One other insect record from the Perth Basin (Riek 1968*a*) is from the Early Jurassic, not Triassic as originally thought.

Several crustaceans have been reported from the Sydney Basin. They include a syncarid, a xiphosuran and a branchiopod from the Hawkesbury Sandstone (Chilton 1929; Riek 1964, 1968*b*), and an isopod from the Ashfield Shale (Chilton 1917).

Conodonts

Reports of conodonts (Column 6 and Figure 27) from the Triassic marine sequences in western marginal basins are limited. McTavish (1970) recorded a Middle or possibly Early Triassic fauna from near the base of the Locker Shale in the Carnarvon Basin. The appearance of the first definitive Triassic conodont biostratigraphy by Sweet et al. (1971) enabled the Western Australian evidence to be assessed more critically, and shortly afterwards McTavish (1973) published what is still the only comprehensive Australian paper. His specimens were obtained from cores cut in the Locker Shale in five hydrocarbon exploration wells in the Carnarvon Basin, and one in the Perth Basin. The last mentioned sample was from the same horizon from which Skwarko & Kummel (1974) subsequently reported Dienerian ammonoids. The presence of many Early Triassic index species demonstrated that the Western Australia conodont faunas extended from the *Neospathodus dieneri* to the *N. timorensis* Zones of Sweet et al. (1971), thus ranging in age from Dienerian to late Spathian. McTavish's conclusions were used by Dolby & Balme (1976) in their account of the Triassic palynology of the Carnarvon Basin, as the basis of their Early Triassic palynostratigraphic zonal scheme.

The presence of conodonts in the Blina Shale, Canning Basin, was noted by Nicoll (1984) but no identifications have been published. Conodonts of definite Middle Triassic age are not known from Western Australia. Jones & Nicoll (1985) reported the first Late Triassic conodonts from a single core sample from the offshore well Sahul Shoals 1 in the Bonaparte Basin. Recent studies of cores from offshore wells and from sea-bottom dredge samples from the North West Shelf have recognised seven conodont zones that span the Norian–Rhaetian (see Nicoll & Foster 1994). The zones, from *Metapolygnathus primitius* to *Misikella posthersteinii*, recognised in western Canada (see Orchard 1983, 1991*a,b*,) and central Europe (Krystyn 1973, 1980, 1988), allow direct correlation with the standard Triassic ammonite zones and, thereby, provide chronologic anchor points for the Australian succession. Most importantly, these conodonts provide independent and more precise time control for the co-occurring spore–pollen and dinocyst assemblages of the North West Shelf, and their palynological correlatives in other parts of Australia. The effect of the conodont studies has been to significantly restrict the age ranges of several Late Triassic palynological zones to the Norian–Rhaetian (see Columns 7–9; Nicoll & Foster 1994).

Foraminifera

Foraminifera have been reported from the Narrabeen and Wianamatta groups in the Sydney Basin (Lovering 1953; Mayne et al. 1974) although these may be inorganic structures (Byrnes & Scheibnerová 1980). In Western Australia, arenaceous forms are known from the Locker Shale (McTavish 1970) and Blina Shale (Tasch & Jones 1979*a*; Heath & Apthorpe 1986) although no detailed studies of these microfaunas exist.

Calcareous Foraminifera have been recovered only from exploration wells, ODP sites and dredgings on the North West Shelf. Apthorpe & Heath (1981) published brief notes on faunas from Barcoo No. 1 Well, Browse Basin, in which they recognised two informal biozones in the Norian. Later, Heath & Apthorpe (1986) provided a more complete account of Anisian foraminiferal faunas in Lawley No. 1 Well, Dampier Subbasin. This age determination was based on palynological rather than microfaunal evidence, although the authors compared the assemblage with those from the Early to Middle Triassic of Spiti.

Diverse Norian foraminiferal assemblages are present in dredge samples from the northern Exmouth Plateau by the Research Vessel *Rig Seismic* (Exon & Williamson 1988; Quilty 1990) and in ODP sites on the Wombat Plateau (Shipboard Scientific Party 1990). These sediments correlate with the uppermost Triassic–basal Jurassic Brigadier Beds, which are developed in reef facies on the northernmost Exmouth Plateau (Williamson et al. 1989). Brenner et al. (1992) have summarised the microfossil studies from the Wombat Plateau and developed a foraminiferal zonal scheme.

Miscellaneous invertebrates

Lingulid brachiopods and echinoids have been reported from the Lower Triassic of the Canning and Carnarvon Basins and the Gympie Block (Runnegar 1969*b*; McTavish 1973) and sponge spicules may be present in the Narrabeen Group (Mayne et al. 1974). Fragments of two nautiloid specimens were recovered from the Kockạtea Shale in Dongara No. 2 Well (Skwarko & Kummel 1974). Otherwise, the only Triassic invertebrates not previously mentioned, are those in the diverse assemblages present in cores from marine and reefal sequences in exploration wells and ODP drill-holes on the North West Shelf (Williamson et al. 1989; Shipboard Scientific Party 1990). They range in age from Rhaetian to, possibly, Early Jurassic and, apart from the foraminifera, nannoplankton and bivalves mentioned above, include corals, echinoderms, bryozoans, sponges and brachiopods.

Vertebrates

Vertebrate remains are more abundant and varied in Australian Triassic sediments than in most other systems. They include diverse labyrinthodonts, a large variety of fish, and smaller numbers of reptiles. Since Banks (1978) published his Triassic correlation chart, vertebrate specialists have been the most prolific contributors to the literature of Australian Triassic palaeontology.

Apart from fish faunas and labyrinthodont remains from the Hawkesbury Sandstone and Wianamatta Group and a palaeoniscoid fish from Leigh Creek (Wade 1953), all these vertebrate remains are of Early Triassic age and come from the Sydney, Bowen, Tasmania, Perth and Canning Basins. They are not biostratigraphically significant, although the faunas may be compared to those in Late Permian to Early Triassic strata in other continents, such as the Beaufort Group of southern Africa and the Indian Panchet Group.

Fish Many fish have been recorded from the Sydney Basin (Egerton 1864; Woodward 1890, 1908; Wade 1935; Hutchinson 1973; Ritchie 1981), and are also known from the Canning (Kemp 1982), Bowen (Turner 1982*c,d*) and Tasmania Basins (Dziewa 1980; Banks et al. 1984). Triassic fish were included in a checklist of fossil fish presented by Long & Turner (1984). Little attempt has been made to compare the Australian fish faunas with those from other continents and they have been dated by reference to other palaeontological evidence.

Amphibians and reptiles In contrast to the other Gondwana continents, amphibians are much more common than reptiles in Australian Triassic sediments. As noted earlier, the identification of the larval brachyopid *Blinasaurus wilkinsoni* (Stephens 1886) from the Gosford Formation in the Sydney Basin, was the first recognised positive evidence of Triassic strata in Australia. Further amphibian remains have been recovered from the Hawkesbury Sandstone and the Wianamatta Group (Watson 1958).

The only amphibian remains that can be dated by direct or indirect reference to ammonoids or conodonts occur in the Kockatea Shale in BMR No. 10 (Beagle Ridge) Bore, Perth Basin, and in the Blina Shale, Canning Basin (Cosgriff 1965, 1969; Cosgriff & Garbutt 1972; Warren 1980).

A rich amphibian, and associated less abundant reptilian, fauna is present in the Arcadia Formation, the highest unit of the Rewan Group in the Bowen Basin (Bartholomai 1979; Thulborn 1979, 1983, 1986; Jupp & Warren 1986; Warren & Hutchinson 1988, 1990). Although the upper part of the Arcadia Formation is unquestionably Early Triassic, it contains no fossils that enable it to be correlated more precisely with the standard system of stages.

The richest Triassic vertebrate horizon in Australia is the Knocklofty Sandstone in the Tasmania Basin (Cosgriff 1974; Camp & Banks 1978; Banks et al. 1984), which was shown by Banks & Naqvi (1967) to be of Griesbachian or Nammalian age.

Molnar (1984*b,c*), Warren (1984) and Thulborn (1986, 1990) reviewed the distribution and affinities of the reptiles, in the general context of the evolution and palaeogeography of Australian vertebrates.

COLUMNS 7–10: PLANT MICROFOSSILS

Acritarchs and dinocysts (Column 7) The phenomenon of the acritarch 'spike' in Griesbachian to Smithian marine strata was discussed above. This is present in the lower parts of the marine Triassic sequences in the Perth, Carnarvon, Canning, and Bonaparte Basins (Balme & Helby 1973; Dolby & Balme 1976; Grenfell, unpublished data), in the Early Triassic of the Cooper Basin (Evans 1966*a*) and less prominently in the Sydney Basin (Helby 1969, 1973). Few published details exist on Australian Triassic acritarchs. Medd (1966) illustrated electron micrographs of specimens from the Kockatea Shale, and Sappal

(1978) recognised four distinct acritarch associations in the same unit, in BMR No. 10 (Beagle Ridge) Bore.

Neglecting a single Silurian record (see Evitt 1985), the oldest undoubted dinocyst species is from probable Late Anisian strata in the Bonaparte Basin (Stover & Helby 1987). All described Australian Triassic species are from the offshore basins of the North West Shelf and they form the basis of the dinocyst zonal scheme that spans, with one discontinuity, the Late Anisian to Early Jurassic (Helby et al. 1987). Column 7 illustrates the Australian dinocyst zonal nomenclature for the Triassic and suggested correlations with the microplankton and the spore and pollen biostratigraphic schemes.

COLUMNS 8–9: PLANT MICROFOSSILS (Fig. 27)

Nannofossils Coccoliths, including the oldest known representatives of the group, have been recorded from several sites of the Ocean Drilling Programme (Shipboard Scientific Party 1990) on the Wombat Plateau, North West Shelf, where drill-holes bottomed in Late Triassic marine sediments of Norian and Carnian age.

Palynomorphs: Spores and pollen Spores and pollen from Australian Triassic terrestrial plants have been comprehensively documented, especially from the intracratonic basins of the north-eastern part of the continent, as a result of the studies of de Jersey and his collaborators. The first account published on Australian Mesozoic palynology (de Jersey 1949) dealt with the Ipswich Coal Measures, although it was not until 13 years later that formal systematic procedures were applied to Australian Triassic miospores (de Jersey 1962). The history of attempts to establish a Triassic zonal scheme based on palynomorphs is summarised by Balme (1990) and de Jersey & Raine (1990). The initial key came with the recognition of Early Triassic marine units in the Canning and Perth Basins (Brunnschweiler 1954; McWhae et al. 1958) and the application of palynology to the elucidation of their relationships. Balme (1964) proposed a twofold division of the Triassic. The lower occurred in marine sequences and was characterised by the *Taeniaesporites* microflora, which was dominated by lycopsid spores and disaccate conifer pollen. This was succeeded by the *Pteruchipollenites* microflora, in which peltasperm pollen was the most conspicuous component.

Evans (1966*a*) was the first to synthesise palynological data from the continental Triassic deposits of the Sydney, Bowen, Cooper and Surat Basins into a biostratigraphic zonal scheme. Because the Permian of these successions is more complete, the palynological break, that in western marginal basins enables marine basal Triassic assemblages to be differentiated from those of the underlying Upper Permian, is not so obvious. The choice of a lower limit for the Triassic therefore becomes somewhat arbitrary and Evans placed it at the base of his Unit Tr1a. Evans regarded this unit and the immediately overlying Unit Tr1b as older than the base of the Kockatea Shale and equated assemblages from his Unit Tr2a , which encompassed the upper part of the Narrabeen Group in the Sydney Basin and the higher parts of the Rewan Group in the Bowen Basin, with the *Taeniaesporites* microflora. In all, Evans recognised eight units in the composite succession that he studied some of which were effectively acme zones, others taxon range and assemblage zones. The highest (Tr3d) was represented in the Moolayember Formation, Bowen Basin, which is today regarded as Middle Triassic (Anisian).

During the 1960s Helby carried out detailed studies of the palynology of the Triassic in the Sydney Basin and although his first formal zonation (Helby 1973) was based on the Sydney Basin succession, he incorporated data from Queensland basins published by previous workers, principally Evans and de Jersey. In most respects, Helby's zonation paralleled that of Evans, except for the horizon selected to mark the base of the Triassic. Helby referred his lowermost

Protohaploxypinus reticulatus (=*microcorpus*) assemblage, which he equated with Tr1a of Evans, to the Permian, because several of the more distinctive taxa present were known from the Chhidru Formation of the Salt Range (Balme 1970). The *P. microcorpus* assemblage occurs in the lower units of the Narrabeen Group, which had been customarily held to be entirely Triassic, and which shows no apparent sedimentary hiatuses within it. It is probable, therefore, that the palynological record within the Narrabeen Group represents an almost complete transitional floral sequence across the Permian–Triassic boundary.

The palynological zonation of the Triassic of the Ipswich and Bowen Basins and the Esk Trough was developed by de Jersey (1975, 1976, 1979; see Balme 1990, Fig. 10). In the first of these papers de Jersey (1975) introduced formal zonal nomenclature and recognised three broad Oppel-zones separated by two significant hiatuses, the older believed to encompass the Ladinian and the later the early and middle Norian. The *Duplexisporites problematicus* Zone, which was represented in the Esk Formation, is equivalent to part of the *Aratrisporites parvispinosus* assemblage unit of Helby but de Jersey's two higher units, designated the *Craterisporites rotundus* Zone (Carnian) and *Polycingulatus crenulatus* Zone (middle Norian to ?Pliensbachian) are younger than any previously defined within the Australian Triassic. The absence of a clear palynological break within the *P. crenulatus* Zone emphasises the difficulty of selecting a suitable boundary between the Triassic and Jurassic in Australian Mesozoic successions.

Two lower zones attributed to de Jersey are interval zones that span the Permian–Triassic boundary. They were based on cores taken from the lower part of the Rewan Group and upper Bandanna Formation, in drill-holes in the western part of the Bowen Basin. Although de Jersey (1979) proposed them only tentatively, they can be correlated with the lower two units of the Sydney Basin sequence. The relationship between the boundary of the '*Tigrisporites*' *playfordii*–*Lunatisporites pellucidus* and *L. pellucidus*–*Aratrisporites wollariensis* Zones and the base of the Rewan Group needs further investigation but it is diachronous (see also Foster 1979, 1982*b*, 1983).

Plant microfossil assemblages from the Triassic of the Sydney, Ipswich and Bowen Basins have an essential unity and strongly Gondwanan aspect. Coeval sequences in the basins of the North West Shelf contain pollen and spore taxa with clear Tethyan affinities. This led Dolby & Balme (1976) to recognise two Australian Triassic floral provinces. The Onslow microflora, which occurs in the Carnarvon Basin, was thought to have derived from mid-latitude plant communities and contains genera well known from the Alpine Triassic and other Tethyan regions. By contrast, assemblages of the Ipswich microflora, characterising the eastern basins, lack Tethyan elements and are dominated by peltasperm pollen grains, just as *Dicroidium* is the most common element in their plant macrofossil floras. They were considered to represent higher latitude regional floras.

The presence of these Tethyan taxa in the Triassic of the offshore Carnarvon Basin, together with the control provided by conodont evidence in the lower part of the sequence, allowed Dolby & Balme (1976) to relate their zonal scheme for western marginal basins to the standard international Triassic stages, with more confidence than had been possible in eastern Australia. Since this zonal scheme was published, further palynological data from the Alpine Triassic (e.g. Visscher & Brugman 1981; Visscher & Van der Zwan 1981) have strengthened arguments for the general correlations that they proposed. As discussed earlier, the most recent conodont study (Nicoll & Foster 1994) provides more precise links between the standard stages and the palynofloral zones of Dolby & Balme, with tie-points from the Nammalian, and latest Carnian to Rhaetian.

Columns 7–9 show the current palynostratigraphic subdivision of the Australian Triassic put forward by Helby et al. (1987). It represents a synthesis of the schemes suggested by Helby (1973), de Jersey (1975, 1976, 1979) and Dolby & Balme (1976) together with their newly defined *Ashmoripollis reducta* Oppel-zone, of Rhaetian to ?Hettangian age, from a section in Ashmore Reef No. 1 well, North West Shelf. Although Helby et al. regard their biostratigraphic units as Oppel-zones their boundaries are more precisely defined, in terms of oldest occurrences of key species, than those of Dolby & Balme's zones. Ranges of selected biostratigraphically important spore and pollen taxa are summarised by Balme (1990, Fig. 10).

Correlation between the biostratigraphic units recognised in western marginal basins and those of the eastern craton has been facilitated to some extent by the discovery of Onslow-type associations in Triassic sediments below the Eromanga Basin (McKellar 1977; de Jersey & McKellar 1981). Nevertheless, the equivalences indicated in Figure 27 should be regarded as reasonable assumptions rather than as demonstrably accurate correlations.

Megaspores (Fig. 27) Lycopsid megaspores are common in the Australian Triassic but have not often been used for biostratigraphic purposes since the work of Dettmann (1961) who analysed floras from the Leigh Creek Coal Measures, South Australia and the Newtown Coal Measures, Tasmania. The latter were assigned a Rhaetian age, although Dettmann did not rule out the possibility that they were older. The former contained forms first described from the Rhaeto–Lias of Germany. Their presence led Dettmann to conclude that at least the upper part of the sequence at Leigh Creek was of the same age.

Helby (1967) described a new species from the Lower Triassic Wollar Sandstone, Goulburn River District, New South Wales and Scott & Playford (1985) analysed a well preserved Early Triassic megaspore association from the upper part of the Rewan Group, Bowen Basin.

Other publications dealing with Australian Triassic megaspores have had a plant anatomical rather than biostratigraphic emphasis. Helby & Martin (1965) described the megaspores of three species of pleuromeiacean lycopod from the Narrabeen Group. *Skilliostrobus australis* Ash is another Early Triassic, heterosporous lycopsid cone that contains *Aratrisporites* microspores (Ash 1979) and *Horstisporites* megaspores. It is known from the Narrabeen Group, Sydney Basin, and the Knocklofty Sandstone, Tasmania Basin.

COLUMN 10: MEGASCOPIC PLANTS (Fig. 27)

Australian Triassic fossil floras, like those of the other Gondwana continents, are dominated by fronds of the Peltaspermales, using that taxon in the sense of Meyen (1987). By far the most common plant remains in the Gondwanan Triassic are bifurcating peltasperm fronds, belonging to the genus *Dicroidium* (=*Thinnfeldia* of earlier Australian workers). It has therefore been designated the *Dicroidium* flora (Gould 1975) and has a similar stratigraphic range to the *Falcisporites* Superzone of Helby et al. (1987), that is, from uppermost Permian to Early Jurassic. White (1986) has provided excellent illustrations of the principal elements of the *Dicroidium* flora, using mainly specimens from the Narrabeen Group, Sydney Basin.

Although macrofloral lists have been published for localities in many Australian sedimentary basins the only comprehensive, illustrated, systematic accounts are based on collections from the Ipswich Basin and Esk Trough (Walkom 1915, 1917*a,b*, 1924, 1928*a*; Jones & de Jersey 1947; Rigby 1977), the Surat Basin (Holmes 1982) and the Sydney Basin (Walkom 1925; Retallack 1977, 1980), with some shorter but important contributions from the Tasmania Basin (Townrow 1965, 1966, 1967*a,b*).

Walkom's studies provided the earliest firm evidence for the Triassic age of sequences in the Ipswich and Sydney Basins but it is only fairly recently that attempts have been made to use plant macrofossils for biostratigraphic subdivision. Townrow (1966) recognised two zones in the section in the Tasmania Basin, and Retallack (1977, 1980) proposed a fourfold zonal subdivision of the Gondwanan Triassic following a detailed analysis and interpretation of eastern Australian plant communities, especially those from the Sydney Basin.

SEQUENCE STRATIGRAPHY

The concepts of sequence and seismic stratigraphy are becoming increasingly influential, particularly in the study of the Mesozoic and Cainozoic geology of the North West Shelf. Sophisticated depositional and tectonostratigraphic models are being developed by exploration geologists and geophysicists but, for the most part, critical data, and their interpretation, remain proprietary. The most detailed relevant studies so far published are those of Kirk (1985) and Boote & Kirk (1989), which treat the Mesozoic rift domain of the Barrow Province, on the western and north-western Australian continental margin. These papers recognise five clastic wedge sequences in the Mesozoic that are essentially tectonically controlled. Initiation of the oldest virtually corresponds with the beginning of Triassic time and the cycle begins with a transgressive sheet sand, resting unconformably on Palaeozoic strata. The basal sand is overlain by pro-delta and fluvial–deltaic sediments, which grade upwards into a thick fluvial sequence. This first cycle was terminated near the end of the Triassic and is unconformably overlain by the sediments of an Early to Middle Jurassic sedimentary cycle. The Kockatea Shale, which is the expression of the Early Triassic marine transgression in the northern Perth Basin, progressively oversteps older Permian strata and rests on Precambrian crystalline rocks, along the flanks of the Northampton Block. This appears to represent the maximum marine encroachment on to the Australian continental craton during the Triassic and to correspond to the 245.5 Ma downlap surface of Haq et al. (1987).

Two Late Triassic sequence boundaries were identified in sections encountered in drill-holes at sites on Leg 122 on the Wombat Plateau (Shipboard Scientific Party 1990). The older was considered to represent the 224 Ma surface at the Carnian–Norian boundary and the younger the 211 Ma event, near the end of the Rhaetian .

It would be anticipated that an increase in oceanic area, resulting from a considerable eustatic rise in sea-level, would be an important factor in the modification of terrestrial floras. This should be reflected by changes in spore–pollen assemblages and some correspondence might be expected between major downlap surfaces and boundaries of palynologically based biostratigraphic units. There is a possibility of such correspondence at the boundaries of the *Protohaploxipinus samoilovichii–T. playfordii* Zones (245.5 Ma), *T. playfordii–Staurosaccites quadrifidus* Zone (238 Ma) and *Samaropollenites speciosus–Minutosaccus crenulatus* Zones (223 Ma), but these boundaries are too imprecisely known to enable firm conclusions to be drawn.

ACKNOWLEDGMENTS

Dr Hugh Grenfell initially undertook the preparation of this review in 1987, and assembled a great deal of bibliographic material before he was obliged to withdraw from the project. We take this opportunity of acknowledging his considerable contribution and to thank him for his careful documentation, which was made available to us. Many others supplied valuable information, and in particular we are indebted to Dr Robin Helby and Dr John Long. A draft of the paper was read by Dr Geoffrey Warrington of the British Geological Survey, Keyworth, UK, who provided important comments and advice. We are especially grateful for his help.

2.8

Jurassic (Charts 8 and 9)

D. Burger and S. Shafik

INTRODUCTION

It was not until several years after the discovery of Jurassic fossils (Mitchell 1838; J.W. Gregory 1849) that their Jurassic age was established, both in eastern Australia (Clarke 1862, 1867; Moore 1870) and Western Australia (Clarke 1867; Moore 1870; Crick 1894; Etheridge 1901, 1910; Chapman 1904*a,b*).

Geological surveys in southern Queensland and northern New South Wales intensified after 1900, with a large proportion of the palaeontological effort being on the study of plant fossils of the coal measures (Walkom 1915, 1917*a,b*, 1919) and plants and fish of the Great Australian Basin (Walkom 1921; Wade 1942). Further work on the Jurassic in Queensland and New South Wales was in part stimulated by the need to understand the hydrogeology and later the hydrocarbon potential of the Great Australian Basin (see Whitehouse 1955; Day 1964) and the Jurassic coal measures of the Surat and Clarence–Moreton Basins (see McElroy 1963).

Systematic geological exploration in Western Australia started near the end of last century, the earliest summary of the geology of the state being given by Maitland (1919). Subsequent surveys are summarised by Brunnschweiler (1954), by which time the known extent of Jurassic rocks had increased considerably (e.g. Teichert 1940*a,b*; Brunnschweiler 1951; Arkell & Playford 1954). Jurassic sediments are now confirmed in the Canning, Perth, Carnarvon and Bonaparte Basins, the North West Shelf, the Great Australian and Clarence–Moreton Basins, and also in the Otway and Gippsland Basins.

This chapter is an updated and shortened version of Burger (1990*a*) and summarises the essential Jurassic palaeontological data from Gondwanan and Laurasian regions. The main Jurassic record in those areas is condensed into two foldout charts, 8 (Hettangian to Bajocian) and 9 (Bathonian to Tithonian). They outline, in columnar mode, the present global standard stages against which the Australian Jurassic is correlated. Columns on similar subjects carry identical numbers in both charts.

Columns 1–19 relate to various aspects of the global record: Columns 1 and 2 include the geochronology, geomagnetic reversal scheme, and standard Tethyan stages; Columns 3–13 include ammonite successions within the Tethyan stages and their Boreal equivalents; and Columns 14–19 include microfloral biostratigraphic schemes for key regions. Columns 20–28 include selected data from the Australian fossil record and relevant biostratigraphic schemes.

COLUMN 1: MAGNETOSTRATIGRAPHY AND GEOCHRONOLOGY (Fig. 28)

Ages of 205 Ma for the Triassic–Jurassic boundary (Odin & Odin 1990), and 141 Ma for the Jurassic–Cretaceous boundary (Bralower et al. 1990), are accepted herein. In the absence of firmer evidence, stage boundaries are dated on the arbitrary assumption that the average duration of individual ammonite zones in the respective type regions is around 1 Ma. The resulting Early and Middle Jurassic ages approximate those suggested by most recent literature.

Larson & Hilde (1975) and Vogt & Einwich (1979) set up a standard geomagnetic reversal diagram for the Late Jurassic and Early Cretaceous (the so-called Keithley sequence, including anomalies M0 to M25) based on studies of geomagnetic lineations from rift zones in the Pacific (south of Hawaii). Recent studies have also logged weaker reversals in the magnetic interval preceding M25 (the 'Jurassic Quiet Zone'), but not enough is known about those (and older) anomalies to unite them into a standard global sequence. Some correlations are discussed below.

With regard to anomalies M18-M25, fossil control has been obtained from the ocean floor for only a few anomalies. Ogg (1983) reported anomaly M19N in DSDP site 534 (northern Atlantic) to be linked with the lower limit of calpionellid Zone B (which correlates with the lower limit of the combined *B. grandis–Berriasella jacobi* ammonite zone). Strata overlying anomaly M25 in DSDP sites 100 and 105 (northwestern Atlantic) include Jurassic (Oxfordian–Kimmeridgian) nannofossils and foraminifera (see Ogg et al. 1984).

Magnetic logging of land-based sedimentary sequences has yielded much more detailed information. Reversal sequences in Jurassic and Cretaceous fossiliferous pelagic limestones in Spain and Italy have been correlated with the M-sequence of anomalies. Some of the most recent results are specified below, and listed in Figure 28.

Channell et al. (1982) logged several Jurassic reversal sequences in Umbria (central Italy), and Ogg et al. (1984) in Sierra Gorda and Carcabuey (southern Spain). Both studies dated M19 to M21

Jurassic Stage	Upper limit Ma	Upper limit chron	Magnetostratigraphic evidence
TITHONIAN	141 (base *grandis*)	M18	Bosso[1], Foza[1]
	142 (base *jacobi*)	M19N	Carcabuey[2], Foza[1,2,3], Xausa[1,2,3]
KIMMERIDGIAN	146	M22N	Xausa[3,4]
OXFORDIAN	151	M25	Umbria[5]
CALLOVIAN	159	JQZ	Xausa[4]
BATHONIAN	165	(BS)	Blake-Bahama Basin[6]
BAJOCIAN	173		
AALENIAN	180		
TOARCIAN	184		
PLIENSBACHIAN	190		
SINEMURIAN	195		
HETTANGIAN	202		
(TRIASSIC/Rhaetian)	205		

1. Ogg & Lowrie (1986)
2. Ogg & others (1984)
3. Channell & others (1987)
4. Channell & others (1982)
5. Lowrie & Ogg (1986)
6. Steiner & others (1985)

20-1/40

Figure 28: Jurassic Stage boundaries: suggested ages and magnetostratigraphic correlations

as Tithonian and M23 and M24 as Kimmeridgian. Channell et al. (1982) placed the Kimmeridgian–Tithonian boundary within M22. They suggested M25 to be of Oxfordian age, and this was confirmed by Ogg et al. (1984), who suggested the Oxfordian–Kimmeridgian boundary lay at or slightly above M25. Lowrie & Ogg (1986), who recalibrated the M-series of anomalies, also placed this boundary within M25.

Channell et al. (1982) placed the Jurassic–Cretaceous (J–K) boundary (i.e. the base of the combined *B. grandis–B. jacobi* ammonite zone) within M19N in Foza, central Italy, and this was confirmed by study of sequences in Bosso in central Italy (Channell & Grandesso 1987) and Sierra de Lugar in southern Spain (see Ogg & Lowrie 1986).

Ogg & Lowrie (1986) gave a detailed analysis of published and unpublished studies of magnetic reversals near the J-K boundary in Spain, Italy, and Hungary. They concluded that the lower limits of the *B. grandis* and the combined *B. grandis–B. jacobi* ammonite zones fall within anomalies M18 and M19N respectively.

Channell et al. (1987) measured several magnetic reversal sequences in Capriolo and Xausa (northern Italy), and confirmed that the Kimmeridgian–Tithonian boundary probably correlates with M22N, and the Tithonian–Berriasian boundary (i.e. the lower limit of calpionellid Zone B) with the lower limit of M18.

Magnetic reversals older than anomaly M25 have been detected in the Pacific and northwestern Atlantic Ocean floors. At present, the only fossil evidence is available from the Blake–Bahama Basin, where bottom strata overlying anomaly M28 are dated Middle Callovian (Gradstein & Sheridan 1983). Older reversals have been logged also in land-based sequences, such as the Oxfordian in northern Spain (Steiner et al. 1985), the Oxfordian–early Kimmeridgian in Germany (Heller 1977, 1978), the Callovian–early Kimmeridgian in Italy, the Sinemurian–Bathonian in Hungary and Italy (Channell et al. 1982), and the Early–basal Middle Jurassic in northern Italy and Switzerland (Horner & Heller 1983). Time relations of these reversals are still uncertain.

COLUMN 2: JURASSIC STAGES (Fig. 28)

The standard timescale for the Jurassic Period includes eleven stages, which are defined by ammonites. The stages are discussed by George et al. (1969), Morton (1974), Ziegler (1974, 1981), Cope et al. (1980*a,b*), and Mégnien & Mégnien (1980).

The authors adhere to the current use of Tithonian as the youngest Jurassic Stage. However, no boundaries have been defined for the Tithonian, and as the ammonite record near the J–K boundary in the key regions of Europe is incomplete, the relationship of the Tithonian to regional stage concepts such as Portlandian and Volgian is unresolved.

Traditionally, the J-K boundary has been drawn in the Tethyan Realm of Europe at the lower limit of the Berriasian, for example the lower limit of the *B. grandis* Zone (Le Hégarat in Cavelier & Roger 1980). In recent years this boundary has been questioned; firstly, because it is not defined sharply enough, and the contact with the preceding Tithonian Stage has not been defined; secondly, because the correlative sequences in the United Kingdom and Russian Platform contain gaps and show distinct Boreal influences (Fig. 29; see following chapter).

COLUMNS 3–13: AMMONITE BIOSTRATIGRAPHY

The standard Jurassic biostratigraphy is based on the ammonites from the Mediterranean province of the Tethyan Realm. Boreal influences are apparent (Hölder 1979); in general terms the English faunas are Boreal–Subboreal, those from France and Germany Submediterranean. Connections with other Tethyan and Boreal–Arctic regions were selective and impede the develop-

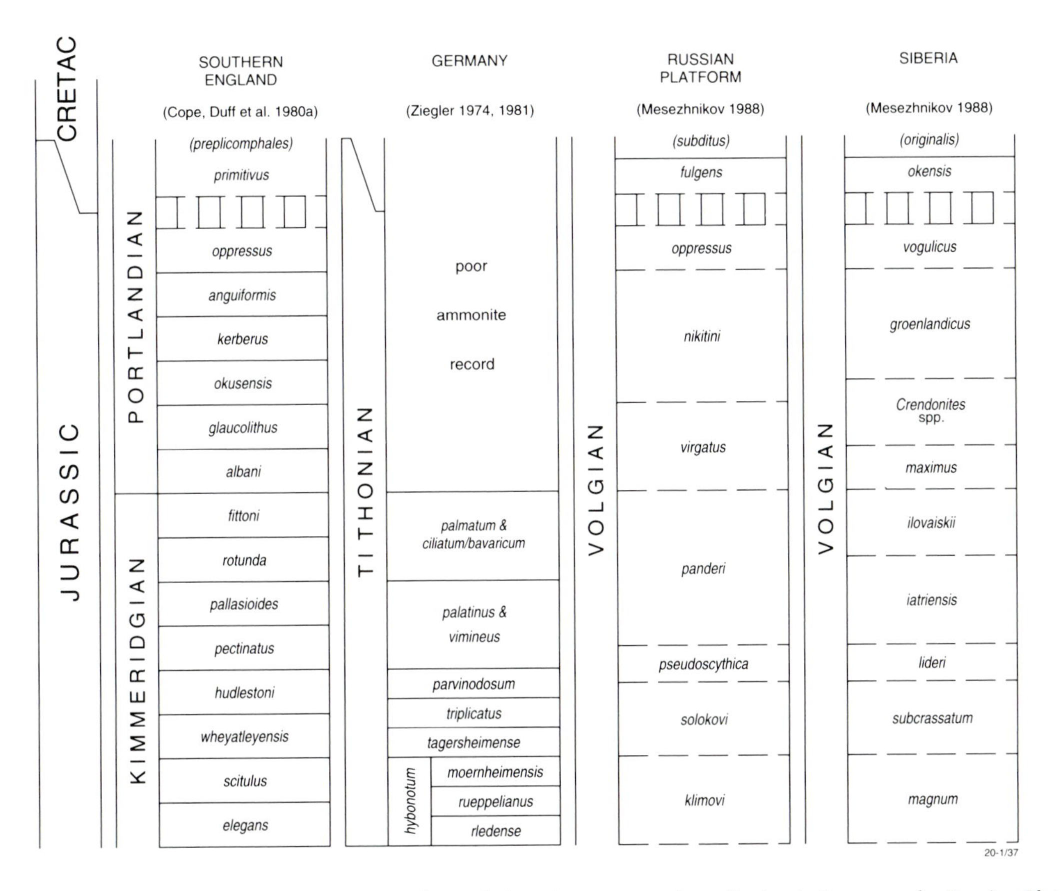

Figure 29: Latest Jurassic ammonite zonal correlations between southern England, Germany, the Russian Platform and Siberia.

ment of a world-wide Jurassic ammonite biostratigraphy. Westernmost Australia and New Zealand were within the eastern Himalayan or Indopacific province (Brinkmann 1959).

At present, the results of the 2nd International Symposium on Jurassic Stratigraphy (Lisbon 1988) are not available. Dr J. Thierry (Université de Bourgogne, Dijon) kindly communicated the proposals for the Middle Jurassic, submitted by him for the French Jurassic Working Group to the International Symposium on Jurassic Stratigraphy (Poitiers 1991). Those proposals have been incorporated in Columns 4 and 5.

Outside Tethyan–Boreal western Europe, Boreal–Arctic and Tethyan ammonite biostratigraphies have been established for the Jurassic (or parts of it) in northern Poland (Kutek et al. 1984), the former USSR, and Siberia (Mesezhnikov 1988). Correlation of sequences in those regions still presents problems (see columns 7–9). Jurassic ammonites are known from western China, but the record is incomplete (Yang 1986). A more marine environment is indicated by the ammonite record of Japan (Sato & Westermann 1991; see Column 10). In North America, a broad ammonite subdivision exists for the Jurassic in the Pacific region only (Imlay 1974; see Column 12). An incomplete zonation for the Middle Jurassic has been erected for western Canada (Westermann 1981).

Due to the late opening of the Indian Ocean the most detailed ammonite record for the Gondwanan Jurassic, from West India (Krishna 1984), does not include the Early Jurassic (Column 11). Early and Middle Jurassic ammonites from South America reveal Tethyan influences (Westermann 1974; Von Hillebrandt 1984; see Column 13). The faunal record of New Zealand includes ammonites, belemnites, and pelecypods, but it has not been subdivided into a zonal system (Stevens & Speden 1978).

COLUMN 14–15: CALCAREOUS NANNOFOSSIL BIOSTRATIGRAPHY

Published Jurassic nannofossil zonal schemes (e.g. Barnard & Hay 1974; Hamilton 1979, 1982; Medd 1982; Bown et al. 1988; Baldanza & Mattioli 1992; Reale et al. 1992) are based on Northern Hemisphere successions. Only those of Bown et al. (1988) and Baldanza & Mattioli (1992) are presented here (Columns 14 and 15). Differences in the stratigraphic ranges of certain species (notably the important *Lotharingius contractus*) in these two zonations are due to their different biogeographic settings: Boreal in northern Europe and Tethyan in Italy (see Cooper 1989). The Toarcian–early Bajocian parts of the two zonal schemes presented in Columns 14 and 15 were discussed by Shafik (1994), who noted that, despite sharing several bioevents, the scheme of Bown et al. (1988) was more applicable to Australian successions.

COLUMNS 16–19: PALYNOLOGY, DINOCYST BIOSTRATIGRAPHY

Studies of marine dinoflagellate cysts have resulted in several widely applicable schemes. The Boreal dinoflagellate record was discussed by Williams (1975) and Bujak & Williams (1977), and a global Jurassic biostratigraphy was proposed by Williams (1977). Sarjeant (1979) and Habib & Drugg (1983) published useful species distribution charts for Middle and Late Jurassic dinoflagellates in the Western (Atlantic) Tethyan Region. Studies covering shorter time-spans are mentioned by Woollam & Riding (1983) and in the comprehensive biostratigraphic summary of Williams & Bujak (1985).

Dinoflagellate records from most of Gondwana are very incomplete. In New Zealand a continuous sequence has been documented only from the latest Jurassic or Heterian upwards (Wilson 1984). At this time a formal biostratigraphy covering almost the entire Jurassic has been developed only for Australia (see below).

COLUMNS 20–28: JURASSIC OF AUSTRALIA

Thick Jurassic sedimentary sequences have been mapped in several basins but, being frequently overlain by younger sequences, they are comparatively poorly known in outcrop.

The Jurassic geological history of Australia, including plate tectonics, basin analysis (stratigraphy, palaeoenvironments), and palaeontology, has been briefly summarised by Bradshaw & Yeung (1992), who compiled the most up-to-date literature lists of seventeen onshore, coastal, and offshore sedimentary basins (Figs 30, 31).

During the Jurassic most of the Australian Plate was above sea-level. The margins of the vast central shield in the east are onlapped by non-marine Jurassic strata, and only the regions in the north (Cape York, Bonaparte Gulf) and west (Perth, Carnarvon, and Canning Basins, North West Shelf) have yielded evidence of marine influence.

COLUMN 20: CALCAREOUS NANNOFOSSILS

Little is known about the Jurassic biostratigraphic succession in the Australian region. Jurassic nannofossils have been documented from the Rowley Terrace, from the ammonite-

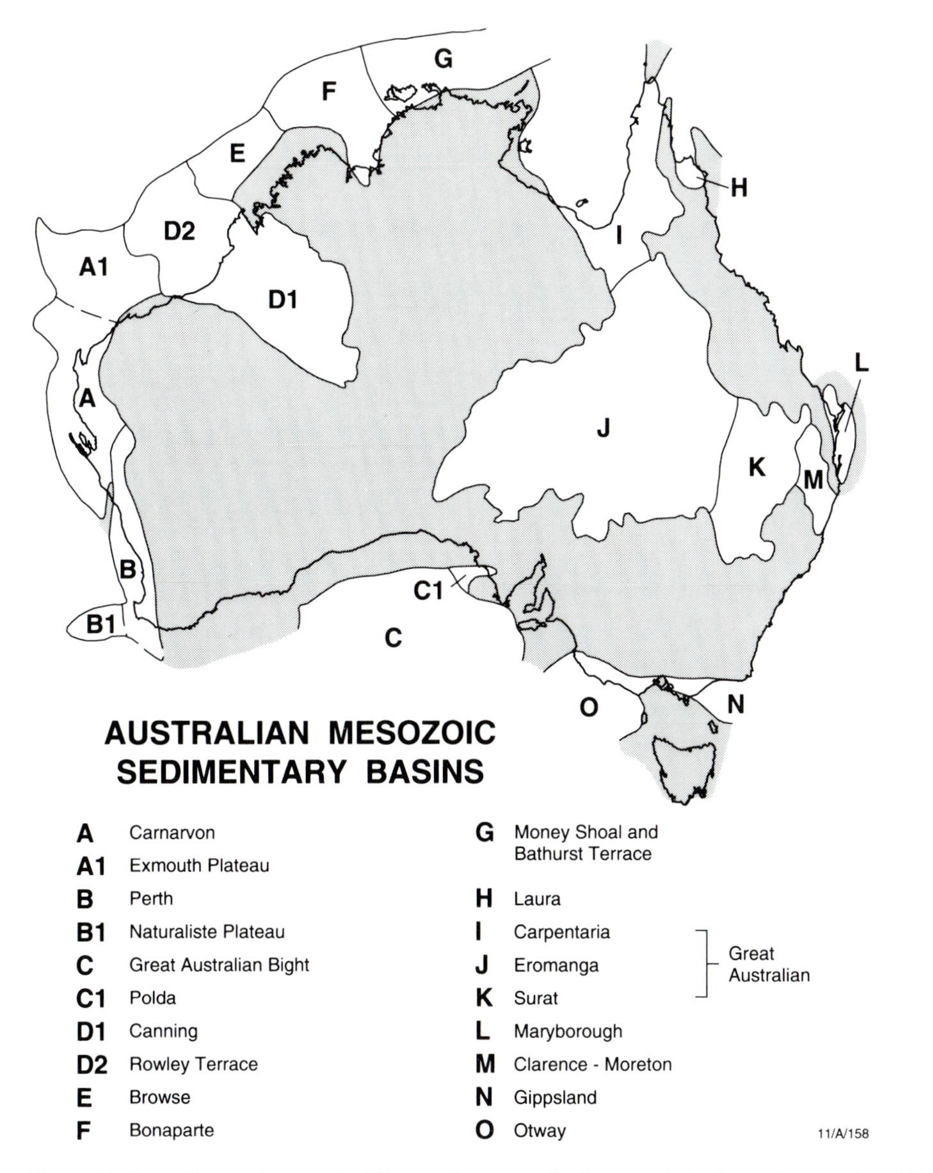

Figure 30: Australian onshore and offshore sedimentary basins containing Jurassic strata (modified from Frakes et al. 1987).

rich Newmarracarra Limestone of the Perth Basin (Shafik 1994), and from DSDP site 261 and ODP site 765 in the Argo Abyssal Plain (Proto-Decima 1974; Dumoulin & Bown 1992).

The nannofossil biostratigraphy of the Late Jurassic–Early Cretaceous sequence of site 765 has been described as problematic, on account of low diversities (associated with poor preservation and barren intervals) and the probability that certain key species are absent or rare, and have truncated ranges (Bown 1992).

The Jurassic levels recorded from the Rowley Terrace are significant, in that they fall within the paralic pre-breakup sequence that was

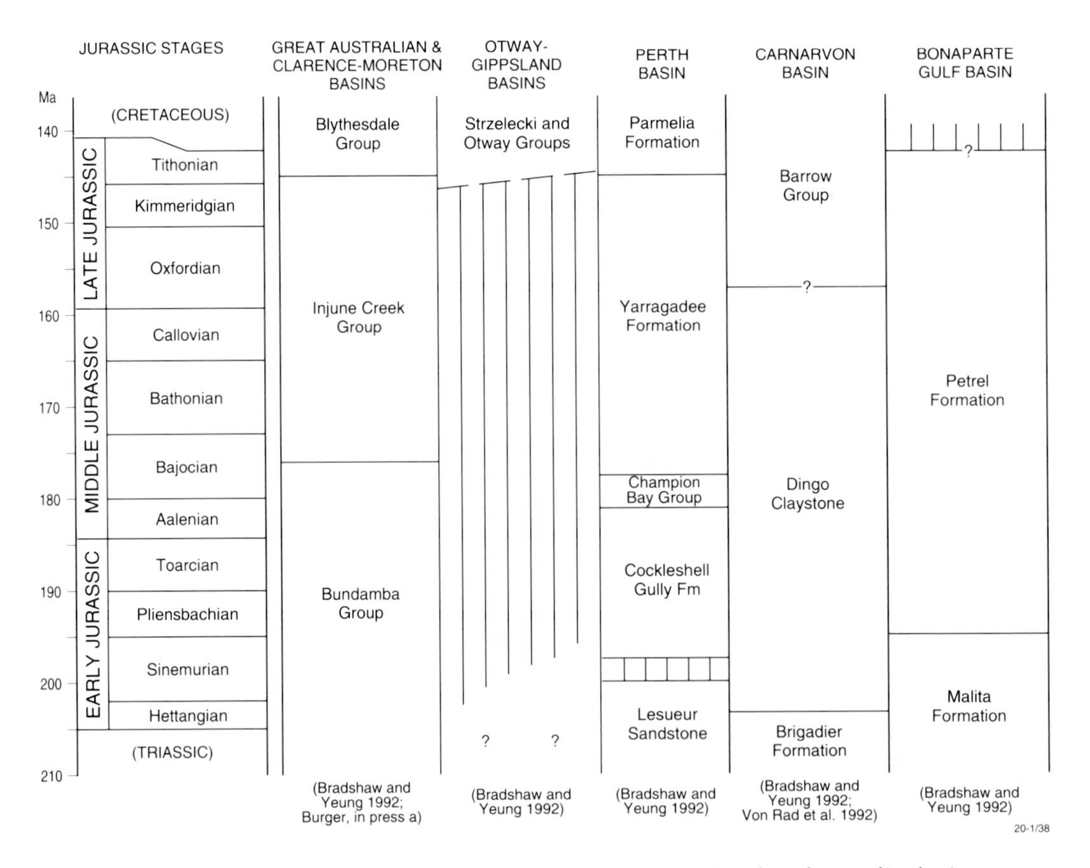

Figure 31: Correlation of principal Jurassic lithostratigraphic successions in selected Australian basins.

deposited during the rifting of Gondwana. Two distinct nannofloras were recorded (Column 20). One suggests an early Toarcian age (*Discorhabdus striatus* Zone of Bown et al. 1988). The other suggest an early Bajocian age (the uppermost *Lotharingius contractus* Subzone of Bown et al. 1988), in addition to a monospecific assemblage (*Watznaueria britannica*) of possible mid-Jurassic age. The Newmarracarra Limestone and early Bajocian Rowley Terrace nannofloras are coeval, being very similar.

The Argo Abyssal Plain nannofloras represent the younger Jurassic (see Dumoulin & Bown, 1992), consistent with the notion that the Argo Abyssal Plain was formed as a result of a breakup event generally Callovian/Oxfordian in age. Two nannofloras are recorded from site 261; one reveals a Tethyan influence, suggesting a late Kimmeridgian to early Tithonian age; the other is Tithonian in age with a lesser Tethyan influence.

COLUMN 21: FORAMINIFERA

Scattered occurrences of agglutinated foraminifera have been reported from the Late Jurassic in the Carnarvon Basin (McWhae et al. 1958). Apthorpe & Heath (1981) set up a zonal system for the Sinemurian–Pliensbachian to middle Bajocian in petroleum exploration wells on the North West Shelf. Agglutinated foraminifera were reported from the Late Jurassic of DSDP sites 259 and 261 at the outer margin of the shelf (Bartenstein 1974; Kuznetsova 1974).

COLUMN 22: PALYNOLOGY, DINOFLAGELLATE CYSTS

Cookson & Eisenack (1958, 1960*b*) first described and reported Jurassic dinoflagellates from Australia. Biostratigraphic studies in many onshore and offshore boreholes in Western Australia (Backhouse 1978, 1988; Wiseman 1980; Helby et al. 1987; and others) have been integrated into a comprehensive zonal scheme for the entire Jurassic by Helby et al. (1987). They dated their zones based on regional faunal evidence, and on comparison with the Jurassic dinoflagellate record from the Western Tethyan Region. In view of the fragmentary evidence and their uncertain conclusions, of which the authors were fully aware, possible alternative ages are here proposed for certain zones (see Column 26).

Additional data have been retrieved from northern Queensland, but are much more restricted. Evans (1966*b*) made the first attempt towards a biostratigraphic analysis of the Late Jurassic in Cape York Peninsula and Papua New Guinea. Burger (1982) and Helby et al. (1987) also investigated latest Jurassic to Early Cretaceous marine rock sequences from the peninsula.

COLUMNS 23–25: PALYNOLOGY, SPORES AND POLLEN

Spores and pollen constitute the principal mediums by which non-marine and marine sedimentary sequences in Australia have been correlated. Early studies by Balme (1957, 1964), de Jersey (1963, 1971, 1975, 1976), Playford & Dettmann (1965), and Evans (1966*a*) have provided the key data from which broad and readily correlatable biostratigraphic schemes have been set up for the Australian Jurassic, which formed an enduring basis for later studies.

Dettmann (1963) and Dettmann & Playford (1969) made detailed studies of latest Jurassic strata from the Great Australian and Otway–Gippsland Basins. Evans (1966*a*) first instituted an informal zonation for the entire Jurassic in the Great Australian Basin. His Early Jurassic intervals were formally defined by Reiser & Williams (1969) and de Jersey (1975, 1976), and the youngest Jurassic interval by Burger (1973, 1989) and Helby et al. (1987). A finer subdivision for the Jurassic of eastern Australia was published by Filatoff & Price (1988). Filatoff (1975) and Backhouse (1978, 1988) presented a formal zonation for the entire Jurassic in the Perth Basin, which was slightly modified by Helby et al. (1987) for continent-wide application.

Zonal schemes and species ranges as published by their authors are set out in Columns 23–25. Several zonal intervals are outlined in more than one scheme, being defined by useful species with pan-Australian distribution. The columns illustrate the different opinions as to ages of various zones. Discrepancies also emerge by comparing ages of spore–pollen and dinocyst sequences. This chapter tries to reconcile those discrepancies by proposing alternative time settings for current zonal schemes that do not violate internal correlations, although several dinocyst zones are defined by rather vague criteria, and are not readily correlated with the parallel spore–pollen sequence. The alternative time settings are argued below, and shown in Column 26.

COLUMN 26: ALTERNATIVE TIME SETTINGS

Dating of palynological zonal sequences in Australia, especially in the Jurassic, is full of uncertainties. The fossil evidence is scarce and open to interpretation. Unavoidably, spore–pollen and dinoflagellate zones have been dated on the basis of separate (if partly overlapping) dossiers of evidence, with the result that in several instances different ages have been deduced for contemporary zonal sequences. This chapter attempts to finds a compromise, by proposing alternative ages for several zones of the pan-Australian

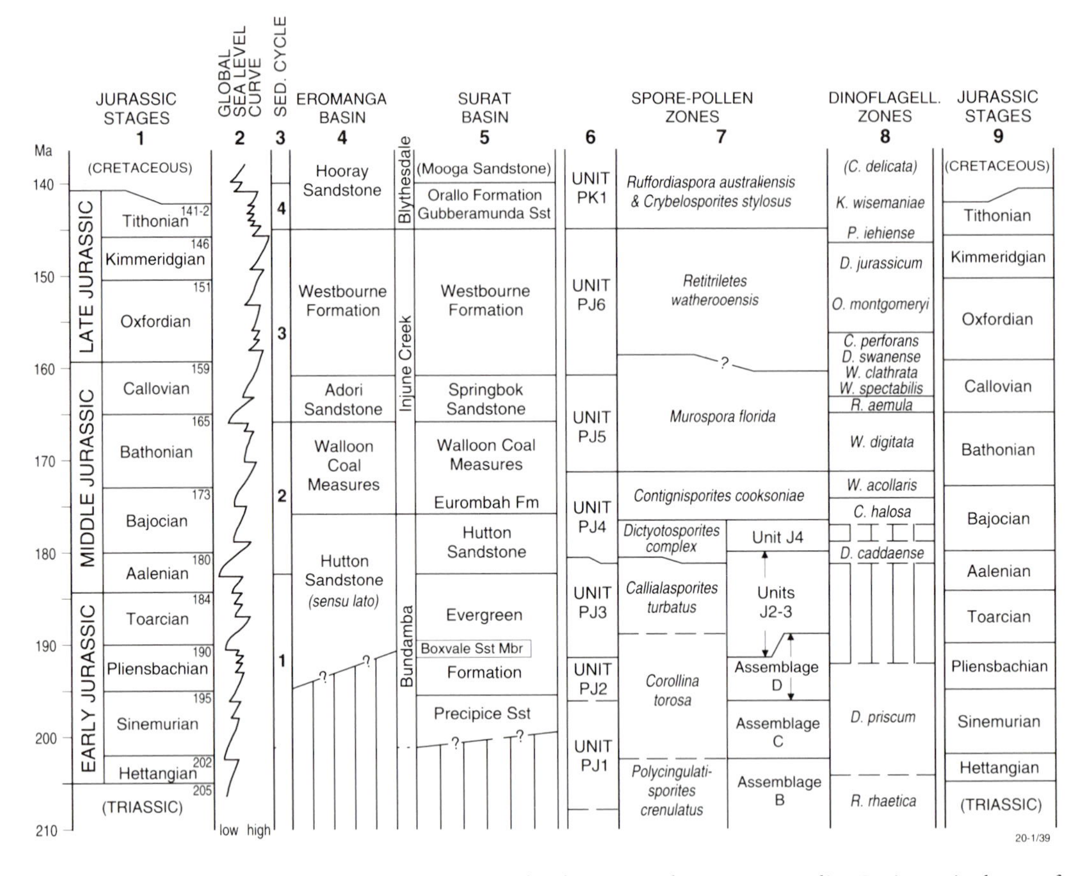

Figure 32: Jurassic palynological zones and associated sediments in the Great Australian Basin: revised ages of sequences argued in text with coastal onlap curve of Haq et al. (1987).

dinoflagellate and spore–pollen biostratigraphies compiled by Helby et al. (1987). The changes in age argued below are of a comparable order of magnitude for both schemes, and time association between them (as given by Helby et al. 1987) is not significantly altered (see also Fig. 32).

Dinoflagellate zones

Several zones in the dinoflagellate scheme of Helby et al. (1987) are not delineated by first appearances of appropriate species (the most consistent and reliable biostratigraphic criterion) but on other palynological events (Fig. 32, Column 8). The precise extent of such zones within larger intervals that are so defined appears somewhat vague. There is room to argue the view that several Middle and Late Jurassic zones may be about one stage older than originally dated. The alternative ages proposed below in items A–H are based in part on a comparison of the Australian record against dinoflagellate evidence from north-western Europe. Long-distance correlations based on stratigraphic ranges of key species carry an element of uncertainty and require independent verification, but attempts made for the Early Cretaceous have yielded encouraging results (Morgan 1980*a*; Burger 1982).

Item A The J–K boundary is generally accepted to fall within the *Pseudoceratium iehiense–Cas-*

siculosphaeridia delicata zonal sequence. However, its exact location is somewhat difficult to place on account of imprecise zonal boundaries within that sequence. Helby et al. (1987) placed the J–K boundary within their *P. iehiense* Zone. On the evidence discussed by Davey (1987) and Burger (in press *b*), it is here placed within the *Kalyptea wisemaniae* Zone.

Item B The *Dingodinium jurassicum* Zone in the Canning Basin is associated with ?mid-Kimmeridgian belemnites (Helby et al. 1987). It includes *Systematophora areolata*, which in Europe and the north-western Atlantic first appears in the late Oxfordian. An Oxfordian age is therefore suggested for the preceding *Omatia montgomeryi* Zone.

Item C This implies that the *Wanaea spectabilis– Cribroperidinium perforans* zonal sequence is not younger than Oxfordian. Its upper *C. perforans* Zone appears to be associated with the earliest appearances of *Leptodinium eumorphum* and *Belodinium dysculum*; in north-western Europe *L. eumorphum* first appears in the early Oxfordian, and *B. dysculum* in the late Oxfordian.

Item D Ammonite evidence from the Carnarvon Basin suggests an (early?) Oxfordian age for the middle part of the *Wanaea clathrata* Zone in the *Wanaea spectabilis–Cribroperidinium perforans* zonal sequence (Helby et al. 1987). Given the imprecise limits of that zone, the evidence is not incompatible with a Callovian age here suggested for the preceding *W. spectabilis* Zone. This zone includes the first appearances of *Scriniodinium crystallinum*, *W. spectabilis*, and *W. clathrata*. All three species first appear within the (early) Callovian of north-western Europe.

Item E *Rigaudella aemula* first appears in the basal Callovian of north-western Europe. In conjunction with item D an early Callovian age is here proposed for the *R. aemula* Zone.

Item F Helby et al. (1987) suggest a Callovian age for the *Wanaea digitata* Zone on the first appearance of *W. digitata* in north-western Europe. The *W. digitata* and preceding *Wanaea acollaris* Zones are here tentatively dated Bathonian on the arguments given in items D and E.

Item G In the Perth Basin the *Dissiliodinium caddaense* Zone is assocated with the Cadda Formation, whose upper part may be correlated with strata containing early Bajocian ammonites (Newmarracarra Limestone, Champion Bay Group). That age implies that the overlying *Caddasphaera halosa* Zone may be at least in part older than Bathonian as dated by Helby et al. (1987). This inference may be drawn also from items D–F, notwithstanding the lacuna between the two zones.

Item H Conodont evidence from the North West Shelf (Nicoll & Foster 1994) suggests that the *Rhaetogonyaulax rhaetica* Zone may extend upwards into the basal Jurassic (see also item J). The index species *R. rhaetica* is also known from the lower Hettangian of the United Kingdom (Woollam & Riding 1983).

Spore–pollen zones

The biostratigraphic schemes of Helby et al. (1987) and Filatoff & Price (1988) are set out in Columns 6 and 7 of Figure 32. Opinions on species ranges diverge and certain zonal boundaries are still being debated. Column 7 of Figure 32 shows Early Jurassic zonal intervals that are defined by first appearances of selected species and have proven readily applicable in north-eastern Australia. De Jersey (1976) defined assemblages B–D from the Clarence–Moreton Basin, and Burger (1976, 1986) defined units J2–4 mostly from the Surat Basin. The alternative ages proposed below (items J–N) are based not only on fossil evidence but also on studies of eustasy (Burger, in press *a,b*).

Item J Evidence for the position of the Triassic–Jurassic (Tr–J) boundary is extremely poor, and de Jersey & Raine (1990) have discussed the problem by extrapolation of data from New Zealand. Combined conodont and spore–pollen data from the North West Shelf suggest that it probably falls within the *Ashmoripollis reducta* Zone (Nicoll & Foster 1994), and thus within the upper *Polycingulatisporites crenulatus* Zone in eastern Australia. De Jersey (1975, 1976) placed the Tr–J boundary within his Assemblage B in the Clarence–Moreton Basin. The coincidence of that assemblage with the *P. crenulatus* Zone also follows from Burger's (in press *a*) suggestion that on their pollen criteria the lower limits of the succeeding *Corollina torosa* Zone and Assemblage C probably coincide.

Item K In the Perth Basin the *Dictyotosporites complex* Zone (*sensu* Filatoff 1975) (e.g. the lower *D. complex* Zone of Helby et al. 1987) is associated with the Cadda Formation and thus dated as early Bajocian (see Item G). Its approximate zonal equivalent in north-eastern Australia is Unit J4 of Burger (1976). This unit is associated with the upper part of the Hutton Sandstone of sedimentary cycle 2, which is linked with an Aalenian–Bajocian episode of low eustatic sea level (Exon & Burger 1981).

Item L In north-eastern Australia the first appearance of the genus *Contignisporites* marks the upper limit of Unit J4. That event falls within the upper Hutton Sandstone–Eurombah Formation of sedimentary cycle 2 and is regarded as most probably Bajocian (McKellar 1974; Price et al. 1985; Burger, in press *b*). In the Perth Basin *C. cooksoniae* first appears in the Yarragadee Formation (Fig. 31). Filatoff (1975) regarded that event as being probably Callovian, and Helby et al. (1987) placed it in the Bathonian. Both age estimates would seem to be too young (see item G), but the possibility exists that the species appeared later in western than in eastern Australia.

Item M The *Retitriletes watherooensis* Zone is associated in the Perth Basin with the Yarragadee Formation (Backhouse 1988), while its correlative Unit PJ6 in north-eastern Australia is associated with the Westbourne Formation of sedimentary cycle 3 (Exon & Burger 1981; Price et al. 1985; Burger 1986). That cycle is here linked with a Callovian to basal Tithonian phase of high eustatic sea-level. Helby et al. (1987) placed the lower limit of the *R. watherooensis* Zone in time against their *Cribroperidinium perforans* dinoflagellate zone, but items C, D, and M would favour a slightly different zonal intercorrelation. Given the imprecise marine zonal boundaries, this discrepancy may be only apparent, but it may also indicate that the index species *R. watherooensis* appeared earlier in eastern than in western Australia.

Item N Spore–pollen sequences traversing the J–K boundary have been documented from many Australian regions. That boundary has been linked with the first appearance of *Ruffordiaspora* and *Cicatricosisporites*, but Dettmann & Playford (1969) and later studies argued that at least one of those types of spores probably appeared in the Late Jurassic. Backhouse (1988) agrees in placing the J–K boundary in Western Australia within the *Biretisporites eneabbaensis* Zone. In the east it probably falls at or close to the boundary between lower and upper *Ruffordiaspora australiensis* Zone, that is, at the first appearance of *Laevigatosporites belfordii* and *Dictyotosporites speciosus* (see Burger 1989).

COLUMN 27: AMMONITES AND BELEMNITES

Ammonites are known from the latest Jurassic of the Canning Basin (Brunnschweiler 1957, 1960), Middle to Late Jurassic of the Carnarvon Basin (McWhae et al. 1958) and the early Middle Jurassic of the Perth Basin (Arkell & Playford 1954; Arkell 1956; Coleman & Skwarko 1967; Hall 1989). Belemnites are known from the latest Jurassic in the Canning Basin (Brunnschweiler 1957, 1960) and the Middle Jurassic (Bajocian) of the Perth Basin (McWhae et al. 1958).

COLUMN 28: VERTEBRATE FAUNAS

The Australian record is poor and fragmented. Freshwater fish have been reported from the Middle to Late Jurassic of Talbragar (Long 1982, 1991; Long & Turner 1984). Amphibians are known from the Early Jurassic Marburg and Evergreen faunas (Warren 1982, 1991; Molnar 1984*a*; Lees 1986*a*). Reptiles are known from Mt Morgan (Bartholomai 1966; Molnar 1982*a,b*, 1991) and the Evergreen fauna (Lees 1986*a*; Molnar 1991), both basal Jurassic, and from the Middle Jurassic Walloon fauna (Molnar 1980, 1982*a,b*, 1984*b*, 1991; Lees 1986*a*).

PLANT MEGAFOSSILS

Plant megafossils have been described in great detail in the Great Australian Basin and Clarence–Moreton Basin by Walkom (1915, 1917*a,b*, 1919, 1921) and Gould (1975, 1980). In Western Australia, identifiable plant megafossils have been reported from the Toarcian and Late Jurassic of the Perth Basin (McWhae et al. 1958). Although locally abundant, the record remains fragmentary, so a megafloral biostratigraphy remains impractical.

BIVALVES AND OTHER INVERTEBRATES

Bivalve molluscs are known from the Late Jurassic of the Carnarvon Basin (Teichert 1940a) and the Canning Basin (Teichert 1940*b*; Brunnschweiler 1957, 1960; Fleming 1959), and the Middle Jurassic of the Perth Basin (Coleman & Skwarko 1967). Other invertebrate faunas are known almost exclusively from Western Australia, and are listed by Quilty (1975).

EUSTASY

Jurassic eustatic sea-level movements and their origins have been discussed by Donovan & Jones (1979), Hallam (1977, 1984), and other authors. A broader-based study, integrating seismic and sequence stratigraphic data to create depositional models, was made by Haq et al. (1987), who published a very detailed (provisional) coastal onlap curve for the Mesozoic and Cainozoic (see Column 2 of Fig. 32).

The record of marine Jurassic faunas and floras in Australia is restricted and fragmentary, and the influence of eustasy in several sedimentary basins has been helpful in dating non-marine sediments. Exon & Burger (1981) and Burger (1986, 1989) linked a repetitive series of cyclic developments in the strata sequence of the Great Australian Basin to Jurassic eustatic sea-level movements. Their conclusions are here modified and shown in Columns 3–5 of Figure 32.

ACKNOWLEDGMENTS

The authors are much indebted to colleagues who have offered valuable criticism — sometimes based on unpublished data — on text and charts of the Jurassic volume in the *BMR Record* series on Australian phanerozoic timescales (Burger 1990*a*). Many kindly sent reprints of recent papers on relevant subjects. He wishes to thank: N.J. de Jersey (Greenbank, Queensland); J.M. Dickins (Canberra, ACT); J.G. Douglas and A.D. Partridge (Melbourne, Vic.); R.L. Armstrong (Vancouver, BC, Canada); P.R. Bown (London, UK); J. Channell (Gainesville, Fla, USA); J.C.W. Cope (Cardiff, Wales); A. von Hillebrandt (Berlin, Germany); W. Lowrie (Zürich, Switzerland); J. Krishna (Varanasi, India); R. Mouterde (Lyon, France); G.S. Odin (Paris, France); J.G. Ogg (West Lafayette, Ind., USA); J.B. Riding and G. Warrington (Keyworth, England); W.A.S. Sarjeant (Saskatoon, Sask., Canada); T. Sato (Tsukuba, Japan); J. Thierry (Dijon, France); and Yang Zunyi (Beijing, China).

2.9

Cretaceous (Charts 10 and 11)

D. Burger and S. Shafik

INTRODUCTION

Although the presence of Cretaceous rocks on both sides of the continent had been suspected for several years (Mitchell 1838; Gregory 1861; Selwyn 1861), it was not until some time later that the Cretaceous age of fossils from Australia was demonstrated (Clarke 1867; Moore 1870).

Subsequent studies recorded Cretaceous rocks and fossils from many more parts of the continent (e.g. Etheridge Snr 1872; Jack & Etheridge 1892; Etheridge Jr 1902, 1907). Cretaceous sedimentary sequences are now known from, among others, the Eucla, Perth, Bonaparte, Carnarvon, Canning, Carpentaria, Eromanga, Surat, Murray, Otway, Gippsland and Bass Basins.

This chapter is an updated and shortened version of Burger (1990*b*) and summarises the main biostratigraphic schemes from Gondwanan and Laurasian regions. The Cretaceous record, including the geochronological timescale and the geomagnetic reversal scheme, is condensed into two foldout charts, 10 (Berriasian to Aptian) and 11 (Albian to Maastrichtian). They outline, in columnar mode, the present correlation of the Australian Cretaceous with the global standard systems (updating Frakes et al. 1987; Bradshaw et al., in press).

Columns 1–20 relate to various aspects of the global record. They include: (i) the geochronology, geomagnetic reversal scheme, and standard Tethyan Stages (Columns 1, 2); (ii) macrofaunal (ammonite, belemnite and bivalve mollusc) biostratigraphic schemes within Tethyan Stages and their Boreal equivalents (columns 3–8); and (iii) microfaunal and microfloral biostratigraphic schemes for various provinces (columns 9–20). Columns 21–37 include selected data from the Australian fossil record and relevant biostratigraphic schemes. Columns on similar subjects carry identical numbers in both charts. However, not all columns appear on both charts.

COLUMN 1: MAGNETOSTRATIGRAPHY AND GEOCHRONOLOGY (Fig. 33)

Standardisation of the sequence of Cretaceous reversals is primarily the result of studies of geomagnetic lineations from rift zones in the Pacific (south of Hawaii) and the northern and southern Atlantic Oceans. Heirtzler et al. (1968) compiled a standard geomagnetic reversal diagram for the Cainozoic and Late Cretaceous, including anomalies 0–34. Larson & Hilde (1975) and Vogt & Einwich (1979) set up a standard reversal diagram for the Late Jurassic and Early Cretaceous, including anomalies M0 to M25 (the numbering proposed by Couillard &

Cretaceous Stage	Upper limit Ma	chron	Magnetostratigraphic evidence
MAASTRICHTIAN	65	92R	Gubbio [1]
CAMPANIAN	73	33N	Gubbio [1, 2]
SANTONIAN	83	34N	Gubbio [1]
CONIACIAN	87	CQZ	Gubbio [1]
TURONIAN	89	CQZ	Gubbio [1]
CENOMANIAN	91	CQZ	Gubbio [1]
ALBIAN	97.5	CQZ	Moria [3]
APTIAN	108	CQZ	Poggio [3], le Guaine [3]
BARREMIAN	115	M0	Valdorbia [3,4,5]
HAUTERIVIAN	123	M7N	Gorgo a Cerbaria Presale [9]
VALANGINIAN	130	M10N	Cismon [9]
BERRIASIAN	135	M14	Caprolo [8], Xausa [8]
(JURASSIC/Tithonian)	141 (base *grandis*)	M18	Bosso [7], Foza [7]
	142 (base *jacobi*)	M19N	Carcabuey [6], Foza [6,7,8], Xausa [6,7,8]

1. Alvarez & others (1977)
2. Lowrie & Alvarez (1981)
3. Lowrie & others (1980)
4. Lowrie & Alvarez (1984)
5. Lowrie & Ogg (1986)
6. Ogg & others (1984)
7. Ogg & Lowrie (1986)
8. Channell & others (1987)
9. Bralower (1987)

20-1/44

Figure 33: Cretaceous Stage boundaries—suggested ages and magnetostratigraphic correlations.

Irving 1975, although perhaps more logical, is not followed). Helsley & Steiner (1969) first recognised a long interval of normal polarity in Cretaceous volcanics of North America (now known as the Cretaceous Quiet Zone, extending between anomalies 34 and M0).

Pioneering work on Neogene volcanics in many parts of the world during the 1960s has demonstrated that ocean floor geomagnetic reversal patterns can be recognised also in suitable land-based sequences (Helsley & Steiner 1969; Alvarez et al. 1977). The advantage of easy accessibility was amply demonstrated by a multidisciplinary study of Palaeogene and Late Cretaceous pelagic limestones near Gubbio (central Italy). Here a sequence of magnetic reversals was logged, of which the oldest could be correlated with sea-floor anomalies 29 to 34, and part of the Cretaceous Quiet Zone. On the evidence of the associated planktonic foraminifera, calpionellids, and nannofossils, this sequence of reversals could be dated Maastrichtian to Cenomanian (Alvarez et al. 1977; Lowrie & Alvarez 1981).

Subsequent studies have correlated magnetic reversal sequences measured from Lower Cretaceous pelagic limestones in central and northern Italy and southern Spain with sea-floor anomalies M0 to M19 (Lowrie et al. 1980; Ogg et al. 1984, 1988; Galbrun 1985; Ogg & Lowrie 1986; Channell & Grandesso 1987; Bralower 1987; Channell et al. 1987). Correlations thus established are set out in the foldout charts and listed in Figure 33.

On the basis of radiometrically dated Early Tertiary and Late Cretaceous ocean-floor anomalies Tarling & Mitchell (1976), Lowrie & Alvarez (1981), Harland et al. (1982), Berggren et al. (1985*a*), and others have extrapolated closely comparable radiometric ages for Late Cretaceous anomalies 29 to 34. Absolute ages of the Aptian–Maastrichtian geomagnetic reversals (Fig. 33) are those of Harland et al. (1982).

At present no accurate ages have been obtained for the Early Cretaceous sea-floor anomalies. As a result, age calibrations by Larson & Hilde (1975), Vogt & Einwich (1979), Harland et al. (1982), and Lowrie & Ogg (1986) have yielded timespans varying between 21 and 29 Ma for the Neocomian. For that reason the estimated ages here adopted are those of Haq & Van Eysinga (1987) for the Early Cretaceous, slightly modified to accept 141 Ma for the Jurassic–Cretaceous boundary, as interpreted by Bralower et al. (1990).

COLUMN 2: CRETACEOUS STAGES (Fig. 33)

The Cretaceous Period is subdivided into twelve stages defined by ammonites (see Muller & Schrenk 1943). At present, several stage boundaries are being reviewed, largely because many stratotype boundaries reflect tectonic events and frequently represent gaps in the record (Hancock 1991). Biostratigraphers are aware of the need for more accurate boundary definitions, if necessary outside the type areas, based not exclusively on ammonites, but also on bivalve molluscs, foraminifera, calpionellids, nannofossils and palynomorphs. Proposals to achieve this were presented at the 1963 *Colloque sur le Crétacé Inférieur* (1965), the 1973 *Colloque sur la limite Jurassique–Crétacé* (1975), and the 1983 Symposium on Cretaceous Stage boundaries (Birkelund et al. 1984), and by Cavelier & Roger (1980).

A satisfactory palaeontological definition for the J–K boundary has never been agreed on, as provincialism in ammonite ecology caused by the latest Jurassic marine regression in Europe has fuelled continuous debates on the relationship between regional Boreal and Tethyan Stages (Volgian, Portlandian, Tithonian, Purbeckian, Berriasian, Ryazanian).

At the 1963 *Colloque sur le Crétacé Inférieur* a recommendation was made to recognise the Berriasian as a separate stage, and was confirmed in 1964 by the Mediterranean Mesozoic Committee in Cassis, France. The accepted J–K boundary in the Tethyan biostratigraphy, that is, the base of the Berriasian, as being the base of the *Berriasella grandis* ammonite zone was upheld by Le Hégarat (Cavelier & Roger 1980, p.96).

At the 1973 *Colloque sur la limite Jurassique–Crétacé* the desirability of adopting the *Berriasella grandis–B. jacobi* interval as the lowermost Cretaceous ammonite zone in the Tethyan Realm was seriously considered as presenting the least disruptive alternative to the traditional J–K boundary (see Yegoyan 1975). This new boundary, which is recognised in southern Spain (Enay & Geyssant 1975), would have an equivalent in the Boreal Realm of the Russia (Fig. 34). It also coincides with the base of the *Calpionella alpina* Zone (calpionellid Zone B of Remane 1978), which has been widely observed, among other regions, in Spain (Allemann et al. 1975; Ogg et al. 1984), Italy (Ogg & Lowrie 1986), and North Africa (Memmi & Salaj 1975), where the standard ammonite succession is not recognised. Both boundaries are indicated in the foldout chart and Figure 34.

The Cretaceous–Tertiary boundary in Europe is generally taken to lie at the Maastrichtian–Danian contact, although several geoscientists would prefer the Danian to be included in the Cretaceous (see Berggren et al. 1985*a*). The type sequences of the Maastrichtian and Danian have been defined respectively in Limburg (the Netherlands) and Stevns Klint (Denmark), and the type boundary has been established at Stevns Klint. The base of the Danian in the sense of Hardenbol & Berggren (1978) lies between the *Abathomphalus mayaroensis* and *Globigerina eugubina* foraminiferal zones.

COLUMNS 3–8: AMMONITE BIOSTRATIGRAPHY

As in the Jurassic, ammonites have been used to subdivide the marine Cretaceous in many regions of the earth. The present standard ammonite biostratigraphy for the Tethyan Cretaceous of north-western Europe (see Column 4) is a compromise based on Birkelund et al. (1984), Robaszynski (1984), Kennedy (1984*a*,*b*, 1986, 1987), and Robaszynski & Amedro (1986). Ammonite occurrences are less frequent in the Late Cretaceous; individual zones are much broader, and the Maastrichtian is usually subdivided by the much more common belemnites (Column 4).

An overall shallowing of the oceans since the Jurassic resulted in sharpening the contrasts between Tethyan, Boreal, and Pacific ammonite provinces. Progressive endemism and impoverishment (even disappearance) of ammonites has complicated world-wide correlations. However, there were frequent interchanges between Boreal and Tethyan faunas (Donze 1973; Owen 1973; Rawson 1973), and these facilitate correlations between the two realms in North America and Eurasia (based on crioceratitid ammonites).

In North America, Early Cretaceous marine deposition was spasmodic; the associated macrofaunas have an impoverished character and contain few ammonites. Marine biostratigraphic schemes for the Gulf region are based largely on bivalve molluscs (*Buchia, Inoceramus, Meleagrinella*) and foraminifera (Jeletzky 1971, 1973; Chamney 1973; Stott 1975).

A large interior sea extended from the Gulf region to north-western Canada during the Albian to Maastrichtian (Williams & Stelck 1975), and study of the rich ammonite faunas has resulted in a detailed zonal scheme established by US and Canadian biostratigraphers (see Chart B). This scheme can be broadly correlated with the European standard zonation and contributes substantially to the geochronology of the Cretaceous. The Albian hoplitid-dominated fauna from Europe spread as far west as Greenland, but is not recognised in the North American interior, where a more or less continuous ammonite record begins with an endemic gastroplinitid fauna, which Owen (1973) regarded as chiefly late Albian, because *Gastroplites cantianus* has been found in the basal *Mortoniceras inflatum* Zone in the United Kingdom.

The Late Cretaceous ammonite sequence is derived from the reviews of Obradovich & Cobban (1975), Stelck (1975), Stott (1975), and Caldwell & North (1984). In the western interior of the USA, the Albian–Cenomanian boundary is placed between the *Neogastroplites maclearni* and *Calycoceras gilberti* Zones (Obradovich & Cobban 1975). Considerable uncertainties remain with regard to the position of the Campanian–Maastrichtian boundary. *Haploscaphites nicolletti* has been found in the early Maastrichtian *Belemnella cimbrica* Zone in Germany, and Obradovich & Cobban (1975) referred to indirect evidence suggesting an earliest Maastrichtian age for the *Baculites reesidei* Zone. However, the authors also pointed out that evidence from planktonic foraminifera suggests that the base of the Maastrichtian may fall below the *Didymoceras stevensoni* Zone in the western Gulf coast, or near the *B. scotti* Zone in New Jersey.

The Cretaceous western interior sea gradually retreated to the south, and the ammonite record ends with the *Berriasella grandis* Zone in Canada and the *Didymoceras cheyennensis* Zone

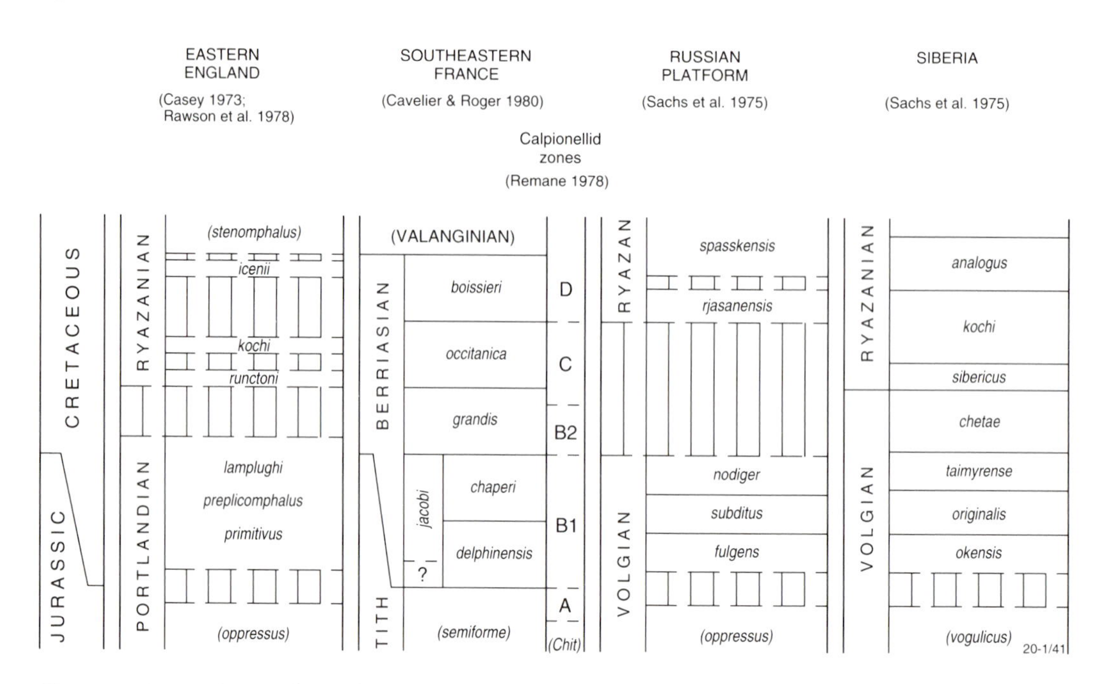

Figure 34: Ammonite zonal correlations at the Jurassic–Cretaceous boundary between England, France, the Russian Platform and Siberia.

in the USA, both zones being acknowledged as early Maastrichtian age (Obradovich & Cobban 1975). The Cretaceous–Tertiary boundary thus lies within a sequence of largely non-marine strata and has been studied in almost a dozen sedimentary basins in the western interior. Coastal lowland and alluvial plain deposits include a diverse reptilian fauna (Russell 1975; Lehman 1987). This classic *Triceratops* fauna has been compared (all or in part) with the late Maastrichtian *Belemnitella junior* and *Belemnella casimirovensis* Zones in Europe (Lanphere & Jones 1978). The demise of this fauna has traditionally been accepted as marking the Cretaceous–Tertiary boundary, and although Sloan et al. (1986) recovered dinosaur remains, together with Palaeocene mammals, in beds above the basal Tertiary 'Z-Coal' overlying the Hell Creek Formation in Montana, Argast et al. (1987) doubted that those remains were *in situ* occurrences.

Brown (1962) suggested that the Cretaceous–Tertiary boundary be drawn at the base of the stratigraphically lowest persistent lignite horizon overlying the highest occurrence of dinosaur remains. North American palynologists have generally followed Brown's definition as changes in the palynological record occur at this level in several places in the western interior (Tschudy & Tschudy 1986). However, the isochronous nature of those changes has not been established beyond doubt (see Berggren et al. 1985*a*).

Palynological changes like those described from North America are unknown in Europe and, as a consequence, the correlative of the Cretaceous–Tertiary boundary in Europe has not yet been accurately pinpointed in North America.

The Boreal Realm in Eurasia presents a more varied and less coherent picture. Despite periodic influxes of Tethyan marine faunal elements (Rawson 1973), which are best displayed in England (Casey 1973), Germany (Kemper 1973*a,b*), and other western European regions, ammonite, belemnite, and bivalve mollusc faunas of strongly endemic character have been used as standard biostratigraphic indicators in all the important geological provinces (Russian

Platform, northern Urals, northern and eastern Siberia). Several biogeographic provinces have been described from China, and Yang (1986) outlined broad Cretaceous ammonoid and bivalve associations. Apparently the detailed ammonite biostratigraphy that has been developed for Japan (kindly communicated by T. Matsumoto; see Column 6) has not been identified in China, suggesting limits to the influence of the Pacific Realm on the East Asian mainland.

Ocean connections existed between Laurasia and Gondwana, permitting faunal interchanges, but the record from Gondwana suggests that they were tenuous on the whole. The Cretaceous of western South America is dominated by the Andean Trough and Magellanean Geosyncline, where the record of ammonites indicates Mediterranean Tethyan influences (Wiedmann 1980). Towards the east (Brazil, Argentina) marine Albian to Maastrichtian sediments with sparse ammonites occur in coastal basins and the eastern continental shelf. In the southern regions of the continent, Early Cretaceous ammonite faunas of the Austral Basin in Patagonia show Boreal, South African, and Caucasian influences, and Late Cretaceous faunas Indo-Pacific affinities (Riccardi 1988). Early Cretaceous ammonites further north (Andean Basin) include Himalayan and Mediterranean elements.

In Africa the record of marine Lower Cretaceous strata is fragmentary. Detailed ammonite biostratigraphic schemes have been published only from Morocco (Wiedmann et al. 1982; Column 7). In both West and South Africa, where nearly-complete Cretaceous sedimentary sequences occur in coastal and offshore regions, microfaunas provide the main regional time framework, although some ammonite control exists in the Algoa Basin and Zululand.

The Cretaceous in India is nowhere fully preserved, but the most complete marine sequences overlying Lower Cretaceous 'Upper Gondwana' strata (frequently in unconformable contact) have been described from the Cauvery, Palar, Godavari–Krishna, and Mahanadi basins on the eastern coast and the Middle and Upper Cretaceous in the north-west (Sastry et al. 1974). A biostratigraphic scheme based on ammonites has been established for the Late Cretaceous in the Trichinopoly District (Sastry et al. 1968; see Column 8).

The biochronostratigraphy of New Zealand covers only the Aptian to Maastrichtian, as the oldest Cretaceous history has been obscured by the Late Mesozoic Rangitata Orogeny. The stages are defined by ammonites and bivalve molluscs (*Aucellina, Inoceramus, Maccoyella*). The present evidence equates the Korangan with the late Aptian, the Motuan with the late Albian, the Ngaterian with the late Albian to early–middle Cenomanian, part of the Arowhanan with the Cenomanian, and part of the Haumurian with the Maastrichtian (Edwards et al. 1988). Positions of the other stages are under consideration.

COLUMNS 9–10: FORAMINIFERAL BIOSTRATIGRAPHY

Aided by the huge influx of data from the subsurface and from the continental shelves and ocean floors, micropalaeontological and palynological research plays an increasingly important role in global correlations. Foraminiferal biostratigraphic schemes (Bolli 1966; Sigal 1977; Van Hinte 1978; Pflaumann & Cepek 1982; Caron 1985) have been published for the Cretaceous in the Atlantic and Mediterranean regions. They have been applied in other parts of the world, but at present no global standard biostratigraphy has been agreed on.

COLUMNS 11–14: CALCAREOUS NANNOFOSSIL BIOSTRATIGRAPHY

Nannofossil biostratigraphic schemes have been devised for the Early Cretaceous of high-latitude regions (north-western Europe, Crux 1989; see Column 24) and mid- to low-latitude regions (e.g. Blake–Bahama Basin, Roth 1983; see Column 24). Neither scheme could be used on material from the Rowley Terrace. However, the Albian high-latitude zonation of Wise & Wind (1977; see Column 24) could be successfully applied in north-eastern and north-western Australia (Shafik, 1985*a*, 1992*a*, unpublished data).

Various events in the nannofossil biostratigraphic schemes set out in Columns 11–14 and 26–28 are coded as follows: FO, FAD indicate earliest appearances (bases of stratigraphic ranges, evolutionary appearances, lowest regular occurrences); LO, LAD denote last appearances (tops of stratigraphic ranges, exits, extinctions, highest regular and common to abundant occurrence, tops of acme occurrences).

COLUMNS 15–20: DINOFLAGELLATE BIOSTRATIGRAPHY

Biostratigraphic studies of marine dinoflagellate cysts have been comprehensively reviewed by Williams & Bujak (1985). Some of the most recently published Tethyan and Boreal zonal schemes are set out in Columns 15–20. Boreal assemblages recorded from Canada and northern Europe reflect phytoplankton provincialism, especially during the Early Cretaceous (Pocock 1976, 1980; Davey 1979, 1982). The European–Mediterranean Tethys record has been analysed in part by Habib & Drugg (1983) and Ogg (1994). They proposed a zonal scheme for the (late) Berriasian to Aptian stratotypes in France, and recognised a virtually identical sequence in the north-western Atlantic. Few of those elements are found in Davey's (1979, 1982) mixed Tethyan–Boreal biostratigraphy for the Early Cretaceous of north-western Europe, in which the base of his *Gochteodinia villosa* Zone coincides with the base of the *Berriasella jacobi* ammonite zone.

Marine sediments of Cretaceous age are rare in China, and the published record is still very fragmentary and localised. Broad biostratigraphic schemes have been proposed only for the Neocomian in the east (Yu 1982) and the Late Cretaceous in the west (Yu & Zhang 1980). In New Zealand, the dinocyst biostratigraphy for the Korangan–Haumurian (late Aptian–Maastrichtian) has been summarised by Wilson (1984).

Correlations of Australian records with the Atlantic Tethyan Cretaceous in western Europe have yielded encouraging results (Morgan 1980*a*; Burger 1982; Helby et al. 1987), and these and other parallel sequential developments (see Williams & Bujak 1985) suggest that an integrated Tethyan Cretaceous biostratigraphy may be set up eventually.

COLUMNS 21-37: CRETACEOUS OF AUSTRALIA

The Cretaceous geological history of Australia, including plate tectonics, stratigraphy, palaeontology, and palaeoenvironments of twenty-three onshore, coastal, and offshore sedimentary basins, has been briefly summarised by Frakes et al. (1987), Dettmann et al. (1992), and Bradshaw et al. (in press). Those authors compiled the most up-to-date lists of geological and palaeontological publications, and their findings are summarised in Figures 35 and 36.

During the Cretaceous the Australian Plate moved comparatively little with regard to the South Pole, its northern rim shifting from 50° to 40° palaeolatitude (Barron et al. 1981), as it separated from the other Gondwana continents. Initial rifting with Antarctica occurred before the Cretaceous, and the sea inundated the rift valleys during the middle Cretaceous. In the west, the Australian and Indian plates separated during the Early Cretaceous. In the east, New Zealand and the Lord Howe Rise broke away during the Santonian (Laird 1981; Johnson & Veevers 1984; Frakes et al. 1987; Bradshaw et al. 1988; Veevers 1988; Von Rad et al. 1992; Buffler 1994; Exon & Colwell 1994).

Although large parts of Australia remained above sea-level, epicontinental seas repeatedly covered the western and north-eastern regions during Early and Middle Cretaceous phases of global high sea-level. By Turonian times most of

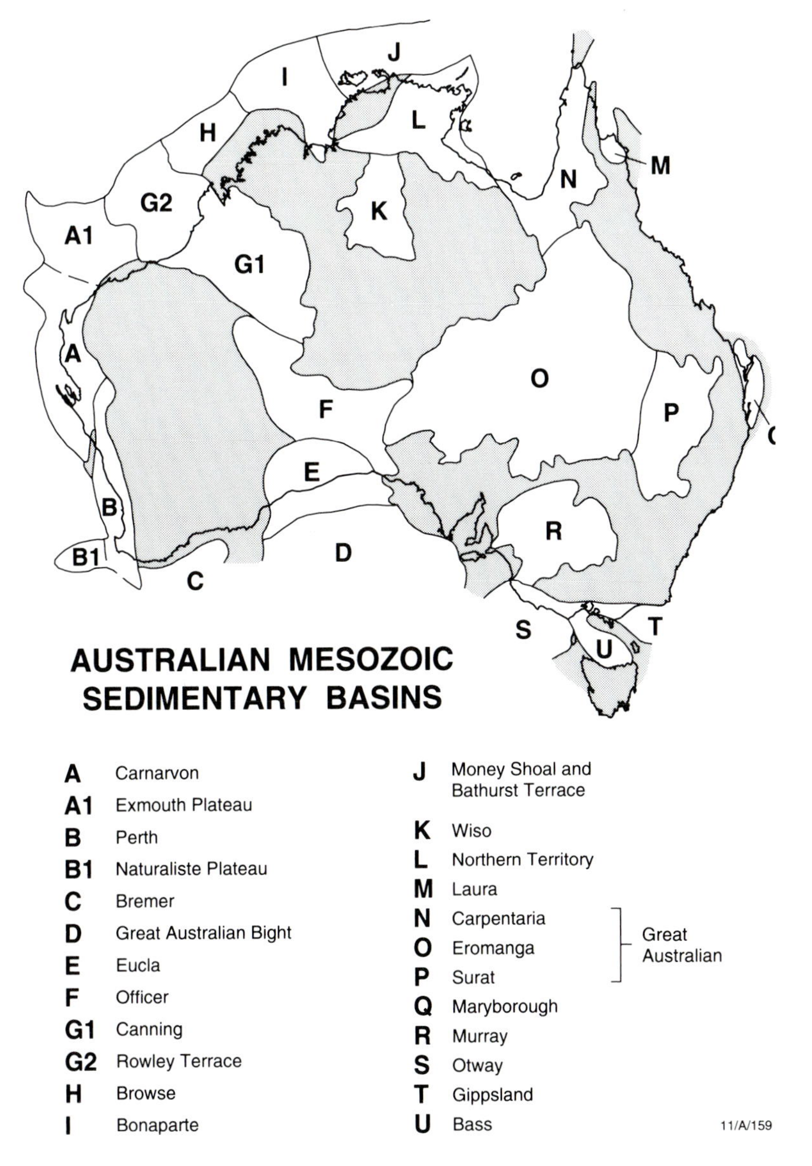

Figure 35: Australian onshore and offshore sedimentary basins containing Cretaceous strata (modified from Frakes et al. 1987).

mainland Australia was above sea-level, and marine sediments are known almost exclusively from offshore and coastal basins on the western and southern margins.

Recent summaries of the Australian Cretaceous palaeontological record have been given in Rich et al. (1991) and Dettmann et al. (1992).

Below are discussed the following fossil groups: ammonites, foraminifera, calcareous nannofossils, dinoflagellates, spores and pollen, megaflora and vertebrates.

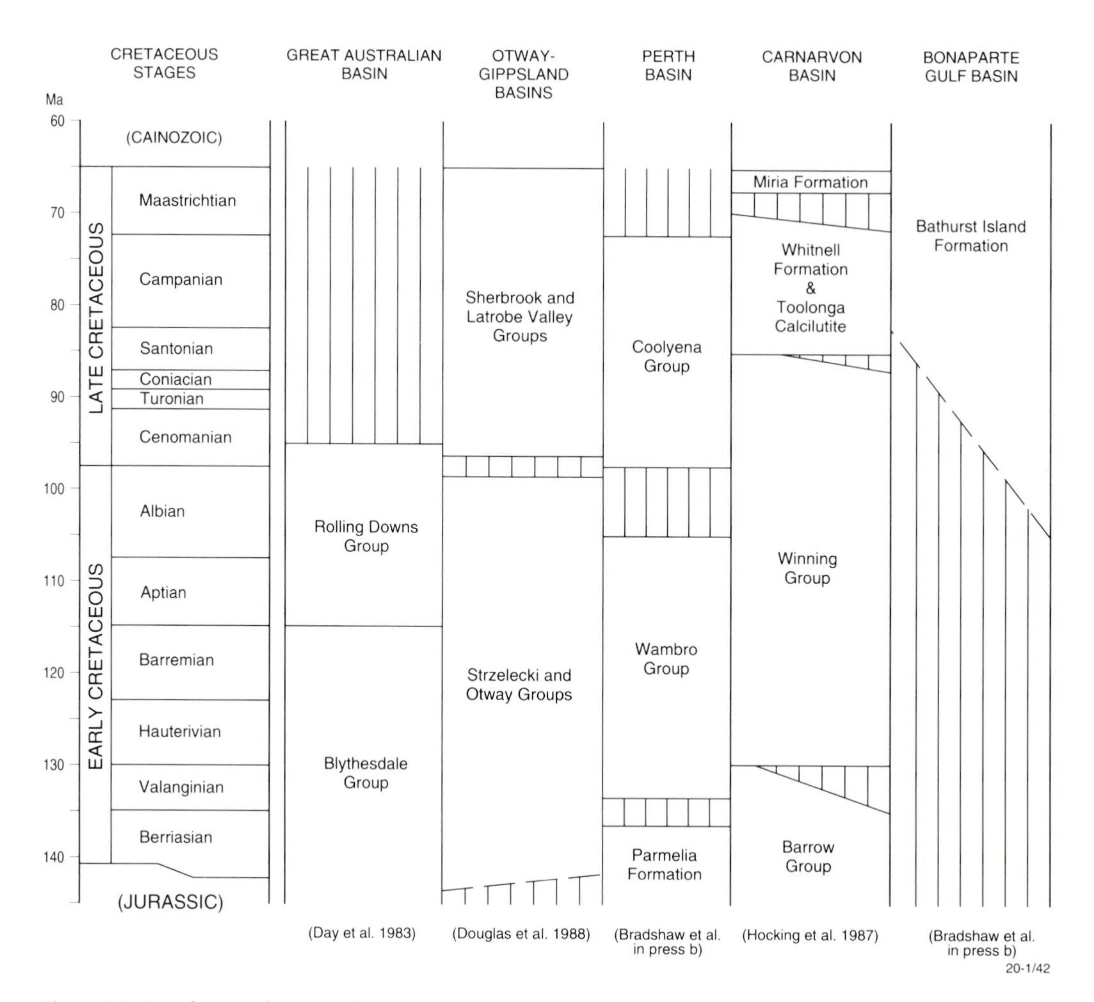

Figure 36: Correlation of principal Cretaceous lithostratigraphic successions in selected Australian basins.

COLUMNS 21–22: AMMONITES

Cretaceous ammonites have been described and reported from the Great Australian Basin (Whitehouse 1955; Reyment 1964; Day 1969, 1974; McNamara 1978, 1980, 1985; Skwarko 1983; and others), and from northern and Western Australia (Brunnschweiler 1959, 1966; Wright 1963; Skwarko 1966, 1983; Henderson & McNamara 1985; Henderson 1990, Henderson et al. 1992; and others). Only the faunas from Bathurst Island and some from the Western Australian Maastrichtian have been directly correlated with the standard Tethyan ammonite succession; the Great Australian Basin fauna is more endemic although it also contains Tethyan elements (Day 1969, 1974; McNamara 1980). Reports from many parts of the continent include more or less isolated ammonite occurrences, and an Australian biostratigraphic sequence has been set up only for the mid-Cretaceous of the Great Australian Basin (Day 1969).

COLUMNS 23–25: FORAMINIFERA

Both benthonic and planktonic foraminifera have been used for biostratigraphic contributions to basin studies, especially in the mid-Cretaceous of north-eastern Australia (Crespin 1963; Ludbrook 1966; Haig 1979; Scheibnerová 1986), and the Middle and Late Cretaceous of the North West Shelf (Wright & Apthorpe 1976; Apthorpe 1979; Quilty 1984*a*; and others). Belford (1958, 1960) and Quilty (1978) described Late Cretaceous agglutinated and calcareous foraminifera from the Perth and Carnarvon Basins. The Late Cretaceous foraminiferal sequence of the Otway Basin has been described by Taylor (1964) and Ludbrook (1971).

COLUMNS 26–28: CALCAREOUS NANNOFOSSIL BIOSTRATIGRAPHY

Cretaceous calcareous nannofossils from several basins marginal to Western Australia and from the epeiric Eromanga Basin were studied by Shafik (1978*a*,*b*, 1985*a*,*b*, 1990*a*, 1994; in Colwell, Graham et al. 1990). Early Cretaceous assemblages were recorded from the Rowley and Carnarvon Terraces, and Late Cretaceous assemblages from the Carnarvon and Perth Basins. In onshore sections of the Carnarvon Basin calcareous nannofossils first appear in the late Turonian. In the western Great Australian Bight Basin, the only known calcareous nannofossils are Maastrichtian in age (Shafik 1990*b*). In the Eromanga Basin Albian nannofossil assemblages are mostly diverse, although usually dominated by only a few species.

So far none of the published zonations can be applied singularly in the western margin of Australia (Bralower & Siesser 1992; Bown 1992; Shafik 1978*a*,*b*, 1994), but reliable biostratigraphic events based on cosmopolitan species have been identified in most areas. Shafik (1990*a*) identified a set of Turonian–Coniacian/early Santonian events in the Carnarvon Basin, and a set of late Coniacian/early Santonian–Campanian events in the Carnarvon and Perth Basins (Column 26).

Cretaceous sections offshore in north-western Australia and along the continent's western and south-western margins were drilled at several DSDP and ODP sites. The most complete nannofossil information came from sites on the Exmouth Plateau and in the adjacent abyssal plains. Sites drilled in the Cuvier and Perth abyssal plains and on the Naturaliste Plateau showed that Albian sediments bearing nannofossils are widespread offshore along the Australian western margin (Thierstein 1974; Bralower & Siesser 1992; Dumoulin & Bown 1992; Mutterlose 1992).

Shafik (1990*a*, 1993) described several nannoprovinces for the Australian region, in particular for the Albian and late Senonian. The co-occurrence of Albian cool-water and warm-water species along the western margin of the continent characterised the Extratropical Nannoprovince (Carnarvon Terrace, Perth Abyssal Plain, Naturaliste Plateau, most of the Papuan Basin, PNG). The coeval Austral Nannoprovince was known from the Eromanga Basin (north-eastern Australia). During the late Campanian–Maastrichtian the Austral Nannoprovince, with abundant cool-water species, existed in the Perth Basin and Great Australian Bight, and the Extratropical Nannoprovince, with a mixture of cool-water and warm-water species, in the Carnarvon Basin and northern Australia. The Tropical Nannoprovince, characterised by abundant warm-water species, extended over most of Papua New Guinea.

COLUMNS 29–32: DINOFLAGELLATE CYSTS

Cookson and her collaborators established a systematic and descriptive foundation with their early studies on the marine Jurassic and Cretaceous phytoplankton in Australia and Papua New Guinea (Deflandre & Cookson 1955; Cookson 1956, 1965; Cookson & Eisenack 1958, 1960*a,b*, 1961, 1962, 1970; Eisenack & Cookson 1960; and many others). Biostratigraphic studies were published from the Carnarvon Basin (Column 29), Perth Basin (Column 30), North West Shelf (Column 32; Backhouse 1978, 1987, 1988; Wiseman 1979; McMinn 1988; and others), and north-eastern Australia (Morgan 1980*a*; Burger 1982).

Helby et al. (1987) proposed the first composite standard zonation for the Australian Cretaceous, based on study of many onshore and offshore boreholes, especially from the south-eastern and north-western Australian and the Papua New Guinea regions (Column 31).

Several recent dinocyst biostratigraphic schemes, and ranges of selected species, are set out against the standard Cretaceous chronology as they were originally published. This demonstrates the differing opinions on range limits of several species, and hence on geological ages of zonal intervals defined by those species. Modified ages and correlations of certain Early Cretaceous zonal intervals are here proposed and argued under Column 33.

COLUMN 33: ALTERNATIVE NEOCOMIAN CORRELATIONS

Slightly modified ages for certain Early Cretaceous dinoflagellate zones, set out in the foldout chart, are proposed here on the following considerations:

- Helby et al. (1987) placed the J–K boundary within their *Pseudoceratium iehiense* Zone, but recent information from Papua New Guinea (Davey 1987) and from the North West Shelf (Claoué-Long et al., in prep.) indicates that this boundary falls within their *Kalyptea wisemaniae* Zone.
- According to Backhouse (pers. comm. June 1989) Control Point CP1 of Wiseman (1979) in the Carnarvon Basin is not older than the *Kaiwaradinium scrutillinum* Zone in the Perth Basin, whose lower limit Backhouse (1987, 1988) placed in the Valanginian.
- The lower limits of the *Phoberocysta lowryi* and *Phoberocysta burgeri* Zones, and Wiseman's (1979) Control Point CP2, are intercorrelated on the first appearance of *Muderongia testudinaria*, and this approximately agrees with Helby et al. (1987). *Muderongia testudinaria* was first described from northern Queensland, where it is accompanied by several other dinoflagellate species that, in the Tethyan northern Atlantic, are not known prior to the Hauterivian (Burger 1982). The range of this species in Australia is therefore assumed not to extend significantly below the Hauterivian.
- Helby et al. (1987) correlated the lower limit of the *Batioladinium jaegeri* Zone of Backhouse (1987) with that of their *Muderongia australis* Zone on the last occurrence of *Canningia reticulata*, *Muderongia testudinaria*, and *Phoberocysta burgeri*. However, first appearances of certain species given by those authors for that part of the sequence suggest that the *B. jaegeri* Zone may correspond approximately in time with the *M. testudinaria* Zone.
- *Odontochitina operculata* and *Ovoidinium cinctum* first appear in the upper part of the *Fromea monilifera* Zone in the Perth Basin (Backhouse 1987, pers. comm. June 1989), and in the *Muderongia australis* Zone (Helby et al. 1987). The authors agree on a Barremian to early Aptian age for the two zones. *O. operculata* first appears near the Hauterivian–Barremian boundary in the Western Tethyan Region, and presumably also in northern Australia (Burger 1982; Helby et al. 1987). For this reason the Hauterivian–Barremian boundary is taken to fall within the basal intervals of the *M. australis* and *F. monilifera* Zones.

COLUMNS 34–35: SPORES AND POLLEN

Broad floral provinces are recognised in the Northern Hemisphere, and they frequently indicate palaeolatitudinal (palaeoclimatic) control, especially during the Early Cretaceous. Both megafloral and spore–pollen records from Australia strongly suggest active interchange of vegetation with adjoining Gondwana continents, and with eastern Laurasia, at least during the Early Cretaceous (Sun 1980; Dettmann 1981; Song et al. 1983). Despite this evidence, intercontinental correlations have met with limited success. The records of Australia and New Zealand also reveal broad similarities but cannot yet be compared in detail (Raine 1984).

Early zonal schemes for eastern Australia (Dettmann 1963; Evans 1966*a*) culminated in the biostratigraphic scheme proposed by Dettmann & Playford (1969). Zonal schemes for the Perth and Carnarvon Basins in Western Australia (Balme 1957, 1964; Backhouse 1978, 1988) cover only the Early Cretaceous, due to the poor spore–pollen record from marine Upper Cretaceous strata (see Column 34).

The pan-Australian biostratigraphy proposed by Helby et al. (1987) is primarily the result of decades of study of hundreds of boreholes drilled in the Otway–Gippsland, Great Australian, and Laura Basins (Burger 1973, 1982; Dettmann 1986; Dettmann & Douglas 1988). The Late Cretaceous zones of that scheme (see Column 35) have been dated on evidence of foraminifera (Taylor 1964; Ludbrook 1971) and dinocysts (Evans 1966*a*; Helby et al. 1987) from the Otway and Bass Basins. The Aptian and Albian zones have been dated from the Great Australian Basin on their association with ammonites (Day 1969, 1974), foraminifera (Playford et al. 1975; Haig 1979), and dinocysts (Evans 1966*a*; Dettmann & Playford 1969; Burger 1980, 1986; Morgan 1980*a*). The Neocomian zones have been dated on their association with invertebrates and dinocysts in the Great Australian and Papuan Basins, and other considerations, such as intercontinental spore correlations and plant migration (Dettmann 1963, 1986; Dettmann & Playford 1969; Burger 1973, 1982, 1986, 1989; Helby et al. 1987). Indirect evidence of eustasy in the non-marine strata of the Great Australian Basin has also given certain clues as to ages of spore and pollen zones (see below).

The Jurassic–Cretaceous boundary is placed within the *Crybelosporites stylosus* Zone, and the lower part of the equivalent *Ruffordiaspora australiensis* Zone (Burger, in press *c*).

COLUMN 36: MEGAFLORAS

Studies of the Cretaceous vegetation have been made primarily in eastern Australia. The Neocomian to Cenomanian megafloral record from the Great Australian Basin is fragmented and restricted largely to poorly preserved leaf and wood impressions, and (locally abundant) silicified wood fragments (Gould 1975). A much richer plant record has been described from the Otway and Gippsland Basins (Douglas 1969, 1973; Cantrill & Webb 1987; Douglas et al. 1988; Taylor & Hickey 1990), and plant remains occur also near the J–K boundary of Tasmania (Tidwell et al. 1989).

COLUMN 37: VERTEBRATE FAUNAS

The record is poor and too episodic for biostratigraphic synthesis. Fishes are well represented in the upper Albian Toolebuc Formation, Great Australian Basin, and in the Aptian Koonwarra Beds, Gippsland Basin (A. Kemp 1991; N. Kemp 1991; Long 1991). Reptiles are well represented in the Late Neocomian to Albian fauna in Victoria, and in the Toolebuc and early–middle Albian Lightning Ridge faunas (Molnar 1991). Birds are known only from isolated feathers (Koonwarra beds) and some skeletal fragments in the late Albian of the Great Australian Basin

(Rich 1991). Amphibians have now been reported from the Early Cretaceous of the Gippsland Basin (Warren 1991). The first known Mesozoic mammal from Australia (jaw and teeth of a monotreme) has recently been discovered in the Lightning Ridge fauna (Rich 1991).

OTHER INVERTEBRATES

Other groups of Cretaceous animals recorded from Australia and not represented in charts A and B have been summarised by Quilty (1975), and include among other taxa: Radiolaria, brachiopods, molluscs (bivalves, gastropods), Crustacea (crabs, ostracods) and other arthropods.

EUSTASY (Fig. 37)

Cretaceous eustatic sea-level movements and their effect on the fossil and environmental records of five continents were first comprehensively investigated by Cooper (1977) and Pitman (1978). Cooper translated the relevant data into a sea-level curve relative to the margins of the ancient Gondwana and Laurasia continents. The earliest attempts to use seismically profiled truncation of strata sequences (primarily in the northern Atlantic region) for the reconstruction of world-wide sea-level movements during the Mesozoic and Cainozoic were made by Vail and collaborators. A broader based study was set up by Haq et al. (1987), who integrated seismic and sequence stratigraphic data from continental margins elsewhere, to create depositional models for the reconstruction of a very detailed Mesozoic and Cainozoic coastal onlap curve (Fig. 37).

Australian workers have studied the evidence of sea-level movements in the Early Cretaceous of the Carnarvon, Perth, and Great Australian Basins, and the Late Cretaceous in Western Australia. Morgan (1980*b*) integrated Aptian–Albian fossil records (in part complemented by lithorecords) into a pan-Australian Cretaceous sea-level curve. It yielded good results for certain basins because, on the whole, there was very little tectonic activity to mask the effects of sea-level movements on sedimentation processes.

Lithological records have been used also to log rising and falling sea-levels. Quilty (1980) observed two Late Cretaceous sedimentary cycles in various basins in Western Australia; they are bounded by widespread unconformities, which he attributed to recurrent eustatic falls of sea-level. Exon & Burger (1981) and Burger (1986) outlined a series of five widespread cycles in the Jurassic and Neocomian non-marine strata sequence of the Great Australian Basin. They argued global sea-level movements to be the most likely cause of those cycles, and linked them with the eustatic curves of Vail et al. (1977) and Cooper (1977).

Various Australian sea-level curves (as drawn by individual authors) are shown in Figure 37. Marine peaks and troughs in different basins do not all coincide. Several of them are evidently of local origin, but the eustatic nature of others is shown by their coincidence with the global curves. In view of their potential value for dating sedimentary sequences they warrant much closer investigation. However, from the study of Matsumoto (1980) it can be inferred that several of the 'global' curves may have been limited largely to the Tethyan Realm, and were, at best, barely felt in Boreal regions.

ACKNOWLEDGMENTS

The authors are indebted to geoscientists from both Australia and overseas who have commented on text and charts of the Cretaceous volume in the *BMR Record* series on Australian Phanerozoic timescales (Burger 1990*b*). Many kindly sent reprints of papers or manuscripts (in press) on relevant subjects. For their assistance, the authors wish to thank: M. Apthorpe and K.J. McNamara (Perth, WA); M.E. Dettmann

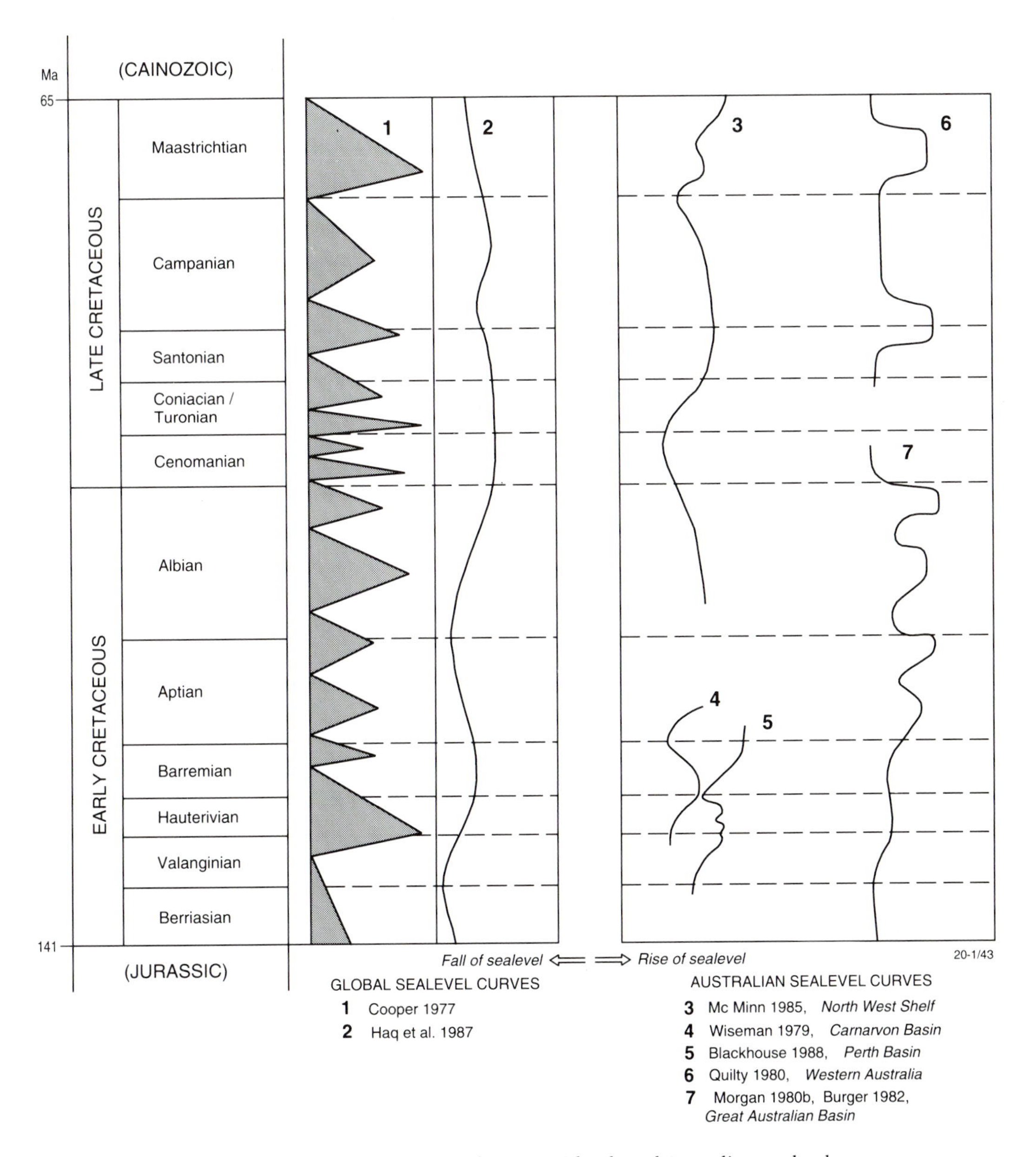

Figure 37: Comparison of proposed global sea-level curves with selected Australian sea-level curves.

and R.E. Molnar (Brisbane, Qld); A.D. Partridge (Melbourne, Vic.); P.G. Quilty (Kingston, Tas.); P. Bengtson (Uppsala, Sweden); T.J. Bralower (Miami, Fla, USA); R.J. Davey (Llandudno, North Wales); B. Galbrun (Paris, France); the late N.F. Hughes (Cambridge, England); W. Lowrie (Zürich, Switzerland); T. Matsumoto, S. Toshimitsu, and M. Noda (Fukuoka, Kyushu,

Japan); D.J. McIntyre (Calgary, Alta, Canada); G.S. Odin (Paris, France); J.G. Ogg (West Lafayette, Ind., USA); K. von Salis Perch-Nielsen (Zürich, Switzerland); F. Robaszynski (Mons, Belgium); G.J. Wilson (Lower Hutt, New Zealand); and Yang Zunyi (Beijing, China).

2.10

Cainozoic (Chart 12)

G. Chaproniere, S. Shafik, E. Truswell, M. Macphail and A. Partridge

INTRODUCTION

The chart is designed to show the interrelationships between zonations based on marine microfossils, macrofossils, or terrestrial faunas, and those based on pollen and spores. These zonal schemes are calibrated, as far as possible, against the timescales for the Palaeogene and Neogene presented in recent compilations by Berggren et al. (in press *a,b*).

The Australian biostratigraphic schemes shown on the chart represent a mixture of zonal concepts. Some of the earlier foraminiferal zones (e.g. Carter 1958*a,b*, 1964; Taylor 1966) are assemblage-zones, as are the palynological zones (Harris 1971, 1985; Stover & Evans 1973; Stover & Partridge 1973;). The mammalian and molluscan assemblages are closely tied to localities, with, in many cases, little understanding of the temporal ranges of the taxa involved. Parts of the foraminiferal biostratigraphic schemes are based on evolving lineages whereas the foraminiferal column of McGowran, and the nannofossil columns of Shafik, are constructed from first and last appearance datums. The advantage of such a presentation of biostratigraphic data, as noted by McGowran (1986) and Shafik (1990*a*), is that it permits ease of correction when advances in knowledge are made; new data can be included within the biostratigraphic system without the formal redefinition of zones and subzones.

The chart represents the most comprehensive attempt yet to relate biostratigraphic schemes based on diverse groups of fossils in the Australian Cainozoic. It owes much, however, to previous regional compilations, such as those of McGowran et al. (1971), Quilty (1972), Abele (1976) and Mallett (1977).

COLUMN 1: TIMESCALE

We have followed the scheme set out by Berggren et al. (in press *a,b*). The correlation between the magnetostratigraphy and the numerical timescale has been updated since the previous compilation (Truswell et al. 1991), which was based on the work of Berggren et al. (1985*a,b*). The period–epoch column and the location of the standard European stages are correlated to the numerical timescale through the magnetic polarity scale (Berggren et al., in press *a,b*).

AUSTRALIAN FORAMINIFERAL BIOSTRATIGRAPHY

The zonal schemes illustrated in this section are arranged broadly in historical order, following their development or application in Australia, superimposed on a geographic arrangement, east to west.

COLUMN 2: INTERNATIONAL FORAMINIFERAL ZONATIONS — LOW-LATITUDE

This compilation uses the planktic foraminiferal low-latitude zonal scheme presented by Berggren et al. (in press *a,b*), and is based on modifications of the Blow (1969, 1979) zonal schemes published previously by Berggren (1973, 1977), Berggren et al. (1985*a,b*) and Berggren & Miller (1988). The Berggren et al. (in press *a,b*) scheme has been correlated with that of Blow (1969, 1979), as modified in part by Kennett & Srinivasan (1983), by using the various bioevents given by Berggren et al. (1985*a,b*; in press *a,b*). A similar scheme based on work by Bolli (1957*a,b*, 1966) is also available for subtropical–tropical areas, and the most recent versions of this (Bolli & Saunders 1985; Toumarkine & Luterbacher 1985) have been incorporated. Since the publication of these two papers it has been realised that some of the bioevents used therein either overlap or coincide. This is reflected in the chart by the modification of the zonal boundaries. Thus, because the first appearances of *Morozovella aragonensis* and *Planorotalites pusilla pusilla* coincide, in the chart they are separated by a diagonal line. The first appearances of *Acarinina pentacamerata* (the index for the base of the *M. aragonensis* Zone) and *Globigerinoides primordius* (index for the base of the *G. primordius* Zone) are now placed below the the FADs of the indices of the overlying zones; these zones are now shown in boxes.

COLUMN 3: INTERNATIONAL FORAMINIFERAL ZONATIONS—SOUTHERN MID-LATITUDE

Because Australia covers a latitudinal range from the tropics to the temperate regions it is appropriate to include a mid-latitude scheme. Berggren et al. (in press *a,b*) proposed schemes for these latitudes, but they have not been included because they encompass only the Miocene to Pleistocene, and are based on Atlantic Ocean sections and so are not pertinent to the Australian region. Instead, a scheme proposed by Jenkins (1985) has been used. This scheme had its origins in the Lakes Entrance Oil Shaft of Victoria (Jenkins 1960), which was substantially modified for use in New Zealand (Jenkins 1965, 1967), before being modified for use over a much broader area. Correlation of this scheme with the low-latitude schemes follows, in part, correlations given by Hornibrook (1978, Tables 7.8, 7.9) and Hornibrook et al. (1989, Figs 5, 6).

COLUMN 4: EAST INDIAN LETTER STAGES

Included is a correlation with the East Indian letter classification, based on 'larger' neritic, tropical foraminifera, following Adams (1984). Adams (1970, 1984) reviewed the development of the letter classification, and provided a correlation of the letter zones with the P and N foraminiferal zones. Some of the correlation details shown on the chart, viz. the lowering of the upper T*e–f* boundary from the base of Zone N9 to within Zone N6, rest on information provided from north-western Australia by Chaproniere (1981*b*, 1984*a,b*). McGowran (1979, 1986) has noted the possibility that changes in biostratigraphic boundaries of provinces based on larger foraminifera are related to shifts in watermass; southern Australian correlations with the standard stages are viewed as extratropical excursions controlled by climate (McGowran 1986, Fig. 4).

COLUMN 5: SOUTH-EASTERN AUSTRALIAN LOCAL STAGES

A terminology based on local stratotypes in south-eastern Australia was first introduced by

Hall & Pritchard (1902) in order to reduce confusion caused by attempts to use European terminology. The stages shown in the column are those in current use in Victoria (Carter 1964; Abele 1976; McGowran et al. 1992; Li & McGowran, in press) and South Australia (Ludbrook & Lindsay 1966; Ludbrook 1973), with the addition of the Willungan Stage as defined by Lindsay (1985). For a history of stage usage reference should be made to Carter (1964) and Abele (1976).

The Janjukian, Balcombian, Kalimnan and Werrikooian were first defined by Hall & Pritchard (1902); Batesfordian was added by Chapman & Singleton (1923); Singleton (1941) added Cheltenhamian and other stages now fallen into disuse, and redefined some of the earlier stages. Crespin (1943*b*) introduced the Mitchellian, Longfordian and Bairnsdalian; these last two she considered as substages of the Balcombian. O.P. Singleton (1954) considered Cheltenhamian and Mitchellian to be time equivalent, a postulate that was supported by Mallett (1977). The relationship of these stages, together with the Kalimnan, was further discussed by Wilkins (1963). The term 'Aldingan', introduced by Hall & Pritchard (1902), was revived by Carter (1964) and redefined by Ludbrook & Lindsay (1966) to restrict it to the Late Eocene sequence typified in the eastern St Vincent Basin.

The name 'Johannian' lacks formal definition, but is used for strata that contain Eocene faunas older than Aldingan (Ludbrook & Lindsay 1966; Ludbrook 1971). 'Wangerripian', a term informally proposed by O.P. Singleton (see Abele 1976), is used for the Middle Paleocene to Early Eocene intervals, typified by the Wangerrip Group in the Otway Basin. Ludbrook (1963, 1973) introduced 'Yatalan', first as a term to describe faunas, later as a stage name to include Pliocene strata between the Kalimnan and Werrikooian in South Australia, replacing the older term 'Adelaidean', and thus avoiding confusion with Precambrian terminology. The relationship between the Werrikooian stratotype and the Pliocene–Pleistocene boundary was reassessed by Singleton et al. (1976).

In his review of the south-eastern Australian Neogene, Carter (1990) provided correlations between the south-eastern Australian local stages and the planktic foraminiferal zonation proposed by Mallett (1978). In this work Carter (1990) considered that the lower part of the Balcombian Stage at its type locality was of equivalent age to the Batesfordian Stage, and so combined the Batesfordian into the lower part of the Balcombian. Li & McGowran (1994) have continued to use both stage names and this usage is continued here. Carter (1990) also did not use the Yatalan Stage because he considered that 'from the evidence available at present, it is not clear whether this Stage name is a synonym of the Werrikooian or whether it should be applied to the earlier phase of the Werrikooian alone or to younger Quaternary deposits' (Carter 1990, p. 191). Because of this uncertainty both names are retained here.

As noted above, there are still questions surrounding the continued use of these stages. Uncertainties remain concerning their correlation with standard European series and stages. The concepts of the stages have varied and in many cases have been extended to cover broader intervals of time than those represented by the stratotypes. Problems of relating stage stratotypes to standard ages are highlighted by Mallett's (1977) analysis of foraminiferal faunas, which asserts that the stratotypes of Bairnsdalian, Cheltenhamian and Mitchellian overlap, and that all fall within the *Globorotalia conomiozea* Zone (Late Miocene).

The continuing usefulness of these stages is subject to debate; although suggestions have been made that their use be discontinued (e.g. Singleton 1967), other authors (e.g. McGowran et al. 1971) have indicated that such a move would be premature, and should only be contemplated when correlations with global stratotypes were firm enough to make local stages redundant. New stages continue to be formalised for local stratigraphic purposes (e.g. Lindsay 1985).

COLUMN 6: CARTER FAUNAL UNITS AND ZONES

Carter (1958*a*,*b*, 1964) established a sequence of faunal units based on distinctive associations or assemblages of foraminifera, including both benthic and planktic forms. The sequence was compiled from outcrop sections in the Longford, Bairnsdale, Lakes Entrance, Aire and Torquay districts of Victoria. The associations first recognised were designated Faunal Units 1–11; these were expressed also as a set of ten named zones. Subsequently, thirteen events based on planktic foraminifera were isolated and used as the basis for intercontinental correlation. The relationship of the faunal units and zones to the P and N zones of Blow (1969) is as indicated by McGowran et al. (1971).

COLUMN 7: LAKES ENTRANCE OIL SHAFT, VICTORIA

The stratigraphy and palaeontology of the Lakes Entrance oil shaft was first described by Crespin (1947). In this work she divided the sequence using the Victorian stage nomenclature current at that time; the location of larger neritic foraminiferal occurrences in this section is based on this work. Subsequently, Jenkins (1958, 1960) identified a sequence of Miocene zones in the Lakes Entrance shaft, East Gippsland, based on planktic foraminiferal bioevents. This was pioneering work at a time when the biostratigraphic usefulness of planktic foraminifera was just being recognised, and formed the basis for the development of a southern temperate zonal scheme based on New Zealand sections. Mallett (1977) pointed out the unsuitability of this sequence as a reference section, noting the condensed nature of the Middle to Late Miocene section there.

COLUMN 8: GIPPSLAND ZONULES

Taylor (1966) established a 'down-sequence' scheme for the Late Eocene to Late Miocene of the offshore Gippsland Basin. It was based on the down-hole appearance of selected taxa, and was elaborated in McGowran et al. (1971) and correlated therein with the P and N zones.

COLUMN 9: SOUTH AUSTRALIAN ZONES

Lindsay (1967, 1969), Ludbrook & Lindsay (1969), and Ludbrook (1971), following the work of Jenkins (1965, 1967) in New Zealand, proposed a zonation based solely on planktic foraminifera, with modifications to suit less open marine conditions. This scheme was used to indicate possible correlations with sequences in New Zealand, East Africa and Trinidad.

These zones are based mainly on bioevents (first and last appearance datums). In the Middle Miocene the sequence used is essentially that of the evolving series leading from *Globigerinoides triloba* s.s. to *Orbulina universa*, first observed by Blow (1956) in Venezuela, and applied to the Australian Miocene by Jenkins (1958, 1960). Modifications to the scheme were expressed by McGowran et al. (1971), who also provided a correlation of the zones with those of Carter (1958*a*,*b*, 1964), Taylor (1966) and Wade (1964).

COLUMN 10: PLANKTIC FORAMINIFERAL EVENTS AND INTERVALS, SOUTHERN AUSTRALIA

This column shows a composite succession of 'local' or Australian biostratigraphic events, constructed from discontinuous sequences at localities from the Naturaliste Plateau to south-western Victoria. The Palaeogene sequence is drawn from McGowran (1978) and McGowran & Beecroft (1985). The full sequence is drawn from

McGowran (1986, 1991), and the indirect basis of the correlation of these events with the P and N zones was demonstrated by McGowran (1986) and Lindsay (1985).

Hollow symbols in the column mark short-lived events in disjunct distributions. It is notable that the biostratigraphic record, particularly in the later part of the Eocene, provides a clear reflection of transgressive events (see Column 30). In this interval, four transgressions, isochronous within the stratigraphic framework shown, are identifiable. McGowran (1986, Fig. 7) refers to these as the Wilson Bluff (P.12), Tortachilla (base P.15), Tuketja (late P.15) and Aldinga (P.17) transgressions; these find a biological expression in ingressions of *Hantkenina* species, and of acmes in *Turborotalia increbescens*. The relationship of the dinoflagellate zones recognised by Harris (1985) to these transgressions is noted below.

COLUMN 11: VICTORIAN ZONES

This zonation (Mallett 1977, 1978) is based on the appearance of Early Miocene to Pleistocene planktic species from the Otway and Gippsland Basins and the Port Phillip Embayment and incorporates a number of internationally recognised datum planes, allowing correlation with Blow's (1969) zones. The numbered 'V' zones on the column right were introduced by Mallett (1977). They are included here as they have been referred to in some publications (e.g. Darragh 1985). Here, these zones have been related to the datum planes of McGowran, and their placement adjusted to fit the Berggren et al. (in press *b*) timescale.

COLUMN 12: LAKES ENTRANCE OIL SHAFT, VICTORIA—PLANKTIC FORAMINIFERAL ASSEMBLAGES

McGowran & Li (1993) and Li & McGowran (1994) studied planktic foraminiferal assemblages over the Miocene part of the Lakes Entrance oil shaft. They recognised a number of assemblages as ecostratigraphic units, based on changes in species diversity and ratios between morphotypes. These variations have been linked to global palaeoclimatic and palaeoceanographic events. Biostratigraphic control is based on a number of bioevents recognised in the sequence.

COLUMN 13: NORTH WEST SHELF ZONATION AND PRINCIPAL FORAMINIFERAL EVENTS

A series of planktic foraminiferal zones for the Paleocene and Eocene of the North West Shelf was established by Wright (1973). Zones T1 to T9 were established because of the difficulty in recognising the international P Zones in the North West Shelf faunas. In his published account, however, and largely to preserve confidentiality, Wright (1977) broadly referred faunas from Scott Reef No. 1 well to the planktic zones of Blow (1969). Zones T10 to T20 are based on more recent work by Heath & Apthorpe (1984) and by Heath (1979), who were also able to recognise Zones N8 to N23 of Blow (1969).

The relationship of the North West Shelf zonation to the 'standard' tropical zonation was further elaborated by Heath & McGowran (1984) and Heath & Apthorpe (1984). One problem is that a few of the P zonal indices appear to have different ranges in the Indian, Atlantic and Pacific Oceans.

This scheme is principally a 'down-hole' one to permit dating during drilling and hence relies heavily on species tops or extinctions. However, some of the zones could not be defined on last appearances and have had to be delineated on first appearances, so require good sidewall control. Such zones include T3, the top of T10, T11a and T16.

COLUMN 14: ASHMORE NO. 1 WELL (BROWSE BASIN) AND NORTH WEST CAPE (CARNARVON BASIN)

This list of foraminiferal events considered important in the Australian and New Zealand region was published by Chaproniere (1984*a*). The placement of some of these events differs from that in the overview of McGowran (1986). Included also was the extinction datum of the larger neritic foraminifer *Lepidocyclina (Eulepidina) badjirraensis.*

COLUMN 15: ODP LEG 122, SITE 761, WOMBAT PLATEAU

Site 761 was drilled on the eastern end of the Wombat Plateau. The Tertiary section was summarised by the Shipboard Scientific Party (1990), and more detail for the Neogene given by Zachariasse (1991). A detailed faunal analysis was not available for the Palaeogene section, but the range of periods of sedimentation during the Paleocene to Oligocene was given. As with other sections in the region, the Early Paleocene rests disconformably on the Late Cretaceous (upper Maastrichtian) (Shipboard Scientific Party 1990). Two intervals (Early Eocene and late Middle Eocene to Early Oligocene) have not been recorded, but whether this is due to sampling or to the presence of hiatuses is unknown. Similar to elsewhere in north-western Australia, the Late Oligocene rests unconformably on the Eocene and contains reworked Eocene faunas. Zachariasse (1991) concluded that most of Zone N.5 and part of the Middle Miocene to Late Pliocene (Zones N.15 to N.21) are missing. The Middle Miocene section is succeeded by an interval containing Late Miocene (Zone N.18) to Early Pliocene (Zone N.19–20) species in association with *Globorotalia truncatulinoides* and *Globigerinoides fistulosus* indicative of the basal part of Zone N.22 (latest Pliocene to earliest Pleistocene). This suggests that the Zone N.18 to N.19–20 forms were reworked into Zone N.22, a feature also recorded by Chaproniere (1991) from the Townsville Trough, offshore Queensland.

COLUMN 16: LARGER NERITIC FORAMINIFERAL DISTRIBUTION

The two columns show the larger neritic foraminiferal distribution for the southern and northern parts of Australia. The distribution for southern Australia is based mainly on the work of McGowran (1979, 1991) with additional details given by Chaproniere (1992), McGowran & Li (1993) and Quilty (1993). The record shows distinct incursions of these forms during intervals of climatic warming. In northern Australia (Palmieri 1971, 1984; McGowran 1979; Chaproniere 1981*a*, 1983, 1984*a,b*; Betzler & Chaproniere 1993) larger neritic forms appear to have made limited incursions during the Eocene, but from the Late Oligocene onwards are often numerous. This distribution is related to the northward movement of Australia during the Tertiary.

OTHER FORAMINIFERAL ZONATIONS

A zonation proposed by Wade (1964), using sequences in the Port Campbell Embayment and the St Vincent and Murray Basins, is not included in the chart. This zonation was based on lineages within *Globigerina, Globigerinoides* and *Globorotalia* spanning the Late Eocene to Middle Miocene. These were related to other Australian biostratigraphic units by McGowran et al. (1971).

In Victoria, data from an unpublished scheme of Nicolls (1968) dealing with faunas above the level of *Orbulina universa* (Middle Miocene) was incorporated by Abele (1976) into a numbered scheme that extended the faunal units of Carter (1958*a,b*) into younger sequences. According to Mallett (1977), this amalgamation involved some redefinition of the Carter units.

CALCAREOUS NANNOFOSSIL BIOSTRATIGRAPHY

The biostratigraphic schemes (zones and/or events) illustrated in this section are arranged broadly in historical order, following their development or application in Australia, superimposed on a geographic arrangement, west to east.

The marine Cainozoic record in Australia lacks continuity, and the nannofossil sequence of events presented below is a set of data from several separate geographic locations pieced together.

COLUMN 17: INTERNATIONAL NANNOFOSSIL ZONATIONS

This compilation uses the 'standard' (temperate to tropical) nannofossil zones of Martini (1971) and the low-latitude (mainly tropical) zones of Bukry (1973, 1975) and Okada & Bukry (1980). Most of the 'standard' and low-latitude zones are based on events marked by first appearances to avoid problems stemming from reworking. However, important boundaries, such as the top of the Eocene or of the Pliocene, are approximated in both zonal schemes by events marked by last appearances. Indeed, most of the zones dividing the Pliocene are based on last appearances.

COLUMN 18: EVENTS ON THE AUSTRALIAN SOUTHERN MARGIN AND NATURALISTE PLATEAU

In the onshore Otway Basin, Shafik (1983) using one section at Browns Creek in the eastern part of the basin, and three in the Gambier Embayment to the west, identified a sequence of biostratigraphic events spanning the interval from the first appearance of *Cyclicargolithus reticulatus* in the Middle Eocene to the last appearance of *Discoaster saipanensis* in the latest Eocene. These nannofossil events were compared to foraminiferal events identified by McGowran (1973, 1978). Shafik (1983) identified two marine ingressions (the *Cyclicargolithus reticulatus* and *Reticulofenestra scissura* ingressions — see also Shafik 1990*b*) well below the base of the calcareous marine section in the Gambier Embayment. Based on offshore material, Shafik (1987, 1992*c*) extended the sequence of biostratigraphic events known in the Otway Basin to levels well within the Oligocene, the youngest being an interval with *Sphenolithus ciperoensis*. Shafik (1987, 1992*c*) also recognised an interval with the key Miocene species *Sphenolithus heteromorphus*.

Further identification of events in the Early to Middle Eocene was made by Shafik (1985*a*) in the sequence on the Naturaliste Plateau, off south-western Australia. In the Great Australian Bight, Shafik (1990*b*) recognised two Early Eocene marine ingressions (the *Discoaster lodoensis* and the Potoroo *D. sublodoensis* ingressions), the younger of which equates with the basal Eocene on the Naturaliste Plateau (Site 264). The nannofossil sequence above these Early Eocene ingressions confirmed the existence of a substantial Oligocene marine record on the Australian margin. This is significant because marine Oligocene sediments are absent from numerous oceanic sections in the Australian region (e.g. Sites 264, 258 on the Naturaliste Plateau) and the south-western Pacific.

Records of the low-latitude, Oligocene species *Sphenolithus distentus* and *S. ciperoensis* on the southern margin of Australia (Shafik 1987, 1990*c*) are important as links with the international zones of Martini (1971) or Okada & Bukry (1980), but their total stratigraphic ranges on the Australian margin may not be the same as in low-latitude sections (Shafik 1992*b*).

Nannofossil data have been used to suggest correlation of local foraminiferal events with the P Zones. For example, Shafik (1981) suggested that the *Hantkenina* interval should be placed high in foraminiferal Zone P.16, rather than high in Zone P.15 as shown here.

COLUMN 19: EVENTS ON THE AUSTRALIAN WESTERN AND NORTH-WESTERN MARGINS

Shafik (1973) presented a nannofossil zonation for the Early Eocene to Middle Oligocene, based on sections in the western and southern continental margins. Subsequently, Shafik (1978*b*) identified the *Heliolithus reidelii* and *Discoaster multiradiatus* (NP 8 and NP 9) Zones of Martini (1971) in the lower Kings Park Shale, Perth Basin. In the younger parts of the same formation (later the Porpoise Bay Formation of Cockbain & Hocking 1989) he identified a cluster of Middle Eocene nannofossil events (correlating with the foraminiferal Zone P.12).

In the offshore Perth Basin, Shafik (1991) identified an almost complete Palaeogene nannofossil record extending from the Early Paleocene (lowest occurrence of *Cruciplacolithus asymmetricus*) to well within the Oligocene (with *Sphenolithus distentus*), and an interval with the key Miocene species *Sphenolithus heteromorphus*. These Oligocene and Miocene records were augmented by detailed study of the nannofossil sequence in Challenger No. 1 Well (Shafik 1992*a*).

Shafik (1990*b*) recorded several assemblages from the Palaeogene sequence on the Carnarvon Terrace (offshore Carnarvon Basin), and on the Exmouth Plateau to the north. Assemblages from above the lowest occurrence of the Late Paleocene *Fasiculithus tympaniformis* to above the lowest occurrence of the Late Miocene *Discoaster quinqueramus* were described from the Rowley Terrace (offshore Canning Basin) by Shafik (1990*c,d,e*).

Events in the Late Oligocene to Early Miocene are partly based on the sequence in Ashmore Reef No. 1 well in the Bonaparte Gulf Basin (Shafik & Chaproniere 1978).

COLUMN 20: ODP LEG 122, SITE 762, EXMOUTH PLATEAU

The Cainozoic sequence recovered at Site 762 contains a few minor hiatuses in the Miocene. Siesser & Bralower (1992) studied this almost complete sequence, and successfully applied Martini's (1971) zonation. However, they had to modify some of the zones, because some marker species were either rare (e.g. *Ellipsolithus macellus, Heliolithus riedelii, Chiasmolithus solitus, Sphenolithus pseudoradians, S. distentus, Discoaster kugleri, Amaurolithus tricorniculatus*), absent (e.g. *Rhabdolithus gladius, Helicosphaera ampliaperta*), or difficult to distinguish consistently (e.g. *Chiasmolithus danicus*, being not always distinguishable from *Cruciplacolithus edwardsii*). The rarity of some of the warm-water species is because the sequence of Site 762 is mostly temperate, not tropical.

PALYNOLOGICAL BIOSTRATIGRAPHY

Biostratigraphic schemes based on spores, pollen and dinoflagellates are concentrated in Australia's south-east. The first published zones were based on spore and pollen distributions in the Otway (Harris 1971) and Gippsland (Stover & Evans 1973; Stover & Partridge 1973) Basins and were erected to fill the stratigraphic needs associated with petroleum exploration in these coastal and offshore basins. Dinoflagellate zonal schemes followed (Partridge 1976; Harris 1985). Within the Murray Basin, an areally extensive basin with a thin sequence of flat-lying sediments, the impetus for fine stratigraphic control came from the need to understand groundwater dynamics within the basin. Zonal schemes in that basin have been varied, and designed to fill local and regional stratigraphic needs (see Columns 22 and 23 below).

There are at present no zonal schemes for northern Australian basins, although thick sequences exist in, for example, the Bonaparte

Gulf Basin. Because of the concentration of palynological reference sections in the south-east of the continent, problems remain in dating sequences within inland basins, which include the large Lake Eyre Basin (Sluiter 1991) and other smaller basins in central Australia (Truswell & Harris 1982; Truswell & Marchant 1986).

One zonal scheme developed in northern Australia was that of Hekel (1972), based on sequences in Queensland coastal and offshore areas. It has not been included on the chart because of difficulties in relating its units to the timescale. Of the five units delineated by Hekel, the basal Unit I could be dated only as possible Paleocene to Middle Oligocene; time control on younger units was considered to be firmer, being established by relationships with foraminifera and nannoplankton. On this basis Hekel suggested that Unit II may be Late Oligocene, Unit III Early Miocene, Unit IV Middle to Late Miocene, and Unit V Pliocene. Foster (1983) reviewed the criteria for the recognition of Hekel's units II and III, and indicated that, on the basis of the ranges of key palynomorphs, as now understood, these units can no longer be used for chronologic or biostratigraphic studies.

COLUMN 21: SPORES AND POLLEN—GIPPSLAND BASIN

The biostratigraphic units of Stover & Evans (1973) and Stover & Partridge (1973) were originally defined in the Gippsland Basin, and were correlated with Taylor's (1966) system of zonules. Partridge (1976) related the Gippsland Basin palynological zones to the 'standard' foraminiferal zones of Berggren (1969), Blow (1969) and Stainforth et al. (1975).

Formally described zones extend from the early Danian–mid-Thanetian lower *Lygistepollenites balmei* Zone to late Early–early Late Miocene *Triporopollenites bellus* Zone. Zones erected for the mid-Late Miocene to Pleistocene by A.D. Partridge are published in name only (e.g. Pole et al. 1993; Macphail & Truswell 1993).

Since 1973 the Gippsland zonation has undergone considerable proprietary revision in two respects. First, intensive drilling has shown many spore and pollen species used to define zone boundaries have extended time distributions. Second, application of sequence stratigraphic concepts has allowed zone boundaries to be interpreted in terms of global eustatic cycles identified by Haq et al. (1987). The correlation presented in Column 21 incorporates these revisions.

Isotopic dating, particularly of basalts overlying, or interbedded with, palyniferous sediments in highland sequences has provided an independent age control on Gippsland Basin zone equivalents recognised there (Owen 1975). These dates accord in a general way with the ages for the zones indicated on the chart, although there are some very minor disparities, reflecting perhaps inaccuracies in the isotopic dates, or ambiguities in the relationship of palyniferous sequences to dated basalts.

The oldest of the zones for which some time control exists is the Upper *Lygistepollenites balmei* Zone. Sub-basaltic sediments at widely separated sites in the Southern Monaro yield assemblages attributable to this zone; ages of 53.2 and 51.0 Ma have been obtained from low in the overlying basalt sequence (Taylor et al. 1990), a relationship that accords well with the ages indicated in the chart. In the Eocene, two dates from the Southern Highlands of New South Wales are relevant. Sediments at Nerriga (Owen 1975) and at Bungonia (Truswell & Owen 1988) have yielded pollen suites referable to the Lower *Nothofagidites asperus* Zone. The pollen-bearing sediments underlie basalts dated as 40–46 Ma for Nerriga and 45–47 Ma for Bungonia. These are at the older end of the age range of the zone as shown.

In the Shoalhaven River, New South Wales, Nott (1992) identified Upper *Nothofagidites asperus* to *Proteacidites tuberculatus* sediments impounded in a lake setting, with basalts damming the lake dated as 29–31 Ma, a date conforming to age range shown. From Tasmania, Macphail & Hill (1994) identified the Lower *P. tuberculatus* Zone in sediments at Wilmot Dam, where overlying basalts are dated as 26.7 Ma. The middle part of the *P. tuberculatus* Zone has been tentatively identified by Pole et al. (1993) in sediments underlying basalts dated as 22.3 Ma (earliest Miocene) at Berwick, Victoria. A similar age relationship was established by Owen (1988) at Kiandra, New South Wales, where freshwater sediments tentatively considered to indicate the middle part of the *P. tuberculatus* Zone are overlain by basalts dated as 21.5–21.7 Ma.

COLUMNS 22–23: SPORES AND POLLEN — MURRAY BASIN

Although palynomorphs are abundant and reasonably diverse in the Murray Basin of south-eastern Australia, assemblages tend to be dominated by a few long ranging taxa. Biostratigraphic marker species are rare, and some species show different time-ranges from those known from adjacent basins such as the Gippsland. For these reasons a number of distinct methods have been used to zone Cainozoic sequences in the Murray Basin. Broadly, these can be classified into those schemes that use qualitative (presence/absence) criteria, and refer to zones established elsewhere, especially in the Gippsland Basin, and schemes that rely on the use of relative pollen frequencies.

The concurrent range-zones or assemblage-zones defined in the Gippsland Basin by Stover & Partridge (1973) have been applied with caution in the Murray Basin. Zones recognised include equivalents of the Lower, Middle and Upper *Nothofagidites asperus*, *Proteacidites tuberculatus* and *Triporopollenites bellus* Zones in boreholes in the Hay–Balranald–Wakool districts in the eastern sector (Martin 1977), and the *P. tuberculatus* and *T. bellus* Zones in the Oakvale corehole in the far west (Truswell et al. 1985). Modified versions of all of these zones were applied by Macphail & Truswell (1989) to subdivide sequences in the central west of the basin. Subsequently, Macphail & Truswell (1993) erected the *Monotocidites galeatus* Zone, based on assemblages in the Late Miocene–Early Pliocene Bookpurnong beds in the same central western part of the Basin.

Two different quantitative methods of zonation have been applied to sequences in the Murray Basin: changes in relative pollen frequencies measured against stratigraphic depth, and changes in the ratios of selected species. Changes in pollen frequencies among major groups were used by Martin (1973, 1984) to define four vegetation 'phases' in probable Plio–Pleistocene sequences in the eastern Murray Basin. There is little time control on these vegetation shifts, and they have not been recognised in the central and western parts of the basin, so that correlation with the *Monotocidites galeatus* Zone of Macphail & Truswell (1993) is speculative.

Another zonation (not shown) based on relative abundance shifts was described by Truswell et al. (1985) for the Oligo–Miocene sequence in the Oakvale borehole. Relative shifts observed in major taxa, including *Nothofagus*, Myrtaceae and Araucariaceae, are probably regionally significant, but are likely to be time-transgressive.

Martin's (1973, 1984) selected ratio method defines zones in terms of the high or low frequencies of selected pollen types expressed as a ratio of the pollen count of major groups. In the column Murray Basin 2, the ratios identified by Martin (as high *Phyllocladidites mawsonii*, *Nothofagidites flemingii* layers, and high Myrtaceae/*Nothofagus* ratios) occur within the Upper *Nothofagidites asperus* to *P. tuberculatus* zonal interval. The method finds application within the Lachlan and Murrumbidgee fans in the central Murray Basin.

COLUMN 24: SPORES AND POLLEN —OTWAY BASIN

The assemblage zones of Harris (1971) were defined in the Otway Basin, and correlated with Australian and intercontinental zones (McGowran et al. 1971). Subsequently, Harris (1985) correlated the Eocene units, that is, the *Cupanieidites orthoteichus* through to the *Sparganiaceaepollenites barungensis* Zones, to the P Zones, following McGowran (1978), and thence to the Palaeogene timescale of Berggren et al. (1985*a*).

Ages assigned to the Otway Basin dinoflagellate and spore–pollen zones presented here incorporate the P Zone correlation of foraminiferal events as modified by McGowran (1989, 1990).

COLUMN 25: DINOFLAGELLATE CYSTS — GIPPSLAND BASIN

The Gippsland Basin assemblage zones of Partridge (1976) have been published in name only, and lack definition in terms of the ranges of their constituent species. Taxonomic assignment of the nominate taxa and age limits of the zones were revised by Partridge (written comm. 1994). The upper three zones in this column are those defined by McMinn (1992) from the Plio–Pleistocene of the offshore Gippsland Basin.

COLUMN 26: DINOFLAGELLATE CYSTS — OTWAY, ST VINCENT AND MURRAY BASINS

The Eocene dinocyst zones of Harris (1985) were defined in the Otway and St Vincent Basins, and were related to McGowran's (1978) foraminiferal correlations, and thence to the Berggren et al. (1985*a,b*) timescale. The relationship of the pattern of eustatic change in the early Tertiary of Australia's southern margin was outlined by Harris (1985), and McGowran (1986), in terms of four named transgressions. Thus, the lower boundaries of the *Achilleodinium biformoides* Zone, and the *Corrudinium incompositum* Zone correspond with the Wilsons Bluff and Tortachilla transgressions respectively; the base of the *Spiniferites ramosus* Zone is below the Aldinga, at the level of a sharp regression.

Dinocysts from the younger Tertiary are poorly known, but Martin (1991) has described assemblages from the western Murray Basin. The oldest part of the succession in the SADME MC63 bore is identified with the *Spiniferites ramosus* Zone; above that, Early to Middle Miocene sequences lack any formal zonation.

Macphail & Truswell (1989), Macphail et al. (1993) and Macphail & Kellett (1993) have analysed dinocyst floras in shallow to marginal marine facies in the central Murray Basin. A number of the index species used in the Gippsland zonation are present, for example, *Corrudinium incompositum* occurs in Upper *Nothofagidites asperus* Zone equivalent facies. Late Miocene to Early Pliocene assemblages are distinct from Oligocene to Middle Miocene assemblages although key species such as *Systematophora placacantha* appear to have extended ranges compared with open marine northern hemisphere sites.

LAND MAMMAL FAUNAL ASSEMBLAGES

The two columns shown are a condensation of the work of Woodburne et al. (1985, 1993), Archer et al. (1989), Rich et al. (1991), and represent Australian land mammal fossil localities. Much of the chronological ordering of the faunas is based on the stage of evolution of selected taxa or on stratigraphic superposition; less commonly, age control is derived from relationship to marine sequences, magnetostratigraphy, or dated volcanic rocks.

COLUMN 27: TASMANIA, VICTORIA, NEW SOUTH WALES

Faunas that can be related to marine rocks include the Wynyard, Beaumaris, Forsyths Bank, Lake Tyers and Hamilton local faunas. Faunas with age constraints established by relationship to dated volcanics include those at Hamilton and Geilston Bay. The oldest Australian mammal faunas are from the Late Oligocene and Early Miocene (Janjukian) of Geilston Bay, Tasmania.

COLUMN 28: SOUTH AUSTRALIA, NORTHERN TERRITORY, QUEENSLAND

The most complete stratigraphic succession of mammalian fossils is in the Lake Eyre Basin, where seven levels of differing age occur, all within the Late Oligocene and Early Miocene. The ages are based on magnetostratigraphy, and have been interpreted as within the interval of Chrons 6Cr to 7Ar (Woodburne et al. 1993). The Riversleigh faunas have been included but the evidence for their age is slim, being based mainly on superposition and biocorrelation (Archer et al. 1989). Faunas with age constraints established by relationship to dated volcanics include those at Bluff Downs.

COLUMN 29: MOLLUSCAN ASSEMBLAGE-ZONES, SOUTH-EASTERN AUSTRALIA

The units shown are those of Darragh (1985). They are based on assemblages from the Murray, Otway, Bass and Gippsland Basins and are related to standard Tertiary timescales by means of the foraminiferal zonations of Taylor (1966), Carter (1964) and Mallett (1977).

COLUMN 30: EOCENE TRANSGRESSIONS, SOUTHERN AUSTRALIA

McGowran (1986) recognised four transgressions, isochronous within the stratigraphic framework shown in Column 10, which he named the Wilson Bluff (Zone P.12), Tortachilla (base Zone P.15), Tuketja (late Zone P.15) and Aldinga (Zone P.17) transgressions. Subsequently, McGowran (1989, 1990) recognised five more transgressions in the Late Paleocene–Early Eocene: the Pebble Point (Zone P.5), Rivernook-A (Zone P.6), Rivernook (Zone P.7), Princetown (Zone P.8 to middle P.9) and Burrungule (late Zone P.9) transgressions.

COLUMN 31: AUSTRALIAN MARINE SEDIMENTARY SEQUENCES

McGowran (1979, Fig. 2) recognised four major sequences separated by hiatuses that formed a broadly isochronous transcontinental pattern in all tectonic environments of continental Australia. The three columns illustrate the variation in length of time represented by hiatuses over the three areas (southern margin, central platform including Carnarvon Basin and North West Shelf; northern platform). McGowran believed that, though eustasy was involved in the generation of these sequences, tectonism also was an important factor underlying the erosion–deposition relationships.

2.11

Quaternary Foraminiferal Biostratigraphy

G.C. Chaproniere

INTRODUCTION

The onshore marine Quaternary record (Fig. 38) is represented mainly by shallow water sequences containing very low diversity calcareous microfossil assemblages. The paucity of planktic foraminiferal assemblages has hampered correlation of these onshore sequences with the marine calcareous microfossil biostratigraphic schemes, with the exception of a few Pliocene and possible Pleistocene records (Mallett 1978, 1982). Recently, with the advent of the AGSO (then BMR) Continental Margins Program, progress has been made towards the establishment of a Quaternary timescale for marine sequences in the Australian area, though much more work needs to be done on the vast collections made by AGSO. It is this new knowledge for the tropical–subtropical offshore marine sequences that is presented here.

NOTES ON THE COLUMNS

The following provides a brief history of the usage of stage and biozone terminology and a source of reference to the biostratigraphy.

COLUMN 1: TIMESCALE

The scheme set out by Berggren et al. (in press *b*) has been followed herein consistent with usage for the Tertiary timescale (above).

PLANKTIC FORAMINIFERAL ZONATIONS

This compilation continues the usage of the international planktic foraminiferal schemes used for the Tertiary timescale (above), which are those that have found general usage globally. It has been traditional to regard the first appearance as FA of *Globorotalia (Truncorotalia) truncatulinoides* marking the Pliocene–Pleistocene boundary in marine sequences. However, it is now known that the bioevent occurs either at or below the base of the Olduvai magnetic event, which is placed within the latest Pliocene.

COLUMN 2: INTERNATIONAL FORAMINIFERAL ZONATIONS, LOW-LATITUDE

The planktic foraminiferal low-latitude zonal scheme used here is that presented by Berggren et al. (in press *b*), which is based on modifications of the Blow (1969, 1979) zonal schemes (Berggren 1973, 1977; Berggren et al. 1985*b*; Berggren & Miller 1988). The Berggren et al. (in press *b*) scheme has been correlated with that of Blow (1969, 1979), as modified by Kennett & Srinivasan (1983), by using the various bioevents given by Berggren et al. (1985*b*; in press *b*). A similar scheme based on the work of Bolli (1957*a,b*,

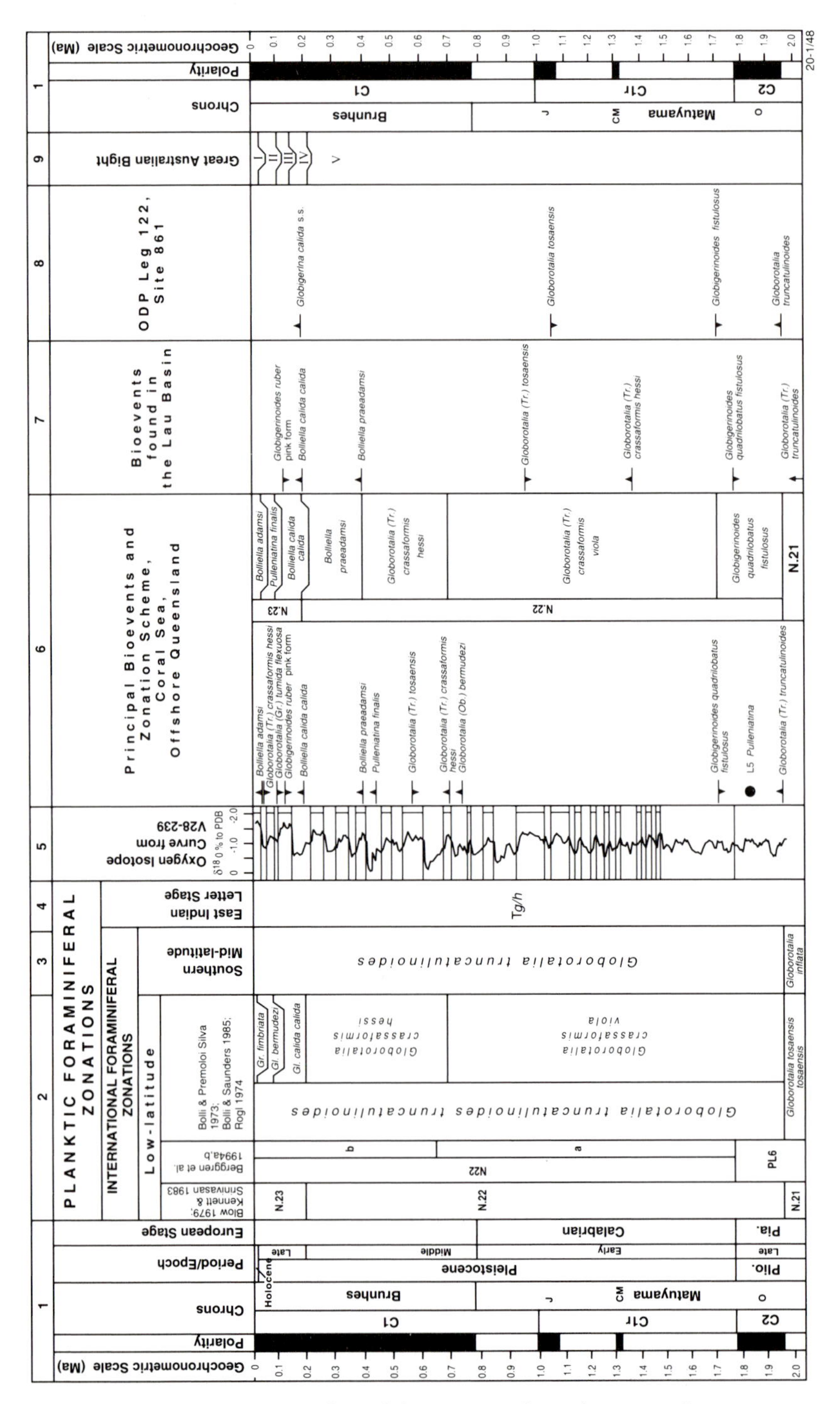

Figure 38: Quaternary timescale and biostratigraphic schemes and events, oxygen isotope curve and magnetostratigraphic scale.

1966) is also available for subtropical–tropical areas, and the most recent versions of this (Bolli & Saunders 1985; Toumarkine & Luterbacher 1985) have been incorporated.

COLUMN 3: INTERNATIONAL FORAMINIFERAL ZONATIONS, SOUTHERN MID-LATITUDE

Because Australia covers a latitudinal range from the tropics to the temperate regions it is appropriate to include a mid-latitude scheme. This is that presented by Jenkins (1985), who was unable to subdivide the single zone that encompasses the Quaternary.

COLUMN 4: EAST INDIAN LETTER STAGES

As for the Tertiary timescale, a column for the East Indian Letter Classification, based on 'larger' neritic tropical foraminifera, following Adams (1984) has been included. Pliocene and Pleistocene neritic larger foraminiferal assemblages have been recorded from northern Australia and have been referred to T*g–h* (Pliocene or Pleistocene) by Palmieri (1974, 1984). Generally these faunas lack planktic forms and so cannot be correlated with any greater accuracy with the timescale.

COLUMN 5: OXYGEN ISOTOPE CURVE

The fluctuations in the ratios of the stable isotopes of oxygen (O^{18}/O^{16}) with time have been used as a chronologic tool since the 1970s. The sequence found in core V28–239 by Shackleton & Opdyke (1976) forms a standard. Correlation between the oxygen isotope scale and the planktic foraminiferal zonal scheme has been achieved for the Middle and Late Pleistocene by Chapronière (1991). Detailed oxygen isotope studies were made by Peerdeman et al. (1993) for ODP Site 820 off the Great Barrier Reef, but no correlation with the planktic foraminiferal zonal scheme has yet been attempted.

COLUMN 6: PRINCIPAL BIOEVENTS AND ZONATION SCHEME, CORAL SEA, OFFSHORE QUEENSLAND

Chaproniere (1991) studied a number of gravity and piston cores from the Townsville Trough and was able to recognise a sequence of biostratigraphic events. Part of the section was able to be successfully correlated with the zonal scheme of Blow (1969) as modified by Bolli & Premoli-Silva (1973). Because a number of the subzonal markers were not present for Zone N.23, as these were restricted to the Atlantic Ocean, it was necessary to use local Indo–western Pacific bioevents to subdivide this interval. Chaproniere (1991) was also able to divide Zone N.22 into four subzones instead of the two suggested by Bolli & Premoli-Silva (1973). The relationship of the bioevents to the magnetostratigraphic scale follows that given by Berggren et al. (1985*b*, in press *b*). Some species, such as *Bolliella praeadamsi* and *Globorotalia (Truncorotalia) crassaformis hessi*, had not previously been correlated against the magnetostratigraphic scale, and the position given in Chaproniere (1991) was based on oxygen isotope stratigraphy.

COLUMN 7: BIOEVENTS FOUND IN THE LAU BASIN

Chaproniere et al. (1994) presented a biostratigraphic and magnetostratigraphic synthesis for a number of holes drilled in the Lau Basin during ODP Leg 135. Magnetostratigraphic interpretations were made for most holes, and the planktic foraminiferal bioevents were able to be related to these. The most important bioevents are shown in Column 7. As can be seen the bioevents within Zone N.22 differ from those given in Column 6, which as noted above, follow those given by Berggren et al. (1985*b*, in press *b*).

Globorotalia (Truncorotalia) truncatulinoides was found to occur near the base of the upper normal interval of Chron C2A (upper Gauss), which is older than that given by Berggren et al. (1985*b*, in press *b*) who suggested the upper part of the upper normal interval of Chron C2. This Gauss record agrees closely with other records of this species from the upper part of the Gauss Chron in the south-western Pacific by Dowsett (1989) and Hills & Thierstein (1989).

The position of the last appearance (LA) of *Globigerinoides fistulosus* was found within the top of the Olduvai magnetic event, a little older than that recorded by Berggren et al. (1985*b*, in press *b*); the sporadic distribution of this species in the Lau Basin sites suggests that this record may not be particularly reliable in that area.

The FA of *Globorotalia (Truncorotalia) crassaformis hessi*, which is recorded from the lower Brunhes Chron, based on oxygen isotope data (Chaproniere 1991), occurs within the upper part of the lower reversed interval of Chron C1r (immediately below the Cobb Mountain magnetic event). The position of the LA of *Globorotalia (Truncorotalia) tosaensis* is somewhat older (immediately above the Jaramillo magnetic event) than that given by Berggren et al. (1985*b*, in press *b*) (within the lower Brunhes).

COLUMN 8: ODP LEG 122, SITE 861

Zachariasse (1991) recognised only the *Globorotalia (Truncorotalia) truncatulinoides* Zone, which he equated with the Pleistocene. He also recognised a *Globigerinoides fistulosus* Zone preceding the *G. (T.) truncatulinoides* Zone, equating this to the Late Pliocene. The *G. fistulosus* Zone was placed at the top of an interval that contains a number of reworked Late Miocene and Early Pliocene species. *G. (T.) truncatulinoides* occurs at the base of this interval, indicating that the *G. (T.) truncatulinoides* Zone ranges over a longer interval than shown by Zachariasse (1991).

Two bioevents, the FA of *Globorotalia (Truncorotalia) truncatulinoides* and the LA of *Globigerinoides fistulosus*, occur at a level below the lowest magnetostratigraphic interpretations (which is somewhere below the Cobb Mountain magnetic event). Hills & Thierstein (1989) suggested that the location of the earliest (between 2.6 and 2.7 Ma) FA of *G. (T.) truncatulinoides* occurs between 20° S and 35° S, and that the age decreases both north and south of this zone, reaching the youngest (1.9–2.0 Ma) at about 15° S and 40° S. Because Site 861 was drilled at a latitude of 16° 44.23' S the age of this bioevent would be expected to be closer to its traditional position, near the base of the Olduvai magnetic event, which has been adopted here; also the LA of *G. fistulosus* has been placed just above the Olduvai magnetic event, because the position appears to be constant (Dowsett 1989; Hills & Thierstein 1989).

The LA of *Globorotalia (Truncorotalia) tosaensis* occurs at the top of the Jaramillo magnetic event, which is somewhat older than that suggested by Berggren et al. (1985*b*, in press *b*), but only slightly older than that recorded by Chaproniere et al. (1994). The FA of *Globigerina calida* s.s. occurs within the upper Brunhes Chron, which is probably at a similar level to that recorded from other sections.

COLUMN 9: GREAT AUSTRALIAN BIGHT

Quaternary planktic foraminiferal biostratigraphic studies for the southern Australian margin have not yet been made, but Almond et al. (1993) present some information for the Late Quaternary. These authors recognised five 'intervals', based on variations in planktic and benthic assemblages. These were correlated with the oxygen isotope curve and cover the interval from stages 1 to 7.

SECTION THREE

Bibliography

Abele, C., 1976. Introduction: Tertiary. pp. 177–91 *in* J.G. Douglas & J.A. Ferguson (eds), q.v.

Adams, C.G., 1970. A reconsideration of the East Indian Letter Classification of the Tertiary. *Bulletin of the British Museum (Natural History), Geology* 19, 85–137.

Adams, C.G., 1984. Neogene larger foraminifera, evolutionary and geological events in the context of datum planes. pp. 47–67 *in* N. Ikebe & R. Tsuchi (eds), *Pacific Neogene Datum Planes. Contributions to Biostratigraphy and Chronology*, University of Tokyo Press.

Agterberg, F.P., 1990. Automated stratigraphic correlation. *Developments in Palaeontology and Stratigraphy* 13, Elsevier, Amsterdam, 424 pp.

Aitchison, J.C., 1988*a*. Early Carboniferous (Tournaisian) Radiolaria from the Neranleigh–Fernvale Beds, Lake Manchester, Queensland, Australia. *Queensland Government Mining Journal* 89, 240–41.

Aitchison, J.C., 1988*b*. Kinderhookian Radiolaria from the Gundahl Complex, Anaiwan terrane, New England Orogen, NSW, Australia. p. 178 *in* R. Schmidt-Effing & A. Braun (eds), *First International Conference on Radiolaria, Abstracts*, Marburg, Germany.

Aitchison, J.C., 1988*c*. Late Palaeozoic radiolarian ages from the Gwydir terrane, New England Orogen, eastern Australia. *Geology* 16(9), 793–95.

Aitchison, J.C., 1988*d*. Radiolaria from the southern part of the New England Orogen, eastern Australia. pp. 49–60 *in* J.D. Kleeman (ed.), *New England Orogen Tectonics and Metallogenesis*, University of New England, Armidale.

Aitchison, J.C., 1989. Discussion: Radiolarian and conodont biostratigraphy of siliceous rocks from the New England Fold Belt. *Australian Journal of Earth Sciences* 36(1), 141–42.

Aitchison, J.C., 1990. Significance of Devonian–Carboniferous radiolarians from accretionary terranes of the New England Orogen, eastern Australia. *Marine Micropaleontology* 15, 365–78.

Aitchison, J.C., 1993*a*. Late Devonian (Frasnian) Radiolaria of the Canning Basin, Western Australia. *Palaeontographica Abt* A 228, 105–28.

Aitchison, J.C., 1993*b*. *Albaillellaria* from the New England Orogen, eastern NSW, Australia. *Marine Micropaleontology* 21, 353–67.

Aitchison, J.C. & Flood, P., 1990. Early Carboniferous radiolarian ages constrain the timing of sedimentation within the Anaiwan terrane, New England Orogen, eastern Australia. *Neues Jahrbuch für Geologie und Paläontologie, Abhandlungen* 180, 1–19.

Alberti, G.K.B., 1984. Beitrag zur Dacryconarida (Tentaculiten) Chronologie des alteren Unter-Devons (Lochkovium und Pragium). *Senckenbergiana lethaea* 65, 27–49.

Alberti, G.K.B., 1993. Dacryoconaride und homoctenide Tentaculiten des Unter- und Mittel-Devons. *Courier Forschungsinstitut Senckenberg* 158, 1–229.

Alberti, H., 1979. Devonian trilobite biostratigraphy. *Special Papers in Palaeontology* 23, 313–24.

Alcock, P.J., 1970. A report on the sedimentology of the Moolayember Formation, Bowen Basin, Queensland. *Bureau of Mineral Resources, Geology & Geophysics, Record* 1970/25, 35 pp.

Allemann, F., Grün, W. & Wiedmann, J., 1975. The Berriasian of Caravaca (Prov. of Murcia) in the subbetic zone of Spain and its importance for defining this stage and the Jurassic–Cretaceous boundary. *Mémoire du Bureau de Recherches Géologiques et Minières* 86, 14–22.

Almond, D., McGowran, B. & Li Qianyu, 1993. Late Quaternary foraminiferal record from the Great Australian Bight and its environmental significance. *Association of Australasian Palaeontologists, Memoir* 15, 417–28.

Alvarez, W., Arthur, M.A., Fischer, A.G., Lowrie, W., Napoleone, G., Premoli-Silva, I. & Roggenthen, W.M., 1977. Upper Cretaceous–Paleocene magnetic stratigraphy at Gubbio, Italy. V. Type section for the Late Cretaceous–Paleocene geomagnetic reversal time scale. *Geological Society of America Bulletin* 88, 383–89.

An Zhisheng, Bowler, J.M., Opdyke, N.D., Macumber, P.G. & Firman, J.B., 1986. Palaeomagnetic stratigraphy of Lake Bungunnia: Plio–Pleistocene precursor of aridity in the Murray Basin, south-eastern Australia. *Paleogeography, Paleoclimatology, Paleoecology* 54, 219–39.

Anan-Yorke, R., 1975. A preliminary report on the palynology of the Merrimbula Group, NSW. *Journal and Proceedings of the Royal Society of New South Wales* 108, 49–50.

Anderson, J.C. & Palmieri, V., 1977. The Fork Lagoons Beds, an Ordovician unit of the Anakie Inlier, central Queensland. *Queensland Government Mining Journal* 78, 260–263.

Andreieff, P., Bellon, H. & Westercamp, D., 1976. Chronométrie et stratigraphie comparée des édifices volcaniques et formations sédimentaires de la Martinique (Antilles françaises). *Bulletin Bureau de Recherches Géologiques et Minières* 4, 335–46.

Andrew, A.S., Hamilton, P.J., Mawson, R., Talent, J.A. & Whitford, D.J., 1994. Isotopic correlation tools in the mid-Palaeozoic and their relation to extinction events. *APEA Journal* 34, 268–76

Anonymous, 1979. Magnetostratigraphic polarity units, a supplementary chapter of the International Subcommission on Stratigraphic Classification, International Stratigraphic Guide. *Geology* 7, 578–83.

Antelo, B., 1969. Hallazgo del género *Protocanites* (Ammonoidea) en el Carbonifero inferior de San Juan. *Ameghiniana* 6, 69–73.

Antevs, E., 1913. The results of Dr Mojberg's Swedish scientific expedition to Australia. *Kungliga Svenska Vetenkapsakademiens Handlingar* 52, 1–6.

Apthorpe, M.C., 1979. Depositional history of the Upper Cretaceous of the North-west Shelf, based upon Foraminifera. *APEA Journal* 19, 74–89.

Apthorpe, M.C. & Heath, R.S., 1981. Late Triassic and Early to Middle Jurassic Foraminifera from the North West Shelf, Australia. *Fifth Australian Geological Convention: Geological Society of Australia, Abstracts* 3, 66.

Archangelsky, S., Azcuy, C.L. & Pinto, I.D., Gonzalez, C.R., Marques Toigo, M., Rösler, O. & Wagner, R.H., 1980. The Carboniferous and Early Permian of the South American Gondwana area: a summary of biostratigraphic information. *Actas II Congres Argentino de Paleontologia y Biostratigrafica y I Congreso Latinamericanos Paleontologia, Buenos Aires, Abril 1978, 4, Simposio Carbonico–Permico*, 257–69.

Archbold, N.W., 1982. Correlation of the Early Permian faunas of Gondwana: Implications for the Carboniferous–Permian boundary. *Journal of the Geological Society of Australia* 29, 267–75.

Archbold, N.W., 1983*a*. Permian marine invertebrate provinces of the Gondwana Realm. *Alcheringa* 7, 59–73.

Archbold, N.W., 1983*b*. Studies on Western Australian Permian brachiopods, 3. The Family Linoproductidae Stehli 1954. *Proceedings of the Royal Society of Victoria* 95(4), 237–54.

Archbold, N.W., 1985. Nineteenth century views on the Australian marine Permian. *Earth Sciences History* 5(1), 12–23.

Archbold, N.W., 1986. Studies on Western Australian Permian brachiopods 6. The genera *Strophalosia* King, 1844, *Heteralosia* King, 1938 and *Echinalosia* Waterhouse, 1967. *Proceedings of the Royal Society of Victoria* 98(3), 97–119.

Archbold, N.W., 1988*a*. Permian brachiopod faunas of the Perth Basin, Western Australia—a study of progressive isolation. *Geological Society of Australia, Abstracts* 21, 46–47.

Archbold, N.W., 1988*b*. Studies on Western Australian Permian brachiopods 8. The Late Permian fauna of the Kirkby Range Member, Canning Basin. *Proceedings of the Royal Society of Victoria* 100, 21–32.

Archbold, N.W., 1993. A zonation of the Permian brachiopod faunas of Western Australia. pp. 313–21 *in* R.H. Findlay, R. Unrug, M.R. Banks & J.J. Veevers (eds), *Gondwana Eight*, Balkema, Rotterdam.

Archbold, N.W. & Dickins, J.M., 1991. Australian Phanerozoic Timescales 6. Permian, a standard for the Permian System in Australia. *Bureau of Mineral Resources, Geology & Geophysics, Record* 1989/36, 1–17.

Archbold, N.W. & Skwarko, S.K., 1988. Brachiopods and bivalves of Kungurian (Late Early Permian) age from the top of the Coolkilya Sandstone, Carnarvon Basin, Western Australia. *Report of the Geological Survey of Western Australia* 23, 1–15.

Archbold, N.W. & Thomas, G.A., 1987. *Fusispirifer* (Spiriferidae, Brachiopoda) from the Permian of Australia and Afghanistan. *Alcheringa* 11, 175–203.

Archbold, N.W., Dickins, J.M. & Thomas, G.A., 1993*b*. Correlations and age of the Western Australian Permian marine faunas. *Bulletin of the Geological Survey of Western Australia* 136, 11–18.

Archbold, N.W., Thomas, G.A. & Skwarko, S.K., 1993*a*. Brachiopoda. *Bulletin of the Geological Survey of Western Australia* 136, 45–51.

Archer, M. & Clayton, G., 1984. *Vertebrate Zoogeography & Evolution in Australasia.* Hesperian Press, Carlisle, WA.

Archer, M., Godthelp, H., Hand, S.J., & Megirian, D., 1989. Fossil mammals of Riversleigh, north-western Queensland: preliminary overview of biostratigraphy, correlation and environmental change. *Australian Zoologist* 25, 29–65.

Argast, S., Farlow, J.O., Gabet, R.M. & Brinkman, D.L., 1987. Transport-induced abrasion of fossil reptilian teeth: implications for the existence of Tertiary dinosaurs in the Hell Creek Formation, Montana. *Geology* 15, 927–30.

Arkell, W.J., 1956. *Jurassic Geology of the World.* Oliver and Boyd Ltd, Edinburgh.

Arkell, W.J. & Playford, P.E., 1954. The Bajocian ammonites of Western Australia. *Royal Society (London), Philosophical Transactions* B 237, 547–605.

Armstrong, J.D., Dear, J.F. & Runnegar, B., 1967. Permian ammonoids from eastern Australia. *Journal of the Geological Society of Australia* 14, 87–97.

Armstrong, R.L., 1978. Pre-Cenozoic Phanerozoic Time Scale—Computer file of critical dates and consequences of new and in-progress decay-constant revisions. pp. 73–91 *in* G.V. Cohee et al., q.v.

Armstrong, R.L., 1982. Late Triassic–Early Jurassic time-scale calibration in British Columbia, Canada. pp. 509–13 *in* G.S. Odin (ed.) q.v.

Ash, S.R.,1979. *Skilliostrobus* gen. nov., a new lycopsid cone from the Early Triassic of Australia. *Alcheringa* 3, 73–89.

Azcuy, C.L. & Jelin, R., 1980. Las Palinozonas del limite Carbonico–Permico en la Cuenca Paganzo. *Actas II Congres Argentino de Paleontologia y Biostratigrafica y I Congreso Latinamericanos Paleontologia, Buenos Aires, Abril 1978, 4, Simposio Carbonico–Permico,* 51–67.

Azcuy, C.L., Sabattini, N. & Taboada, A.C., 1991 [imprint 1990]. Advances in the Lower Carboniferous zonation of Argentina. *Courier Forschungsinstitut Senckenberg* 130, 207–10.

Baag, C. & Helsley, C.E., 1974*a*. Remanent magnetization of a 50 m core from the Moenkopi Formation, Western Colorado. *Geophysical Journal, Royal Astronomical Society* 37, 245–62.

Baag, C. & Helsley, C.E., 1974*b*. Evidence for penecontemporaneous magnetization of the Moenkopi Formation. *Journal of Geophysical Research* 79, 3308–20.

Baars, D.L., 1990. Permian chronostratigraphy in Kansas. *Geology* 18, 687–90

Baars, D.L., Maples, C.G., Ritter, S.M. & Ross, C.A., 1992. Redefinition of the Pennsylvanian–Permian Boundary in Kansas, Mid-Continent USA. *International Geology Review* 34(10), 1021–25.

Backhouse, J., 1978. Palynological zonation of the Late Jurassic and Early Cretaceous sediments of the Yarragadee Formation, central Perth Basin, Western Australia. *Geological Survey of Western Australia Report* 7, 52 pp.

Backhouse, J., 1987. Microplankton zonation of the Lower Cretaceous Warnbro Group, Perth Basin, Western Australia. *Association of Australasian Palaeontologists Memoir* 4, 205–26.

Backhouse, J., 1988. Late Jurassic and Early Cretaceous palynology of the Perth Basin, Western Australia. *Geological Survey of Western Australia Bulletin* 135, 233 pp.

Bagas, L., 1988. Geology of Kings Canyon National Park. *Northern Territory Geological Survey Report* 4, 21 pp.

Baillie, P.W., Banks, M.R. & Rickards, R.B., 1978. Early Silurian graptolites from Tasmania. *Search* 1, 46–47.

Baksi, A.K., Hsu, V., McWilliams, M.O. & Farrar, E., 1992. ^{40}Ar/^{39}Ar dating of the Brunhes–Matuyama geomagnetic field reversal. *Science* 256, 356–57.

Baldanza, A. & Mattioli, E., 1992. Biozonazione a nannofossili calcarei del Giurassico inferiore-medio della Provincia Mediterranea (Dominio Tetideo): revisione ed ampliamento. *Paleopelagos* 2, 69–77.

Balfe, P.E., 1982. Permian stratigraphy of the Springsure–Arcturus Downs area. *Queensland Government Mining Journal* 83(965), 133–59.

Balme, B.E., 1957. Spores and pollen grains from the Mesozoic of Western Australia. *CSIRO Fuel Research, Coal Section, Technical Communication* 25, 48 pp.

Balme, B.E., 1964. The palynological record of Australian pre-Tertiary floras. pp. 49–80 *in* L.M. Cranwell (ed.), *Ancient Pacific Floras, Proceedings Tenth Pacific Science Congress,* University of Hawaii Press, Honolulu .

Balme, B.E.,1970. Palynology of Permian and Triassic strata in the Salt Range and Surghar Range, West Pakistan. pp. 305–453 *in* B. Kummel & C. Teichert (eds), Stratigraphic Boundary Problems: Permian and Triassic of West Pakistan.

Department of Geology, University of Kansas, Special Publication 4.

Balme, B.E., 1979. Palynology of Permian–Triassic boundary beds at Kap Stosch, East Greenland. *Meddelelser om Grønland* 200(6), 5–37.

Balme, B.E., 1980*a*. Palynology and the Carboniferous–Permian boundary in Australia and other Gondwana continents. *Palynology* 4, 43–55.

Balme, B.E.,1980*b*. Palynology of Late Permian–Early Triassic strata in Greenland and Alaska, and its plant geographic implications. *5th International Palynological Conference, Cambridge 1980, Abstracts*, 25.

Balme, B.E., 1988. Miospores from Late Devonian (early Frasnian) strata, Carnarvon Basin, Western Australia. *Palaeontographica* B209, 109–66.

Balme, B.E., 1990. Australian Phanerozoic Timescales: Triassic. *Bureau of Mineral Resources, Geology & Geophysics, Record* 1989/37, 28 pp.

Balme, B.E. & Helby, R.J., 1973. Floral modifications at the Permian–Triassic boundary in Australia. *Canadian Society of Petroleum Geologists Memoir* 2, 433–44.

Banks, M.R., 1978. Correlation chart for the Triassic System of Australia. *Bureau of Mineral Resources, Geology & Geophysics, Bulletin* 156C, 1–39.

Banks, M.R., 1988. The base of the Silurian System in Tasmania. *Bulletin of the British Museum of Natural History, Geology* 43, 191–94.

Banks, M.R. & Baillie, P.W., 1989. Late Cambrian to Devonian. pp. 182–237 *in* C.F. Burrett & E.L. Martin, q.v.

Banks, M.R., & Burrett, C.F., 1980. A preliminary Ordovician biostratigraphy of Tasmania. *Journal of the Geological Society of Australia* 26, 363–76.

Banks, M.R. & Clarke, M.J., 1987. Changes in the geography of the Tasmania Basin in the Late Paleozoic. pp. 1–14 *in* G.D. McKenzie (ed.) Gondwana six: stratigraphy, sedimentology and paleontology. *American Geophysical Union, Geophysical Monograph* 41.

Banks, M.R. & Naqvi, I.H.,1967. Formations close to the Permo–Triassic boundary in Tasmania. *Papers and Proceedings of the Royal Society of Tasmania* 101, 17–30.

Banks, M.R., Cosgriff, J. & Kemp, N.,1984. A Tasmanian Triassic stream community. pp. 291–298 *in* M. Archer & G. Clayton (eds), q.v.

Barnard, T. & Hay, W.W., 1974. On Jurassic coccoliths: a tentative zonation of the Jurassic of southern England and north France. *Eclogae Geologicae Helvetiae* 67 (2), 563–85.

Barrick, J.E. & Klapper, G., 1976. Multielement Silurian (late Llandoverian–Wenlockian) conodonts of the Clarita Formation, Arbuckle Mountains, Oklahoma, and phylogeny of *Kockelella. Geologica et Palaeontologica* 10, 59–100.

Barron, E.J., Harrison, C.G.A., Sloan, J.L. & Hay, W.W., 1981. Paleogeography, 180 million years ago to the present. *Eclogae Geologicae Helvetiae* 74, 433–70.

Bartenstein, H., 1974. Upper Jurassic–Lower Cretaceous primitive arenaceous foraminifera from DSDP sites 259 and 261, eastern Indian Ocean. *Initial Reports of the Deep Sea Drilling Project* 27, 683–95.

Bartholomai, A., 1966. The discovery of plesiosaurian remains in freshwater sediments in Queensland. *Australian Journal of Science* 28, 437–38.

Bartholomai, A., 1979. New lizard-like reptiles from the Early Triassic of Queensland. *Alcheringa* 3, 225–34.

Bassett, M.G., 1979. 100 years of Ordovician geology. *Episodes* 2, 18–21.

Bassett, M.G., 1985. Towards a 'common language' in stratigraphy. *Episodes* 8, 87–92.

Bassett, M.G., Lane, P.D. & Edwards, D. (eds), 1991. The Murchison Symposium, Proceedings of an international conference on the Silurian System. *Special Papers in Palaeontology* 44, 397 pp.

Bassett, M.G., Cocks, L.R.M., Holland, C.H., Rickards, R.B. & Warren, P.T., 1975. The type Wenlock Series. *Institute of Geological Science, Report* 75/13, 1–19.

Baxter, J.W. & Brenckle, P.L., 1982. Preliminary statement on Mississippian calcareous foraminiferal successions of the Midcontinent (USA) and their correlation to western Europe. *Newsletters on Stratigraphy* 11(3), 136–53.

Baxter, J.W. & Von Bitter, P.H., 1984. Conodont succession in the Mississippian of southern Canada. pp. 253–64 *in* P.K. Sutherland & W.L. Manger (eds), 1984*b*, q.v.

Becker, R.T., 1988. Ammonoids from the Devonian–Carboniferous Boundary in the Hasselbach Valley (Northern Rhenish Slate Mountains). pp. 193–213 *in* G. Flajs, R. Feist & W. Ziegler (eds), q.v.

Becker, R.T., 1993. Anoxia, eustatic changes, and Upper Devonian to lowermost Carboniferous global ammonoid diversity. *Systematics Special Volume* 47, 105–64.

Becker, R.T. & House, M.R., 1994. International Devonian goniatite zonation, Emsian to Givetian, with new records from Morocco. *Courier Forschungsinstitut Senckenberg* 169, 79–135.

Becker, R.T., House, M.R. & Kirchgasser, W.T., 1993. Devonian goniatite biostratigraphy and timing of facies movements in the Frasnian of the Canning Basin, Western Australia. pp. 293–321 *in* E.A. Hailwood & R.B. Kidd (eds), High Resolution Stratigraphy. *Geological Society Special Publication* 70.

Becker R.T., House, M.R., Kirchgasser, W.T. and Playford, P.E., 1991. Sedimentary and faunal changes across the Frasnian/Fammenian boundary in the Canning Basin of Western Australia. *Historical Biology*, 5, 183–96.

Becker, R.T., Feist, R., Flajs, G., House, M.R. & Klapper, G., 1989. Frasnian–Famennian extinction events in the Devonian at Coumiac, southern France. *Compte Rendu de l'Académie des Sciences, Paris, Série II*, 309, 259–66.

Becker, R.T., Bless, M.J.M., Brauckmann, C., Friman, L., Higgs, K.D., Keupp, H., Korn, D., Langer, W., Paproth, E., Racheboeuf, P., Stoppel, D., Streel, M. & Zakowa, H., 1984. Hasselbachtal, the section best displaying the Devonian–Carboniferous boundary beds in the Rhenish Massif (Rheinisches Schiefergebirge). *Courier Forschungsinstitut Senckenberg* 67, 181–91.

Belford, D.J., 1958. Stratigraphy and micropalaeontology of the Upper Cretaceous of Western Australia. *Geologische Rundschau* 47, 629–47.

Belford, D.J., 1960. Upper Cretaceous foraminifera from the Toolonga Calcilutite and Gingin Chalk, Western Australia. *Bureau of Mineral Resources, Geology & Geophysics, Bulletin* 57, 198 pp.

Belford, D.J., 1968. Occurrence of the genus *Draffania* Cummings in Western Australia. *Bureau of Mineral Resources, Geology & Geophysics, Bulletin* 92, 49–52.

Belford, D.J., 1970. Upper Devonian and Carboniferous Foraminifera, Bonaparte Gulf Basin, northwestern Australia. *Bureau of Mineral Resources, Geology & Geophysics, Bulletin* 108(1), 1–38.

Belka, Z., & Groessens, E., 1986. Conodont succession across the Tournaisian–Viséan boundary beds at Salet, Belgium. *Bulletin de la Société Géologique de Belgique* 95, 257–80.

Bengtson, S., 1986. Siliceous microfossils from the Upper Cambrian of Queensland. *Alcheringa* 10, 195–216.

Bengtson, S., Conway Morris, S., Cooper, B.J., Jell, P. A. & Runnegar, B.N., 1990. Early Cambrian fossils from South Australia. *Association of Australasian Palaeontologists, Memoir* 9, 364 p.

Benson, W.N., 1921. A census and index of the Lower Carboniferous Burindi fauna. *Records of the Geological Survey of New South Wales* 10, 12–74.

Benson, W.N., 1922. Materials for the study of the Devonian palaeontology of Australia. *Records of the Geological Survey of New South Wales* 10, 83–204.

Beres, J. & Soffel, H., 1985. Magnetostratigraphie und Anisotropie der magnetischen Suszeptibilität von Proben aus der Forschungsbohrung Obernsees zwischen Bohrmeter 1341,15 und 1170,40 (Perm–Trias Übergang). *Geologica Bavarica* 88, 153–61.

Berggren, W.A., 1969. Cainozoic chronostratigraphy, planktonic foraminiferal zonation and the radiometric time scale. *Nature* 224, 1072–75.

Berggren, W.A., 1973. The Pliocene time scale: calibration of planktonic foraminifera and calcareous nannoplankton zones. *Nature* 243, 391–97.

Berggren, W.A., 1977. Atlas of Palaeogene planktonic foraminifera. Some species of the genera, *Subbotina, Planorotalites, Morozovella, Acarinina* and *Truncorotaloides*. pp. 205–300 *in* A.T.S. Ramsay (ed.), *Oceanic Micropalaeontology*, Academic Press, London.

Berggren, W.A., & Miller, K.G., 1988. Paleogene tropical planktonic foraminiferal biostratigraphy and magnetobiochronology. *Micropaleontology* 34, 362–80.

Berggren, W.A., Kent, D.V. & Flynn, J.J., 1985*a*. Paleogene geochronology and chronostratigraphy. pp. 141–95 *in* N.J. Snelling (ed.), q.v.

Berggren, W.A., Kent, D.V. & Van Couvering, J.A., 1985*b*. Neogene geochronology and chronostratigraphy. pp. 211–50 *in* N.J. Snelling (ed.), q.v.

Berggren, W.A., Kent, D.V., Flynn, J.J. & Van Couvering, J.A., 1985*c*. Cenozoic geochronology. *Geological Society of America Bulletin* 96, 1407–18.

Berggren, W.A., Kent, D.V., Obradovich, J.D. & Swisher III, C.C., 1992. Towards a revised Paleogene geochronology. pp. 29–45 *in* D.R. Prothero & W.A. Berggren (eds), *Eocene–Oligocene Climatic and Biotic Evolution*, Princeton University Press, Princeton, NJ.

Berggren, W.A., Kent, D.V., Swisher III, C.C. & Miller, K.G., in press *a*. A revised Paleogene geochronology and chronostratigraphy. *In* W.A. Berggren & D.V. Kent (eds), Geochronology, time scales and stratigraphic correlation. *Society for Sedimentary Geology (SEPM) Special Publication.*

Berggren, W.A., Kent, D.V., Swisher III, C.C. & Miller, K.G., in press *b*. A revised Neogene geochronology and chronostratigraphy. *In* W.A. Berggren & D.V. Kent (eds), Geochronology, time scales and stratigraphic correlation. *Society for Sedimentary Geology (SEPM) Special Publication.*

Bergström, S.M., 1971. Conodont biostratigraphy of the Middle and Upper Ordovician of Europe and eastern North America. *Geological Society of America Memoir* 127, 83–162.

Bergström, S.M., 1973. Ordovician conodont biostratigraphy. pp. 47–58 *in* A. Hallam (ed.), *Atlas of Palaeobiogeography*, Elsevier, Amsterdam.

Bergström, S.M. & Mitchell, C.E., 1990. Trans-Pacific graptolite faunal relations: the biostratigraphic position of the base of the Cincinnatian Series (Upper Ordovician) in the standard Australian graptolite zone succession. *Journal of Paleontology* 64, 992–97.

Betzler, C.G. & Chaproniere, G.C.H., 1993. Paleogene and Neogene larger foraminifers from the Queensland Plateau: biostratigraphy and environmental significance. *Proceedings of the Ocean Drilling Program, Scientific Results* 133, 51–66.

Beyrich, E., 1854. Über die Stellung des hessischen Tertiärbildungen. *Monatsberichte der Königlich Preussischen Academie der Wissenschaften* (1854), 640–66.

Bigsby, J.J., 1868. *Thesaurus siluricus. The Flora and Fauna of the Silurian Period. With Addenda (from Recent Acquisitions)*, John van Voorst, London, iii + 214 pp.

Birkelund, T., Hancock, J.M., Hart, M.B., Rawson, P.F., Remane, J., Robaszynski, F., Schmid, F. & Surlyk, F., 1984. Cretaceous stage boundaries—proposals. *Geological Society of Denmark Bulletin* 33, 3–20.

Bischoff, G.C.O., 1986. Early and Middle Silurian conodonts from mid-western New South Wales. *Courier Forschungsinstitut Senckenberg* 89, 1–337.

Bischoff, G.C.O. & Prendergast, E.I., 1987. Newly discovered Middle and Late Cambrian fossils from the Wagonga Beds of New South Wales, Australia. *Neues Jahrbuch für Geologie und Paläontologie, Abhandlung* 175(1), 39–64.

Bittner, A., 1892. Was ist Norisch? *Jahrbuch Geologischen Reichsanstalt* 42, 387–96.

Blatchford, T., 1927. Geology of portions of the Kimberley Division, with special reference to the Fitzroy Basin and the possibilities of the occurrence of mineral oil. *Bulletin of the Geological Survey of Western Australia* 93, 1–56.

Blow, W.H., 1956. Origin and evolution of the foraminiferal genus *Orbulina* d'Orbigny. *Micropaleontology* 2, 57–70.

Blow, W.H., 1969. Late Middle Eocene to Recent planktonic foraminiferal biostratigraphy. pp. 199–421 *in* P. Brönnimann & H.H. Renz (eds), *Proceedings of the First International Conference on Planktonic Microfossils* I.

Blow, W.H., 1979. *The Cainozoic Globigerinida*, vols 1–3, 1–1413, E.J. Brill, Leiden.

Bogoslovskaya, M.F. & Pavlova, E.E., 1988. On the development of the ammonoids of the family Spirolegoceratidae. *Paleontologicheskii Zhurnal* 1988(2), 111–14.

Bolli, H.M., 1957*a*. Planktonic Foraminifera from the Oligocene–Miocene Cipero and Lengua formations of Trinidad, BWI. *United States National Museum Bulletin* 215, 97–131.

Bolli, H.M., 1957*b*. The genera *Globigerina* and *Globorotalia* in the Paleocene–Lower Eocene Lizard Springs Formation of Trinidad, BWI. *United States National Museum Bulletin* 215, 61–81.

Bolli, H.M., 1966. Zonation of Cretaceous to Pliocene marine sediments based on planktonic Foraminifera. *Asociacion Venezolana de Geologia, Mineria y Petroleo, Boletin Informativo* 9, 1–32.

Bolli, H.M. & Premoli-Silva, I., 1973. Oligocene to Recent planktonic Foraminifera and stratigraphy of the Leg 15 sites in the Caribbean Sea. *Initial Reports of the Deep Sea Drilling Project* 15, 475–97.

Bolli, H.M. & Saunders, J.B., 1985. Oligocene to Holocene low latitude planktic Foraminifera. pp. 155–262 *in* H.M. Bolli, J.B. Saunders & K. Perch-Nielson (eds), *Plankton Stratigraphy*, Cambridge University Press, Cambridge.

Boote, D.R.D. & Kirk, R.B., 1989. Depositional wedge cycles on evolving plate margin, Western and northwestern Australia. *American Association of Petroleum Geologists Bulletin* 73, 216–43.

Boucot, A.J., 1975. *Evolution and Extinction Rate Controls*, Elsevier, Amsterdam, 427 pp.

Boucot, A.J., 1990. Silurian and pre-Upper Devonian bio-events. pp. 125–32 *in* E.G. Kauffman & O.H. Walliser (eds), *Extinction events in earth history*, Springer-Verlag.

Boulard, C., 1993. Biochronologie quantitative; concepts, méthodes et validité. *Documents des Laboratoires de Géologie, Lyon* 128, 259 pp.

Bouroz, A. & Doubinger, J., 1977. Report on the Stephanian–Autunian boundary and on the contents of Upper Stephanian and Autunian in their stratotypes. pp. 147–69 *in* V.M. Holub & R.H. Wagner (eds), *Symposium on Carboniferous Stratigraphy*, Ústředního ústavu geologického.

Bouroz, A., Einor, O.L., Gordon, M., Meyen, S.V. & Wagner, R.H., 1978. Proposals for an international chronostratigraphic classification of the Carboniferous. *Compte Rendu 8ème Congrès International de Stratigraphie et de Géologie du Carbonifere, Moskva, 1975, 1: General Problems of the Carboniferous Stratigraphy*, 'Nauka', Moscow, 36–69.

Bown, P.R., 1992. New calcareous nannofossil taxa from the Jurassic–Cretaceous boundary interval of Sites 765 and 261, Argo Abyssal Plain. *Proceedings of the Ocean Drilling Program, Scientific Results* 123, 369–79.

Bown, P.R., Cooper, M.K.E. & Lord, A.R., 1988. A calcareous nannofossil biozonation scheme for the early to mid Mesozoic. *Newsletters on Stratigraphy* 20 (2), 91–114.

Bowring, S.A., Grotzinger, J.P., Isachsen, C.E., Knoll, A.H., Pelechaty, S.M. & Kolosov, P., 1993. Calibrating rates of Early Cambrian evolution. *Science* 261, 1293–98.

Bradshaw, J., Nicoll, R.S. & Bradshaw, M.T., 1990. The Cambrian to Permo–Triassic Arafura Basin, northern Australia. *APEA Journal* 30, 107–27.

Bradshaw, M.T. & Yeung, M., 1992. *Palaeogeographic Atlas of Australia. Volume 8: Jurassic*, Bureau of Mineral Resources, Canberra.

Bradshaw, M.T., Wells, A.T. & O'Brien, P.E., 1994. Revision of the stratigraphic position of the Towallum Basalt, Clarence–Moreton Basin—Implications for Australian Jurassic biostratigraphy. *Bureau of Mineral Resources, Geology & Geophysics, Bulletin* 241, 138–143.

Bradshaw, M.T., Burger, D., Yeung, M. & Beynon, R.M., in press. *Palaeogeographic Atlas of Australia. Volume 9: Cretaceous*, Australian Geological Survey Organisation, Canberra.

Bradshaw, M.T., Yeates, A.N., Beynon, R.M., Brakel, A.T., Langford, R.P., Totterdell, J.M. & Yeung, M., 1988. Palaeogeographic evolution of the North West Shelf region. pp. 29–54 *in* P.G. Purcell & R.R. Purcell (eds), q.v.

Brakel, A.T., 1983. Correlation between the Comet Platform and the Denison Trough, Bowen Basin. pp. 295–302 *in Permian Geology of Queensland*, Geological Society of Australia, Queensland Division.

Bralower, T.J., 1987. Valanginian to Aptian calcareous nannofossil stratigraphy and correlation with the upper M-sequence magnetic anomalies. *Marine Micropaleontology* 11, 293–310.

Bralower, T.J. & Siesser, W.G., 1992. Cretaceous calcareous nannofossil biostratigraphy of sites 761, 762 and 763, Exmouth and Wombat plateaus, north-west Australia. *Proceedings of the Ocean Drilling Program, Scientific Results* 122, 529–56.

Bralower, T.J., Ludwig, K.R., Obradovich, J.D. & Jones, D.L., 1990. Berriasian (Early Cretaceous) radiometric ages from the Grindstone Creek Section, Sacramento Valley, California. *Earth and Planetary Science Letters* 98, 62–73.

Braun, A., 1990. Oberdevonische Radiolarien aus Kieselschieter-Gerollen des unteren Maintales bei Frankfurt. *Monatshefte Geologisches Jahrbuch Hessen* 118, 5–27.

Braun, A., 1991. Radiolarien aus dem Unter-karbon Deutschlands. *Courier Forschungsinstitut Senckenberg* 133, 1–177.

Braun, A. & Gusky, H.-J., 1991. Kieselige Sedimentgesteine des Unterkarbon im Rhenoherzynikum—eine Bestandsaufnahme. *Geologica et Palaeontologica* 25, 57–77.

Braun, A. & Schmidt-Effing, R., 1993. Biozonation, diagenesis and evolution of radiolarians in the Lower Carboniferous of Germany. *Marine Micropaleontology* 21, 369–83.

Brenckle, P.L., 1991 [imprint 1990]. Foraminiferal division of the Lower Carboniferous/Mississippian in North America. *Courier Forschungsinstitut Senckenberg* 130, 65–78.

Brenckle, P.L. & Groves, J.R., 1986. Calcareous foraminifers from the Humboldt Oolite of Iowa: Key to Early Osagean (Mississippian) correlations between eastern and western North America. *Palaios* 1, 561–81.

Brenckle, P.L., Lane, H.R. & Collinson, C., 1974. Progress toward reconciliation of Lower Mississippian conodont and foraminiferal zonations. *Geology* 2, 433–37.

Brenckle, P.L., Marshall, F.C., Waller, S.F. & Wilhelm, M.H., 1982. Calcareous microfossils from the Mississippian Keokuk Limestone and adjacent formations, upper Mississippi Valley: Their meaning for North American and intercontinental correlation. *Geologica et Palaeontologica* 15, 47–88.

Brenner, W., Bown, P.R., Bralower, T.J., Crasquin-Soleau, S., Dèpêche, F., Dumont, T., Martini, R., Siesser, W.G. & Zaninetti, L., 1992. Correlation of Carnian to Rhaetian palynological, foraminiferal, calcareous nannofossil and ostracode biostratigraphy, Wombat Plateau. *Proceedings of the Ocean Drilling Program, Scientific Results* 122, 487–95.

Briden, J.C. & Mullan, A.J., 1984. Superimposed Recent, Permo–Carboniferous and Ordovician palaeomagnetic remanence in the Builth Volcanic Series. *Earth and Planetary Science Letters* 69, 413–21.

Briggs, D.E.G. & Rolfe, W.I.D., 1983. New Concavicarida (new order: ?Crustacea) from the Upper Devonian of Gogo, Western Australia, and the palaeoecology and affinities of the group. *Special Papers in Palaeontology* 30, 249–76.

Briggs, D.J.C., 1987. Permian productid zones of New South Wales. *Advances in the Study of the Sydney Basin. Proceedings of the Twenty-first Symposium, Department of Geology, University of Newcastle*, 135–42.

Briggs, D.J.C., 1989. Can macropalaeontologists, palynologists and sequence stratigraphers ever agree on eastern Australian Permian correlations? *Advances in the Study of the Sydney Basin. Proceedings of the Twenty-third Symposium, Department of Geology, University of Newcastle*, 29–35.

Briggs, D.J.C., 1993*a*. Chronostratigraphic correlation of Australian Permian depositional sequences. *Advances in the Study of the Sydney Basin. Twenty-seventh Newcastle Symposium, Department of Geology, University of Newcastle, Publication* 545, 51–58.

Briggs, D.J.C., 1993*b*. Permian depositional sequences of the Sydney–Bowen Basin. *Advances in the Study of the Sydney Basin. Twenty-seventh Newcastle Symposium, Department of Geology, University of Newcastle, Publication* 545, 247–54.

Brinkmann, R., 1959. *Abriss der Geologie. Vol. 2: Historische Geologie*, Ferdinand Enke Verlag, Stuttgart.

Brinkmann, R., 1960. *Geologic Evolution of Europe* [translated by J.E. Sanders], Enke, Stuttgart, and Hafner, New York, 161 pp.

Brongniart, A., 1829. *Tableau des Terrains qui Composent L'écorce du Globe ou Essai sur la Structure de la Partie Connue de la Terre*, Paris, 435 pp.

Brower, J.C. & Bussey, D.T., 1985. A comparison of five quantitative techniques for biostratigraphy. pp. 279–306 *in* F.M. Gradstein et al. (eds), q.v.

Brown, H.Y.L., 1895. Report on Northern Territory explorations. *South Australia Parliamentary Paper* 82, 1–30.

Brown, I.A., 1948. Lower Ordovician brachiopods from Junee district, Tasmania. *Journal of Paleontology* 22(1), 35–39.

Brown, I.A., 1949. Occurrence of the brachiopod genus *Plectodonta* Kozlowski at Bowning, New South Wales. *Journal and Proceedings of the Royal Society of New South Wales* 82, 196–201.

Brown, R.W., 1962. Paleocene flora of the Rocky Mountains and Great Plains. *United States Geological Survey Professional Paper* 375, 119 pp.

Brownlaw, R.L.S. & Jell, J.S., 1994. Coral biostratigraphy of the Sadler and Pillara limestones, Canning Basin, Western Australia. *Australasian Palaeontological Convention 1994, Abstracts & Programme*, p. 22.

Brunnschweiler, R.O., 1951. Notes on the geology of Dampier Land, north-western Australia. *Australian Journal of Science* 14, 6–8.

Brunnschweiler, R.O., 1954. Mesozoic stratigraphy and history of the Canning Desert and Fitzroy Valley, Western Australia. *Journal of the Geological Society of Australia* 1, 35–54.

Brunnschweiler, R.O., 1957. The geology of Dampier Peninsula, Western Australia. *Bureau of Mineral Resources, Geology & Geophysics, Report* 13, 19 pp.

Brunnschweiler, R.O., 1959. New Aconaeceratinae (Ammonoidea) from the Albian and Aptian of

Australia. *Bureau of Mineral Resources, Geology & Geophysics, Bulletin* 54, 19 pp.

Brunnschweiler, R.O., 1960. Marine fossils from the Upper Jurassic and the Lower Cretaceous of Dampier Peninsula, Western Australia. *Bureau of Mineral Resources, Geology & Geophysics, Bulletin* 59, 52 pp.

Brunnschweiler, R.O., 1966. Upper Cretaceous ammonites from the Carnarvon Basin of Western Australia. *Bureau of Mineral Resources Geology & Geophysics, Bulletin* 58, 58 pp.

Buch, L. von, 1839. *Über den Jura in Deutschland*, Der Königlich Preussischen Akademie der Wissenschaften, Berlin, 87 pp.

Buchan, K.L. & Hodych, J.P., 1989. Early Silurian palaeopole for redbeds and volcanics of the King George IV Lake area, Newfoundland. *Canadian Journal of Earth Sciences* 26, 1904–17.

Buckland, W., 1818. Appended table. *In* W. Phillips, *Selection of Facts Arranged so as to Form an Outline of the Geology of England and Wales*. London.

Buffler, R.T., 1994. Geological history of the eastern Argo Abyssal Plain based on ODP drilling and seismic data. *AGSO Journal of Australian Geology and Geophysics* 15, 157–64.

Bujak, J.P. & Williams, G.L., 1977. Jurassic palynostratigraphy of offshore eastern Canada. pp. 321–29 *in* F.M. Swain (ed.), *Stratigraphic Micropaleontology of Atlantic Basin and Borderlands*, Elsevier, Amsterdam.

Bukry, D., 1973. Low-latitude coccolith biostratigraphic zonation. *Initial Reports of the Deep Sea Drilling Project* 15, 685–703.

Bukry, D., 1975. Coccolith and silicoflagellate stratigraphy, north-western Pacific Ocean, Deep Sea Drilling Project Leg 32. *Initial Reports of the Deep Sea Drilling Project* 32, 677–701.

Bulman, O.M.B., 1963. On *Glyptograptus dentatus* (Brongniart) and some allied species. *Palaeontology* 6, 665–89.

Burek, P.J., 1970. Magnetic reversals: their application to stratigraphic problems. *American Association of Petroleum Geologists Bulletin* 54, 1120–39.

Burger, D., 1973. Spore zonation and sedimentary history of the Neocomian, Great Artesian Basin, Queensland. *Geological Society of Australia Special Publication* 4, 87–118.

Burger, D., 1976. Appendix 5: Palynological observations in the Surat Basin. pp. 51–55 *in* A.R. Jensen et al., A guide to the geology of the Bowen and Surat basins in Queensland. *Excursion Guide no. 3C for the 25th International Geological Congress, Sydney*.

Burger, D., 1980. Palynological studies in the Lower Cretaceous of the Surat Basin, Australia. *Bureau of Mineral Resources, Geology & Geophysics, Bulletin* 189, 106 pp.

Burger, D., 1982. A basal Cretaceous dinoflagellate suite from north-eastern Australia. *Palynology* 6, 161–92.

Burger, D., 1986. Palynology, cyclic sedimentation, and palaeoenvironments in the Late Mesozoic of the Eromanga Basin. *Geological Society of Australia Special Publication* 12, 53–70.

Burger, D., 1988. Early Cretaceous environments in the Eromanga Basin; palynological evidence from GSQ Wyandra-1 corehole. *Association of Australasian Palaeontologists Memoir* 5, 173–86.

Burger, D., 1989. Stratigraphy, palynology, and palaeoenvironments of the Hooray Sandstone, eastern Eromanga Basin, Queensland and New South Wales. *Queensland Department of Mines Report* 3.

Burger, D., 1990*a*. Australian Phanerozoic timescales: 8. Jurassic. Biostratigraphic charts and explanatory notes. *Bureau of Mineral Resources, Geology & Geophysics, Record* 1989/38.

Burger, D., 1990*b*. Australian Phanerozoic time scales. Volume 9. Cretaceous. Geochronological charts and explanatory text. *Bureau of Mineral Resources, Geology & Geophysics, Record* 1989/39.

Burger, D., 1994. Past and present palynological studies in the Clarence–Moreton Basin, Queensland and New South Wales. *Bureau of Mineral Resources, Geology & Geophysics, Bulletin* 241, 164–80.

Burger, D., 1994. Palynology of the Walloon Coal Measures, Kangaroo Creek Sandstone, and Grafton Formation, Clarence–Moreton Basin, Queensland and New South Wales. *Bureau of Mineral Resources, Geology & Geophysics, Bulletin* 241, 181–88.

Burger, D., in press. Mesozoic palynomorphs from the North West Shelf, offshore Western Australia. *Palynology*.

Burr, T., 1846. *Remarks on the Geology and Mineralogy of South Australia*, Andrew Murray, Adelaide, 32 pp.

Burrett, C.F., 1979. *Tasmanognathus*: a new Ordovician conodontophorid genus from Tasmania. *Geologica et Palaeontologica* 13, 31–38.

Burrett, C.F. & Goede, A., 1987. Mole Creek—a geological and geomorphological field guide. *Geological Society of Australia (Tasmanian Division) Guidebook* 1, 1–21.

Burrett, C.F. & Martin, E.L. (eds), 1989. Geology and mineral resources of Tasmania. *Geological Society of Australia, Special Publication* 15, 574 pp.

Burrett, C.F., Stait, B. & Laurie, J., 1983*a*. Trilobites and microfossils from the Middle Ordovician of Surprise Bay, southern Tasmania, Australia. *Association of Australasian Palaeontologists, Memoir* 1, 177–93.

Burrett, C., Stait, B. & Laurie, J., 1983*b*. The deep, the dark and the dirty—an Ordovician shelf to basin transition in southern Tasmania, Australia. *Geological Society of Australia, Abstracts* 9, 198–99.

Byrnes, A. & Scheibnerová, V.,1980. The inorganic nature of marine microfossils from the Wianamatta Group. *Micropaleontology* 26 (3), 323–26.

Caldwell, W.G.E. & North, B.R., 1984. Cretaceous stage boundaries in the southern Interior Plains of Canada. *Geological Society of Denmark Bulletin* 33, 57–69.

Calver, C.R. & Castleden, R.H.,1981. Triassic basalts from Tasmania. *Search* 12, 40–41.

Calver, C.R., Clarke, M.J. & Truswell, E.M., 1984. The stratigraphy of a Late Palaeozoic borehole section at Douglas River, eastern Tasmania. A synthesis of marine macroinvertebrates and palynological data. *Royal Society of Tasmania Papers and Proceedings* 118, 137–67.

Camp, C.L & Banks, M.R., 1978. A proterosuchian reptile from the Early Triassic of Tasmania. *Alcheringa* 2, 143–58.

Campbell, K.S.W., 1953. The fauna of the Permo–Carboniferous Ingelara Beds of Queensland. *Department of Geology, University of Queensland Papers* 4 (1), 1–43.

Campbell, K.S.W., 1955. *Phricodothyris* in New South Wales. *Geological Magazine* 92, 378–84.

Campbell, K.S.W., 1956. Some Carboniferous productid brachiopods from New South Wales. *Journal of Paleontology* 30, 463–80.

Campbell, K.S.W., 1957. A Lower Carboniferous brachiopod–coral fauna from New South Wales. *Journal of Paleontology* 31, 34–98.

Campbell, K.S.W., 1961. Carboniferous fossils from the Kuttung rocks of New South Wales. *Palaeontology* 4(3), 428-74.

Campbell, K.S.W., 1962. Marine fossils from the Carboniferous glacial rocks of New South Wales. *Journal of Paleontology* 36(1), 38–52.

Campbell, K.S.W. & Durham, G.J., 1970. A new trinucleid trilobite from the Upper Ordovician of New South Wales. *Palaeontology* 13, 573–80.

Campbell, K.S.W. & Engel, B.A., 1963. The faunas of the Tournaisian Tulcumba Sandstone and its members in the Werrie and Belvue synclines, New South Wales. *Journal of the Geological Society of Australia* 10(1), 55–122.

Campbell, K.S.W. & McKellar, R.G., 1969. Eastern Australian Carboniferous invertebrates: sequence and affinities. pp. 77–119 *in* K.S.W. Campbell (ed.) *Stratigraphy and Palaeontology*, ANU Press, Canberra.

Campbell, K.S.W. & McKelvey, B.C., 1972. The geology of the Barrington district, NSW. *Pacific Geology* 5, 7–48.

Campbell, K.S.W., Brown, D.A. & Coleman, A.R., 1983. Ammonoids and the correlation of the Lower Carboniferous rocks of eastern Australia. *Alcheringa* 7(2), 75–123.

Cande, S.C. & Kent, D.V., 1992*a*. A new geomagnetic polarity time scale for the Late Cretaceous and Cenozoic. *Journal of Geophysical Research* 97, 13917–51.

Cande, S.C. & Kent, D.V., 1992*b*. Ultrahigh resolution marine magnetic anomaly profiles: a record of continuous paleointensity variations. *Journal of Geophysical Research* 97, 15075–83.

Cantrill, D.J. & Webb, J.A., 1987. A reappraisal of *Phyllopteroides* Medwell (Osmundaceae) and its stratigraphic significance in the Lower Cretaceous of eastern Australia. *Alcheringa* 11, 59–85.

Carey, S.W., 1937. The Carboniferous sequence in the Werrie Basin. *Proceedings of the Linnean Society of New South Wales* 62, 341–76.

Carey, S.W. & Browne, W.R., 1938. Review of the Carboniferous stratigraphy, tectonics, and palaeogeography of New South Wales and Queensland. *Journal and Proceedings of the Royal Society of New South Wales* 71, 591–614.

Caron, M., 1985. Cretaceous planktic Foraminifera. pp. 17–86 *in* H.M. Bolli, J.B. Saunders & K. Perch-Nielsen (eds), *Plankton Stratigraphy*, Cambridge University Press.

Carter, A.N., 1958*a*. Tertiary Foraminifera from the Aire District, Victoria. *Geological Survey of Victoria, Bulletin* 55, 1–76.

Carter, A.N., 1958*b*. Pelagic Foraminifera in the Tertiary of Victoria. *Geological Magazine* 95, 297–304.

Carter, A.N., 1964. Tertiary Foraminifera from Gippsland, Victoria and their stratigraphical significance. *Geological Survey of Victoria, Memoir* 23, 1–54.

Carter, A.N., 1990. Time and space events in the Neogene of south-eastern Australia. pp. 183–93 *in* R. Tsuchi (ed.), *Pacific Neogene Events, Their Timing, Nature and Interrelationship*, University of Tokyo Press, Tokyo.

Cas, R., 1983. Palaeogeographic and tectonic development of the Lachlan Fold Belt, south-eastern Australia. *Geological Society of Australia Special Publication* 10, 104 pp.

Cas, R.A.F. & VandenBerg, A.H.M., 1988. Ordovician. pp. 5–102 *in* J.G.Douglas & J.A. Ferguson (eds), q.v.

Casey, R., 1973. The ammonite succession at the Jurassic–Cretaceous boundary in eastern England. 193–266 *in* R. Casey & P.F. Rawson (eds), The Boreal Lower Cretaceous. *Geological Journal Special Issue* 5.

Cavelier, C. & Roger, J. (eds), 1980. Les étages français et leurs stratotypes. *Mémoire du Bureau de Recherches Géologiques et Minières* 109, 295 pp.

Cawood, P.A., 1976. Cambro–Ordovician strata, northern New South Wales. *Search* 7, 317–18.

Chaloner, W.G. & Sheerin, A., 1979. Devonian macrofloras. *Special Papers in Palaeontology* 23, 145–61.

Chamney, T.P., 1973. Micropalaeontological correlation of the Canadian Boreal Lower Cretaceous. pp. 19–40 *in* R. Casey & P.F. Rawson (eds), The Boreal Lower Cretaceous. *Geological Journal Special Issue* 5.

Chang, K.H., 1975. Unconformity-bounded stratigraphic units. *Bulletin of the Geological Society of America* 86, 1544–52.

Channell, J.E.T. & Erba, E., 1992. Early Cretaceous polarity chrons CM0 to CM11 recorded in northern Italian land sections near Brescia. *Earth and Planetary Science Letters* 108, 161–179.

Channell, J.E.T. & Grandesso, P., 1987. A revised correlation of Mesozoic polarity chrons and calpionellid zones. *Earth and Planetary Science Letters* 85, 222–40.

Channell, J.E.T., Bralower, T.J. & Grandesso, P., 1987. Biostratigraphic correlation of Mesozoic polarity chrons CM1 to CM23 at Capriolo and Xausa (Southern Alps, Italy). *Earth and Planetary Science Letters* 85, 203–21.

Channell, J.E.T., Ogg, J.G. & Lowrie, W., 1982. Geomagnetic polarity in the Early Cretaceous and Jurassic. *Philosophical Transactions of the Royal Society of London* A 306, 137–46.

Channell, J.E.T., Lowrie, W., Pialli, P. & Venturi, F., 1984. Jurassic magnetic stratigraphy from Umbrian (Italian) land sections. *Earth and Planetary Science Letters* 68, 309–25.

Chapman, F., 1904*a*. On some Foraminifera and Ostracoda from Jurassic (Lower Oolite) strata near Geraldton, Western Australia. *Proceedings of the Royal Society of Victoria* 16, 185–206.

Chapman, F., 1904*b*. On a collection of Upper Palaeozoic and Mesozoic fossils from Western Australia and Queensland in the National Museum, Melbourne. *Proceedings of the Royal Society of Victoria* 16, 306–35.

Chapman, F., 1904*c*. New or little-known Victorian fossils in the National Museum. 3. Some Palaeozoic Pteropoda. *Proceedings of the Royal Society of Victoria* 16, 336–42.

Chapman, F., 1907. On some fossils from Silurian limestones, Dolodrook Valley, Mt Wellington, Victoria. *Victorian Naturalist* 24, 34.

Chapman, F., 1908*a*. Preliminary notes on a collection of trilobite remains from Dolodrook River, N. Gippsland. *Proceedings of the Royal Society of Victoria* 21, 268–69.

Chapman, F., 1908*b*. Report on fossils. *Records of the Geological Survey of Victoria* 2(4), 207–11.

Chapman, F., 1911. New or little-known Victorian fossils in the National Museum. Part 12: On a trilobite fauna of Upper Cambrian age (Olenus Series) in NE Gippsland, Victoria. *Proceedings of the Royal Society of Victoria* 23, 305–24.

Chapman, F., 1917*a*. On the occurrence of fish remains and a *Lingula* in the Grampians, western Victoria. *Records of the Geological Survey of Victoria* 4 (1), 83–86.

Chapman, F., 1917*b*. Report on Cambrian fossils from Knowsley East, near Heathcote. *Records of the Geological Survey of Victoria* 4(1), 87–89.

Chapman, F., 1929. On some trilobites and brachiopods from the Mount Isa district, northwestern Queensland. *Proceedings of the Royal Society of Victoria* 41, 206–16.

Chapman, F. & Singleton, F.A., 1923. The Tertiary deposits of Australia. *Proceedings of the Pan Pacific Science Congress* 1, 985–1024.

Chaproniere, G.C.H., 1981*a*. Late Oligocene to Early Miocene planktic Foraminiferida from Ashmore Reef No. 1 Well, north-west Australia. *Alcheringa* 5, 103–31.

Chaproniere, G.C.H., 1981*b*. Australian mid-Tertiary larger foraminiferal associations and their bearing on the East Indian Letter Classification. *BMR Journal of Australian Geology & Geophysics* 6, 145–51.

Chaproniere, G.C.H., 1983. Tertiary larger foraminiferids from the north-western margin of the Queensland Plateau, Australia. *Bureau of Mineral Resources, Geology & Geophysics, Bulletin* 217, 31–57.

Chaproniere, G.C.H., 1984*a*. Oligocene and Miocene larger Foraminiferida from Australia and New Zealand. *Bureau of Mineral Resources, Geology & Geophysics, Bulletin* 188, 1–98.

Chaproniere, G.C.H., 1984*b*. The Neogene larger foraminiferal sequence in the Australian and New Zealand regions, and its relevance to the East Indian Letter Stage Classification. *Palaeogeography, Palaeoclimatology, Palaeoecology* 46, 25–35.

Chaproniere, G.C.H., 1991. Pleistocene to Holocene planktic foraminiferal biostratigraphy of the Coral Sea, offshore Queensland, Australia. *BMR Journal of Australian Geology & Geophysics* 12, 195–221.

Chaproniere, G.C.H., 1992. The distribution and development of Late Oligocene and Early Miocene reticulate globigerines in Australia. *Marine Micropaleontology* 18, 279–305.

Chaproniere, G.C.H., Styzen, M.J., Sager, W.W., Nishi, H., Quinterno, P. & Abrahamsen, N., 1994. Late Neogene biostratigraphic and magnetostratigraphic synthesis, Leg 135. *Proceedings of the Ocean Drilling Program, Scientific Results* 135, 857–77.

Chatterjee, S. & Hotton, N.,1986. The paleoposition of India. *Journal of South East Asian Earth Sciences* 1 (3), 145–89.

Chatterton, B.D.E., 1971. Taxonomy and ontogeny of Siluro–Devonian trilobites from near Yass, New South Wales. *Palaeontographica, Abt A* 137(1–3), 1–108, 1–24.

Chatterton, B.D.E. & Campbell, K.S.W., 1980. Silurian trilobites from near Canberra and some related forms from the Yass Basin. *Palaeontographica, Abt* A 167(1–3), 77–119, 1–16.

Chatterton, B.D. & Wright, A.J., 1986. Silicified Early Devonian trilobites from Mudgee, New South Wales. *Alcheringa* 10, 279–96.

Chatterton, B.D.E., Johnson, B.D. & Campbell, K.S.W., 1979. Silicified Lower Devonian trilobites from New South Wales. *Palaeontology* 22, 799–837.

Chen, X.Y. & Barton, C.E., 1991. Onset of aridity and dune-building in central Australia: sedimentological and magnetostratigraphic evidence from Lake Amadeus. *Paleogeography, Paleoclimatology, Paleoecology* 84, 55–73.

Cheng, Y.-N., 1986. Taxonomic studies on Upper Paleozoic Radiolaria. *National Museum of Natural Science, Special Publications* 1, 311 pp.

Chernykh, V.V. & Reshetkova, N.P., 1987. *The Biostratigraphy and Conodonts from Carboniferous–Permian Boundary Deposits of the West Slope of Middle and South Urals*, Urals Science Centre, Academy of Science, Sverdlovsk, 1–50.

Chilton, C.L., 1917. A fossil isopod belonging to the freshwater genus *Phreatoicus. Journal and Proceedings of the Royal Society of New South Wales* 51, 365–88.

Chilton, C.L., 1929. On a fossil shrimp from the Hawkesbury Sandstone. *Journal and Proceedings of the Royal Society of New South Wales* 62, 366–68.

Chivas, A.R., De Deckker, P., Nind, M., Thiriet, D. & Watson, G., 1986. The Pleistocene palaeoenvironmental record of Lake Buchanan: an atypical Australian playa. *Palaeogeography, Palaeoclimatology, Palaeoecology* 54, 131–52.

Chlupáč, I. & Kukal, Z., 1986. Reflection of possible global Devonian events in the Barrandian area, CSSR. pp. 171–79 *in* O.H. Walliser (ed.), *Global Bio-events*, Springer, Berlin.

Chlupáč, I., Kríž, J. & Schönlaub, H.P., 1980. Field trip E: Silurian and Devonian conodont localities of the Barrandian. pp. 147–80 *in* H.P. Schönlaub (ed.), Second European conodont symposium

(ECOS II), guidebook, abstracts. *Abhandlungen der Geologischen Bundesanstalt* 35.

Churkin, M.C., Carter, C. & Johnson, B., 1977. Subdivision of Ordovician and Silurian time scale using accumulation rates of graptolitic shale. *Geology* 5, 452–56.

Claesson, K.C., 1979. Early Palaeozoic geomagnetism of Gotland. *Geologiska Föreningens i Stockholm Förhandlingar* 101, 149–55.

Claoué-Long, J.C. Jones, P.J. & Roberts, J., 1991. Precise zircon dating of Palaeozoic stratigraphy. *Terra Abstracts.* European Union of Geosciences (EUG) VI, Strasbourg.

Claoué-Long, J.C., Jones, P.J. & Roberts, J., 1993. The age of the Devonian–Carboniferous boundary. *Annals Société de Géologique de Belgique* 115(2), 531- 49.

Claoué-Long, J.C., Jones, P.J. & Truswell, E.M., 1989. Calibration of the Phanerozoic Time Scale: a pilot study towards accurate dating of Phanerozoic biozones. *In 1989 BMR Research Symposium: Extended Abstracts,* Bureau of Mineral Resources, Geology & Geophysics, Canberra, 7 pp.

Claoué-Long, J.C., Compston, W., Roberts, J. & Fanning, C.M. (in press). Two Carboniferous ages: a comparison of SHRIMP zircon dating with conventional zircon ages and $^{40}Ar/^{39}Ar$ analysis. *In* W.A. Berggren & D.V. Kent (eds), *Geochronology, Time Scales and Stratigraphic Correlation, SEPM Special Publication.*

Claoué-Long, J.C., Jones, P.J., Roberts, J. & Maxwell, S., 1990. SHRIMP determination of the age of the Devonian–Carboniferous boundary. *In* 7th International Conference on Geochronology, Cosmochronology and Isotope Geology, Canberra, 1990. *Geological Society of Australia, Abstracts* 27, 20.

Claoué-Long, J.C., Jones, P.J., Roberts, J. & Maxwell, S., 1992. The numerical age of the Devonian–Carboniferous boundary. *Geological Magazine* 129, 281–91.

Claoué-Long, J.C., Zhang Zichao, Ma Guogan & Du Shaohua, 1991. The age of the Permian–Triassic boundary. *Earth and Planetary Science Letters* 105, 182–90.

Clarke, E. de C., 1937. Correlation of the Carboniferous and Permian formations of Australia. II. Western Australia. *Report of the Australian and New Zealand Association for the Advancement of Science* 23, 427–30.

Clarke, M.J., 1987. Late Permian (late Lymington = ?Kazanaian) brachiopods from Tasmania. *Alcheringa* 11, 261–89.

Clarke, M.J., 1989. Lower Parmeener Supergroup. pp. 295–309 *in* C.F. Burrett & E.L. Martin, q.v.

Clarke, M.J., 1990. Late Palaeozoic (Tamarian; Late Carboniferous–Early Permian) cold-water brachiopods from Tasmania. *Alcheringa* 14, 53–76.

Clarke, M.J., 1992. Hellyerian and Tamarian (Late Carboniferous–Lower Permian) invertebrate faunas from Tasmania. *Tasmanian Geological Survey Bulletin* 69, 1–52.

Clarke, M.J. & Baillie, P.W., 1984. Geological Atlas 1:55,000 series. Sheet 77 (8512N), Maria, *Tasmania Department of Mines, Geological Survey Explanatory Report.*

Clarke, M.J. & Banks, M.R., 1975. The stratigraphy of the Lower Permo–Carboniferous parts of the Parmeener Supergroup, Tasmania. pp. 453–467 *in* K.S.W. Campbell (ed.), *Gondwana Geology,* ANU Press, Canberra.

Clarke M.J. & Farmer, N., 1976. Biostratigraphic nomemclature for Late Palaeozoic rocks in Tasmania. *Proceedings of the Royal Society of Tasmania* 110, 91–109.

Clarke, W.B., 1862. On the occurrence of Mesozoic and Permian faunae in eastern Australia. *Quarterly Journal of the Geological Society* 18, 244–47.

Clarke, W.B., 1867. On marine fossiliferous Secondary formations in Australia. *Quarterly Journal of the Geological Society* 23, 7–12.

Clayton, G., Loboziak, S., Streel, M., Turnau, E. & Utting, J., 1991 [1990 imprint]. Palynological events in the Mississippian (Lower Carboniferous) of Europe, North Africa and North America. *Courier Forschungsinstitut Senckenberg* 130, 79–84.

Cockbain, A.E., 1974. Triassic conchostracans from the Kockatea Shale. *Geological Survey of Western Australia, Annual Report for 1973,* 104–6.

Cockbain, A.E., 1980. Permian ammonoids from the Carnarvon Basin—A review. *Geological Survey of Western Australia, Annual Report for 1979,* 100–105.

Cockbain, A.E. & Hocking, R.M., 1989. Revised stratigraphic nomenclature in Western Australian Phanerozoic basins. *Geological Survey of Western Australia, Record* 1989/15.

Cockbain, A.E. & Playford, P.E., 1989. The Devonian of Western Australia: a review. pp. 743–54 *in* N.J. McMillan, A.F. Embry & D.J. Glass (eds), vol. 1, q.v.

Cocks, L.R.M., 1985. The Ordovician–Silurian Boundary. *Episodes* 8, 98–100.

Cocks, L.R.M. & Nowlan, G.S., 1993. New left hand side for correlation diagrams. *Silurian Times—a Newsletter of the Silurian Subcommission* 1, 6–8.

Cocks, L.R.M., Toghill, P. & Ziegler, A.M., 1970. Stage names within the Llandovery Series. *Geological Magazine* 107, 79–87.

Cocks, L.R.M., Woodcock, N.H., Rickards, R.B., Temple, J.T. & Lane, P.D., 1984. The Llandovery Series of the type area. *British Museum (Natural History), Geology, Bulletin* 38(3), 131–82.

Cohee, G.V., Glaessner, M.F. & Hedberg, H.D. (eds), 1978. Contributions to the Geological Time Scale. *American Association of Petroleum Geologists, Studies in Geology* 6, 388 pp.

Coleman, P.J. & Skwarko, S.K., 1967. Lower Triassic and Middle Jurassic fossils at Enanty Hill, Mingenew, Perth Basin, Western Australia. *Bureau of Mineral Resources, Geology & Geophysics, Bulletin* 92, 197–215.

Collinson, C., Rexroad, C.B. & Thompson, T.L., 1971. Conodont zonation of the North American Mississippian. *Geological Society of America Memoir* 127, 353–94.

Collinson, C., Scott, A.J. & Rexroad, C.B., 1962. Six charts showing biostratigraphic zones and correlations based on conodonts from the Devonian and Mississippian rocks of the upper Mississippi Valley. *Illinois Geological Survey Circular* 328, 1–32.

Colloque sur la Limite Jurassique–Crétacé (Lyon & Neuchatel 1973), 1975. *Mémoire du Bureau de Recherches Géologiques et Minières* 86, 393 pp.

Colloque sur le Crétacé Inférieur (Lyon, 1963), 1965. *Mémoire du Bureau de Recherches Géologiques et Minières* 34, 840 pp.

Colwell, J.B., Graham, T.L. & Others, 1990. Stratigraphy of Australia's NW continental margin (Project 121–26)—Post-cruise report for BMR Survey 96. *Bureau of Mineral Resources, Geology and Geophysics, Record* 1990/85, 1–123.

Combaz, A. & Peniguel, G., 1972. Étude palynostratigraphie de l'Ordovicien dans quelques sondages du Bassin de Canning (Australie Occidentale). *Bulletin Centre Recherche Pau SNPA* 6, 121–67.

Compston, W. & Williams, I.S., 1992. Ion probe ages for the British Ordovician and Silurian stratotypes. pp. 59–67 *in* B.D. Webby & J.R. Laurie (eds), q.v.

Compston, W. Williams, I.S., Kirschvink, J.L., Zhang Zichao & Ma Guogan, 1992. Zircon U-Pb ages for the Early Cambrian time-scale. *Journal of the Geological Society, London* 149, 171–84.

Condon, M.A., 1954. Progress report on the stratigraphy and structure of the Carnarvon Basin, Western Australia. *Bureau of Mineral Resources, Geology & Geophysics, Report* 15, 163 pp.

Condon, M.A., 1967. The geology of the Carnarvon Basin, Western Australia, Part 2: Permian stratigraphy. *Bureau of Mineral Resources, Geology & Geophysics, Bulletin* 77, 1–191.

Conil, R., 1969. General discussion of Subcommission on Carboniferous Stratigraphy VIII. Lower boundary of the Viséan. *Compte Rendu 6ème Congrès International de Viséen Stratigraphie et Géologie du Carbonifère, Sheffield, 1967*, 1, 188–9.

Conil, R., Groessens, E. & Pirlet, H., 1976. Nouvelle charte stratigraphique du Dinantien type de le Belgique. *Annales de la Société Géologique du Nord* 96, 363–71.

Conil, R., Longerstaey, P.J. & Ramsbottom, W.H.C., 1989. Matériaux pour l'étude micropaléontologique du Dinantien de Grande-Bretagne. *Mémoires de l'Institut Géologique de Louvain*, XXX, 187 pp.

Conil, R., Austin, R.L., Lys, M. & Rhodes, F.H.T., 1969. La limite des étages Tournaisien et Viséen au stratotype de l'assise de Dinant. *Bulletin de la Société Belge de Géologie* 77 (1), 39–69.

Conil, R., Groessens, E., Laloux, M. & Poty, E., 1989. La limite Tournaisien/Viséen dans la région-type, *Annales de la Société Géologique de Belgique*, 112 (1), 177–89.

Conil, R., Dreesen, R., Lentz, M.-A., Lys, M. & Plodowski, G., 1986. The Devono–Carboniferous transition in the Franco–Belgian Basin with reference to Foraminifera and brachiopods. *Annales de la Société Géologique de Belgique* 109, 19–26.

Conil, R., Groessens, E., Laloux, M., Poty, E. & Tourneur, F., 1991 [1990 imprint]. Carboniferous guide Foraminifera, corals, and conodonts in the Franco–Belgium and Campine basins: their potential for widespread correlation. *Courier Forschungsinstitut Senckenberg* 130, 15–30.

Conkin, J.E. & Conkin, B.M., 1968. A revision of some Upper Devonian Foraminifera from Western Australia. *Palaeontology* 11, 601–09.

Conway Morris, S., 1988. Radiometric dating of the Precambrian–Cambrian boundary in the Avalon Zone. *New York State Museum and Geological Survey Bulletin* 463, 53–58.

Conway Morris, S., 1989. South-eastern Newfoundland and adjacent areas (Avalon Zone). pp. 7–39 *in* J.W. Cowie & M.D. Brasier (eds), *The Precambrian–Cambrian Boundary*, Clarendon Press, Oxford.

Conybeare, W.D. & Phillips, W., 1822. *Outlines of the Geology of England and Wales, with an Introductory Compendium of the General Principles of that Science, and Comparative Views of the Structure of Foreign Countries, Part I*, 470 pp., William Phillips, London.

Cookson, I.C., 1956. Additional microplankton from Australian Late Mesozoic and Tertiary sediments. *Australian Journal of Marine and Freshwater Research* 7, 183–91.

Cookson, I.C., 1965. Cretaceous and Tertiary microplankton from south-eastern Australia. *Proceedings of the Royal Society of Victoria* 78, 85–93.

Cookson, I.C. & Eisenack, A., 1958. Microplankton from Australian and New Guinean Upper Mesozoic sediments. *Proceedings of the Royal Society of Victoria* 70, 19–79.

Cookson, I.C. & Eisenack, A., 1960*a*. Microplankton from Australian Cretaceous sediments. *Micropaleontology* 6, 1–18.

Cookson, I.C. & Eisenack, A., 1960*b*. Upper Mesozoic microplankton from Australia and New Guinea. *Palaeontology* 2, 243–61.

Cookson, I.C. & Eisenack, A., 1961. Upper Cretaceous microplankton from the Belfast No. 4 bore, south-western Victoria. *Proceedings of the Royal Society of Victoria* 74, 69–76.

Cookson, I.C. & Eisenack, A., 1962. Additional microplankton from Australian Cretaceous sediments. *Micropaleontology* 8, 485–507.

Cookson, I.C. & Eisenack, A., 1970. Cretaceous microplankton from the Eucla Basin, Western Australia. *Proceedings of the Royal Society of Victoria* 83, 137–58.

Cooper, B.J., 1980. Toward an improved Silurian conodont biostratigraphy. *Lethaia* 13, 209–27.

Cooper, B.J., 1981. Early Ordovician conodonts from the Horn Valley Siltstone, Central Australia. *Palaeontology* 24, 147-83.

Cooper, B.J., 1984. Historical perspective: Australia's first geology book, 1846. *Geological Survey of South Australia, Quarterly Geological Notes*, April 1984, 90, 2.

Cooper, B.J., 1986. A record of Ordovician conodonts from the Warburton Basin, South Australia. *Geological Survey of South Australia, Quarterly Geological Notes* 100, 8–14.

Cooper, J.A., Jenkins, R.J.F., Compston, W., & Williams, I.S., 1990. Ion microprobe U-Pb zircon dating within the Lower Cambrian of South Australia. *In* 7th International Conference on Geochronology, Cosmochronology and Isotope Geology, Canberra. *Geological Society of Australia, Abstracts* 27, 21.

Cooper, J.A., Jenkins, R.J.F., Compston, W. & Williams, I.S., 1992. Ion probe zircon dating of a mid Early Cambrian tuff in South Australia. *Journal of the Geological Society, London*, 149, 185–92.

Cooper, M.K.E., 1989. Nannofossil provincialism in the Late Jurassic–Early Cretaceous (Kimmeridgian to Valanginian) period. pp. 223–46 *in* J.A. Crux & S.E. van Heck (eds), *Nannofossils and their Applications, Proceedings of the International Nannofossil Association Conference, London, 1987*, Ellis Horwood Ltd, Chichester.

Cooper, M.R., 1977. Eustasy during the Cretaceous: its implications and importance. *Palaeogeography, Palaeoclimatology, Palaeoecology* 22, 1–60.

Cooper, R.A., 1973. Taxonomy and evolution of *Isograptus* Moberg in Australasia. *Palaeontology* 16, 45–115.

Cooper, R.A., 1979. Sequence and correlation of Tremadoc graptolite assemblages. *Alcheringa* 3, 7–19.

Cooper, R.A. & Grindley, G.W. (eds) 1982. Late Proterozoic to Devonian sequences in south-eastern Australia, Antarctica and New Zealand and their correlation. *Geological Society of Australia Special Publication* 9, 103 pp.

Cooper, R.A. & Lindholm, K., 1990. A precise worldwide correlation of early Ordovician graptolite sequences. *Geological Magazine* 127, 497–525.

Cooper, R.A. & McLaurin, A.N., 1974. *Apiograptus* gen. nov. and the origin of the biserial graptoloid rhabdosome. *Special Papers in Palaeontology* 13, 75–86.

Cooper, R.A. & Ni Yunan, 1986. Taxonomy, phylogeny, and variability of *Pseudisograptus* Beavis. *Palaeontology* 29, 313–63.

Cooper, R.A. & Stewart, I.R., 1979. The Tremadoc graptolite sequence of Lancefield, Victoria. *Palaeontology* 22, 767–97.

Corbett, K.D., 1975. Preliminary report on the geology of the Red Hills–Newton Creek area, West Coast Range, Tasmania. *Tasmanian Department of Mines Technical Report* 19, 11–25.

Corbett, K.D. & Banks, M.R., 1973. Ordovician stratigraphy of the Florentine Valley Synclinorium, south-west Tasmania. *Papers and Proceedings of the Royal Society of Tasmania* 109, 111–20.

Cosgriff, J.W., 1965. A new genus of Temnospondyli from the Triassic of Western Australia. *Journal of the Royal Society of Western Australia* 48, 65–90.

Cosgriff, J.W., 1969. *Blinasaurus* a brachyopid genus from Western Australia and New South Wales. *Journal of the Royal Society of Western Australia* 62, 65–88.

Cosgriff, J.W., 1974. Lower Triassic Temnospondyli of Tasmania. *Geological Society of America, Special Papers* 149, 1–134.

Cosgriff, J.W., 1984. The temnospondyl labyrinthodonts of the earliest Triassic. *Journal of Vertebrate Palaeontology* 4, 30–46.

Cosgriff, J.W. & Garbutt, N.K., 1972. *Erythrobatrachus noonkanbahensis* a trematosaurid species from the Blina Shale. *Journal of the Royal Society of Western Australia* 55, 5–18.

Couillard, R. & Irving, E., 1975. Palaeolatitude and reversals: evidence from the Cretaceous Period. *Geological Association of Canada Special Paper* 13, 21–29.

Cowie, J.W., 1992. Two decades of research on the Proterozoic–Phanerozoic transition: 1972–1991. *Journal of the Geological Society, London* 149, 589–92.

Cowie, J.W. & Bassett, M.G., 1989. IUGS 1989 Global Stratigraphic Chart with geochronometric and magnetostratigraphic calibration. *Episodes* 12 (2), supplement.

Cowie, J.W. & Glaessner, M.F., 1975. The Precambrian–Cambrian boundary: A symposium. *Earth Science Reviews* 11, 209–51.

Cowie, J.W. & Harland, W.B., 1989. Chronometry. pp. 186–198 *in* J.W. Cowie & M.D. Brasier (eds), *The Precambrian–Cambrian Boundary*, Clarendon Press, Oxford.

Cowie, J.W. & Johnson, M.R.W., 1985. Late Precambrian and Cambrian geological time-scale. pp. 47–64 *in* N.J. Snelling (ed.), q.v.

Cowie, J.W., Ziegler, W. & Remane, J., 1989. Stratigraphic commission accelerates progress, 1984 to 1989. *Episodes* 12, 79–83.

Cox, J.C., 1880. Notes on the Moore Park borings. *Proceedings of the Linnean Society of New South Wales* 5, 273–80.

Crespin, I., 1943*a*. Conodonts from Waterhouse Range, central Australia. *Transactions of the Royal Society of South Australia* 67, 231–33.

Crespin, I., 1943*b*. The stratigraphy of the Tertiary marine rocks in Gippsland, Victoria. Department of Supply and Shipping, Australia, *Palaeontological Bulletin* 4, 1–101.

Crespin, I., 1947. A summary of the stratigraphy and palaeontology of the Lakes Entrance Oil Shaft, Gippsland, Victoria. *Bureau of Mineral Resources, Geology & Geophysics, Report* 1947/69, 1–5.

Crespin, I., 1963. Lower Cretaceous arenaceous Foraminifera of Australia. *Bureau of Mineral Resources, Geology & Geophysics, Bulletin* 66, 110 pp.

Crick, G.C., 1894. On a collection of Jurassic Cephalopoda from Western Australia—obtained by Harry Page Woodward, FGS, Government Geologist—with description of the species. *Geological Magazine* NS, Decade IV, 1, 385–93, 433–41.

Crux, J.A., 1989. Biostratigraphy and palaeontological applications of Lower Cretaceous nannofossils from northwestern Europe. pp. 143–211 *in* J.A. Crux & S.E. van Heck (eds), *Nannofossils and their Applications, Proceedings of the International Nannofossil Association Conference, London, 1987*, Ellis Horwood Ltd, Chichester.

Cunningham, A. 1825. *In* Baron Field (ed.), *Geographical Memoirs on New South Wales*, John Murray, London, 504 pp.

Dachroth, W., 1976. Gesteinsmagnetische Marken im Perm Mitteleuropas. *Geologisches Jahrbuch* E 10, 3–63.

Dachroth, W., 1988. Gesteinsmagnetischer Vergleich Permischer Schichtenfolgen in Mitteleuropa. *Zeitschrift für Geologischen Wissenschaften* 16, 959–68.

Daily, B., 1956. The Cambrian in South Australia. pp. 91–147 *in* J. Rodgers (ed.), El Sistema Cambrico, su paleogeografia y el problema de su base. 2. *20th Session International Geological Congress, Mexico.*

Daily, B., 1963. The fossiliferous Cambrian succession on Fleurieu Peninsula, South Australia. *Records of the South Australian Museum* 14, 570–601.

Daily, B., 1966. See Harris, W.K. & Daily, B.

Daily, B., 1972. The base of the Cambrian and the first Cambrian faunas. *Centre for Precambrian Research, University of Adelaide, Special Paper* 1, 13–41.

Daily, B., 1975. See Cowie, J.W. & Glaessner, M.F.

Daily, B., 1976*a*. New data on the base of the Cambrian in South Australia. *Izvestiya Akademii Nauk SSSR, Seriya Geologicheskaya* 3, 45–52.

Daily, B., 1976*b*. The Cambrian of the Flinders Ranges. *25th Session International Geological Congress, Sydney, Excursion Guidebook* 33A, 15–19.

Daily, B. & Jago, J.B., 1975. The trilobite *Lejopyge* Hawle & Corda and the Middle–Upper Cambrian boundary. *Palaeontology* 18, 527–50.

Daintree, R., 1872. Notes on the geology of the colony of Queensland. *Quarterly Journal of the Geological Society of London* 28, 271–317.

Darragh, T.A., 1985. Molluscan biogeography and biostratigraphy of the Tertiary of southeastern Australia. *Alcheringa* 9, 83–116.

Davey, R.J., 1979. The stratigraphic distribution of dinocysts in the Portlandian (latest Jurassic) to Barremian (Early Cretaceous) of north-west Europe. *American Association of Stratigraphic Palynologists Contribution* Series 5B, 49–81.

Davey, R.J., 1982. Dinocyst stratigraphy of the latest Jurassic to Early Cretaceous of the Haldager No. 1 borehole, Denmark. *Danmarks Geologiske Undersøgelse* B 6, 3–57.

Davey, R.J., 1987. Palynological zonation of the Lower Cretaceous, Upper and uppermost Middle Jurassic in the north-western Papuan Basin of Papua New Guinea. *Geological Survey of Papua New Guinea Memoir* 13, 77 pp.

David, T.W.E., 1889. Cupriferous tuffs of the passage beds between the Triassic Hawkesbury Series and the Permo–Carboniferous coal-measures of New South Wales. *Report of the Australasian Association for the Advancement of Science* 1, 275–90.

David, T.W.E., 1932. *Explanatory Notes to Accompany a New Geological Map of the Commonwealth of Australia*, Commonwealth Council for Scientific and Industrial Research, Australasian Medical Publishing Co. Ltd, Sydney, 177 pp.

David, T.W.E., 1950. *The Geology of the Commonwealth of Australia.* W.R. Browne (ed.), vol. 1, Edward Arnold, London, 747 pp.

David, T.W.E & Pittman, E.F., 1883. On the occurrence of *Lepidodendron australe* in the Devonian rocks of New South Wales. *Records of the Geological Survey of New South Wales* 3, 195–200.

David, T.W.E. & Pittman, E.F. 1899. Palaeozoic radiolarian rocks of New South Wales. *Quarterly Journal of the Geological Society of London* 55, 16–37.

Davydov, V.I., 1984. The zonal subdivision of the Upper Carboniferous in south-western Darvaz. *Bulletin of the Moscow Society of Naturalists, Geological Series* 59 (3), 41–57.

Davydov, V.I., 1988. On the problems and principles of stage and zonal division of the Upper Carboniferous. *Annual of the All-Union Paleontological Society* 31, 256–69.

Davydov, V.I., Barskov, I.S., Bogoslovskaya, M.F., Leven, E.Y., Popov, A.V., Akhmetshina, L.Z. & Kozitskaya, R.I., 1992. The Carboniferous–Permian Boundary in the former USSR and its correlation. *International Geology Review* 34(9), 889–906.

Day, R.W., 1964. Stratigraphy of the Roma–Wallumbilla area. *Geological Survey of Queensland Publication* 318, 23 pp.

Day, R.W., 1969. The Lower Cretaceous of the Great Artesian Basin. pp. 140–73 *in* K.S.W. Campbell (ed.), *Stratigraphy and Palaeontology*, ANU Press, Canberra.

Day, R.W., 1974. Aptian ammonites from the Eromanga and Surat basins, Queensland. *Geological Survey of Queensland Publication* 360, *Palaeontological Paper* 34, 19 pp.

Day, R.W., Whitaker, W.G., Murray, C.G., Wilson, I.H. & Grimes, K.G., 1983. Queensland geology: a companion volume to the 1: 2 500 000 scale geological map (1975). *Geological Survey of Queensland, Publication* 383, 1–194.

Dear, J.F., 1968. The geology of the Cania district. *Publications of the Geological Survey of Queensland* 330, 1–27.

Dear, J.F., 1972. Preliminary biostratigraphic subdivision of the Permian brachiopod faunas in the northern Bowen Basin and their distribution throughout the basin. *Geological Survey of Queensland, Report* 49, 1–19.

Dear, J.F., McKellar, R.G. & Tucker, R.M., 1971. Geology of the Monto 1:250 000 sheet area. *Geological Survey of Queensland, Report* 46, 1–124.

Debrand-Passard, S., Enay, R. & Rioult, M. (eds), 1980. Jurassique supérieur. *Mémoire du Bureau de Recherches Géologiques et Minières* 101, 195–253.

Debrenne, F. & Gravestock, D.J., 1990. Archaeocyatha from the Sellick Hill Formation and Fork Tree Limestone on Fleurieu Peninsula, South Australia. pp. 290–309 *in* J.B. Jago & P.S. Moore (eds), The evolution of the Late Precambrian and Early Palaeozoic rift complex: Adelaide Geosyncline. *Geological Society of Australia Special Publication* 16.

Deflandre, G. & Cookson, I.C., 1955. Fossil microplankton from Australian Late Mesozoic and Tertiary sediments. *Australian Journal of Marine and Freshwater Research* 6, 242–313.

de Jersey, N.J., 1949. Principal microspores in the Ipswich coals. *Papers of the Department of Geology, University of Queensland* 3, 1–8.

de Jersey, N.J., 1962. Triassic spores and pollen grains from the Ipswich Coalfield. *Geological Survey of Queensland, Publication* 307, 1–18.

de Jersey, N.J., 1963. Jurassic spores and pollen grains from the Marburg Sandstone. *Geological Survey of Queensland, Publication* 13, 18 pp.

de Jersey, N.J., 1966. Devonian spores from the Adavale Basin. *Geological Survey of Queensland, Publication* 334, *Palaeontological Paper* 3, 1–28.

de Jersey, N.J., 1970. Triassic miospores from the Blackstone Formation, Aberdare Conglomerate and Raceview Formation. *Geological Survey of Queensland, Publication* 348, *Palaeontological Paper* 22, 1–41.

de Jersey, N.J., 1971. Triassic miospores from the Tivoli Formation and Kholi Sub-Group. *Geological Survey of Queensland, Publication* 353, *Palaeontological Paper* 28, 1–40.

de Jersey, N.J., 1975. Miospore zones in the Lower Mesozoic of south-eastern Queensland. pp. 159–72 *in* K.S.W. Campbell (ed.), *Gondwana Geology, Proceedings of the Third Gondwana Symposium, Canberra, Australia, 1973*, ANU Press, Canberra.

de Jersey, N.J., 1976. Palynology and time relationships in the lower Bundamba Group (Moreton Basin). *Queensland Government Mining Journal* 77, 461–65.

de Jersey, N.J., 1979. Palynology of the Permo-Triassic transition in the western Bowen Basin. *Geological Survey of Queensland, Publication* 374, *Palaeontological Paper* 46, 1–39.

de Jersey, N.J. & McKellar, J.L., 1981. Triassic palynology of the Warang Sandstone (northern Galilee Basin) and its phytogeographic implications. pp. 31–37 *in* M.M. Cresswell & P. Vella (eds), *Gondwana Five*, Balkema, Rotterdam.

de Jersey, N.J. & Raine, J.I., 1990. Triassic and earliest Jurassic miospores from the Murihiku Supergroup, New Zealand. *New Zealand Geological Survey, Palaeontological Bulletin* 62, 164 pp.

De Koninck, L.G., 1877. Recherches sur les fossiles Paléozoiques de la Nouvelle-Galles du Sud (Australie). *Mémoires de Société Royale Science Liège*, series 2,6, and 7. (Translation by T.W.E. David, 1898. *Memoir, Geological Survey of New South Wales, Palaeontology* 6.)

Delepine, G., 1941. On upper Tournaisian goniatites from New South Wales, Australia. *Annals and Magazine of Natural History*, series 11, 7, 386–95.

Denmead, A.K., 1964. Notes on marine macrofossils with Triassic affinities from Maryborough Basin, Queensland. *Australian Journal of Science* 27(4), 117.

Dennison, J.M., 1985. Devonian eustatic fluctuations in Euramerica: discussion. *Geological Society of America Bulletin* 96, 1595–97.

Derycke, C., 1992. Microrestes de Sélaciens et autres vertébrés du Dévonien supérieur du Maroc. *Muséum National d'Histoire naturelle, Bulletin* (4), 14, 15–61.

De Souza, H.A.F., 1982. Age data from Scotland for the Carboniferous time scale. pp. 455–60 *in* G.S. Odin (ed.), q.v.

Dettmann, M.E., 1961. Lower Mesozoic megaspores from Tasmania and South Australia. *Micropaleontology* 7(1), 71–89.

Dettmann, M.E., 1963. Upper Mesozoic microfloras from south-eastern Australia. *Proceedings of the Royal Society of Victoria* 77, 1–138.

Dettmann, M.E., 1981. The Cretaceous flora. pp. 357–75 *in* A. Keast (ed.), *Ecological biogeography of Australia*, Junk, the Hague.

Dettmann, M.E., 1986. Early Cretaceous palynoflora of subsurface strata correlative with the Koonwarra Fossil Bed, Victoria. *Association of Australasian Palaeontologists Memoir* 3, 79–110.

Dettmann, M.E. & Douglas, J.G., 1988. Palaeontology. pp. 164–76 *in* J.G. Douglas & J.A. Ferguson (eds), q.v.

Dettmann, M.E. & Playford, G., 1969. Palynology of the Australian Cretaceous: a review. pp. 174–210 *in* K.S.W. Campbell (ed.), *Stratigraphy and Palaeontology*, ANU Press, Canberra.

Dettmann, M.E., Molnar, R.E., Douglas, J.G., Burger, D., Fielding, C., Clifford, H.T., Francis, J., Jell, P., Rich, T., Wade, M., Rich, P.V., Pledge, N., Kemp, A. & Rozefelds, A., 1992. Australian Cretaceous terrestrial faunas and floras: biostratigraphic and biogeographic implications. *Cretaceous Research* 13, 207–62.

Deutsch, E.R., 1980. Magnetism of the Mid-Ordovician Tramore Volcanics, SE Ireland, and the question of a wide Proto-Atlantic Ocean. *Journal of Geomagnetism and Geoelectricity, Supplement* III, 32, 77–98.

Deutsch, E.R. & Somayajulu, C., 1973. Palaeomagnetism of Ordovician ignimbrites from Killary Harbour, Eire. *Earth and Planetary Science Letters* 7, 337–45.

Dickins, J.M., 1963. Permian pelecypods and gastropods from Western Australia. *Bureau of Mineral Resources, Geology & Geophysics, Bulletin* 63, 203 pp.

Dickins, J.M., 1968. Correlation of the Permian of the Hunter Valley, New South Wales, and the Bowen Basin, Queensland. *Bureau of Mineral Resources, Geology & Geophysics, Bulletin* 80, 27–44.

Dickins, J.M., 1969. Permian marine macrofossils from the Springsure Sheet area. *Bureau of Mineral Resources, Geology & Geophysics, Report* 123, Appendix 1, 72–94.

Dickins, J.M., 1976. Correlation Chart for the Permian System of Australia. *Bureau of Mineral Resources, Geology & Geophysics, Bulletin* 156B, 1–26.

Dickins, J.M., 1983. The Permian Blenheim Subgroup of the Bowen Basin and its time relationships. pp. 269–74 *in Permian Geology of Queensland*, Geological Society of Australia, Queensland Division.

Dickins, J.M., 1984. Evolution and climate in the Upper Palaeozoic. pp. 317–27 *in* P. Brenchley (ed.), *Fossils and Climate*, John Wiley & Sons.

Dickins, J.M., 1985. Late Palaeozoic glaciation. *BMR Journal of Australian Geology & Geophysics* 9, 163–69.

Dickins, J.M., 1989. Youngest Permian marine macrofossil fauna from the Bowen and Sydney basins, eastern Australia. *BMR Journal of Australian Geology & Geophysics* 11, 63–79.

Dickins, J.M. & Campbell, H.J., 1992. Permo–Triassic boundary in Australia and New Zealand. pp. 175–78 *in* W.C. Sweet, Yang Zunyi, J.M. Dickins & Yin Hongfu (eds), *Permo–Triassic Events in the Eastern Tethys—Stratigraphy, Classification, and Relations with Western Tethys*, Cambridge University Press, Cambridge.

Dickins, J.M. & Malone, E.J., 1968. Regional subdivision of the Back Creek Group (= Middle Bowen beds), Bowen Basin, Queensland. *Queensland Government Mining Journal* 69, 292–301.

Dickins, J.M. & Malone, E.J., 1973. Geology of the Bowen Basin, Queensland. *Bureau of Mineral Resources, Geology & Geophysics, Bulletin* 130, 1–154.

Dickins, J.M. & McTavish, R.A., 1963. Lower Triassic marine fossils from the Beagle Ridge (BMR 10) Bore, Perth Basin, Western Australia. *Journal of the Geological Society of Australia* 10(1), 123–40.

Dickins, J.M. & Shah, S.C., 1981. Permian paleogeography of peninsular and Himalayan India and the relationship with the Tethyan region. pp. 79–83 *in* M.M. Cresswell & P. Vella (eds.), *Gondwana Five*, Balkema, Rotterdam.

Dickins, J.M. & Shah, S.C., 1987. The relationship of the Indian and Western Australian Permian marine faunas. pp. 15–20 *in* G.D. McKenzie (ed.), Gondwana Six. *American Geophysical Union, Geophysical Monograph* 41.

Dickins, J.M., Gostin, V.A. & Runnegar, B., 1969. The age of the Permian sequence in the southern part of the Sydney Basin. pp. 211–15 *in* K.S.W. Campbell (ed.), *Stratigraphy and Paleontology*, ANU Press, Canberra.

Dickins, J.M., Archbold, N.W., Thomas, G.A. & Campbell, H.J., 1989. Mid-Permian correlation. pp. 185–98 *in* Jin Yugan & Li Chun (eds), *Compte Rendu 11ème Congrès International de Stratigraphie et de Géologie du Carbonifère* 2.

DiVenere, V.J. & Opdyke, N.D., 1988. A Mississippian magnetic stratigraphy: Maringouin Peninsula, New Brunswick. *EOS* 69, 340.

DiVenere, V.J. & Opdyke, N.D., 1990. Paleomagnetism of the Maringouin and Shepody formations, New Brunswick: a Namurian magnetic stratigraphy. *Canadian Journal of Earth Sciences* 27, 803–10.

DiVenere, V.J. & Opdyke, N.D., 1991*a*. Magnetic polarity stratigraphy and Carboniferous paleopole positions from the Joggins section, Cumberland structural basin, Nova Scotia. *Journal of Geophysical Research* 96, 4051–64.

DiVenere, V.J. & Opdyke, N.D., 1991*b*. Magnetic polarity stratigraphy in the uppermost Mississippian Mauch Chunk Formation, Pottsville, Pennsylvania. *Geology* 19, 127–30.

Dodds, B., 1949. Mid-Triassic Blattoidea from the Mount Crosby insect bed. *Papers of the Department of Geology, University of Queensland* 3(10), 1–11.

Dolby, J.H. & Balme, B.E., 1976. Triassic palynology of the Carnarvon Basin, Western Australia. *Review of Palaeobotany and Palynology* 22, 105–68.

Donovan, D.T. & Jones, E.J.W., 1979. Causes of world-wide changes in sea-level. *Journal of the Geological Society of London* 136, 187–92.

Donze, P., 1973. Ostracod migrations from the Mesogean to Boreal provinces in the European Lower Cretaceous. pp. 155–160 *in* R. Casey & P.F. Rawson (eds), The Boreal Lower Cretaceous. *Geological Journal Special Issue* 5.

Douglas, J.G., 1969. The Mesozoic floras of Victoria. Parts 1 and 2. *Geological Survey of Victoria Memoir* 28, 310 pp.

Douglas, J.G., 1973. The Mesozoic floras of Victoria. Part 3. *Geological Survey of Victoria Memoir* 29, 185 pp.

Douglas, J.G. & Ferguson, J.A. (eds), 1976. *Geology of Victoria*, 1st edition. Geological Society of Australia Special Publication 5, 528 pp.

Douglas, J.G. & Ferguson, J.A. (eds), 1988. *Geology of Victoria*, 2nd edition, Geological Society of Australia Victorian Division, Melbourne, 663 pp.

Douglas, J.G., Abele, C., Benedek, S., Dettmann, M.E., Kenley, P.R., Lawrence, C.R., Rich, T.H.V. & Rich, P.V., 1988. Mesozoic. pp. 213–50 *in* J.G. Douglas & J.A. Ferguson (eds), q.v.

Douglass, D.N., 1988. Palaeomagnetics of Ringerike Old Red Sandstone and related rocks, southern Norway: implications for pre-Carboniferous separation of Baltica and British terranes. *Tectonophysics* 148, 11–27.

Douglass, R.C., 1987. Fusulinid biostratigraphy and correlations between the Appalachian and Eastern interior basins. *US Geological Survey Professional Paper* 1451, 1–95.

Dowsett, H.J., 1989. Application of the graphic correlation method to Pliocene marine sequences. *Marine Micropaleontology* 14, 3–32.

Draper, J.J., 1983. Origin of pebbles in mudstones in the Denison Trough. pp. 305–16 *in Permian Geology of Queensland*, Geological Society of Australia, Queensland Division.

Draper, J.J., Palmieri, S.M., Parfrey, S.M. & Rigby, J.F., 1990. Early Permian faunas from the Cracow area, South-eastern Bowen Basin, Queensland. *Geological Society of Australia, Tenth Australian Geological Convention, Abstracts*, 52–53.

Druce, E.C., 1969. Devonian and Carboniferous conodonts from the Bonaparte Gulf Basin, northern Australia and their use in international correlation. *Bureau of Mineral Resources, Geology & Geophysics, Bulletin* 98, 242 pp.

Druce, E.C., 1970*a*. Frasnian conodonts from Mount Morgan, Queensland. *Bureau of Mineral Resources, Geology & Geophysics, Bulletin* 108, 75–84.

Druce, E.C., 1970*b*. Lower Carboniferous conodonts from the northern Yarrol Basin, Queensland. *Bureau of Mineral Resources, Geology & Geophysics, Bulletin* 108, 91–105.

Druce, E.C., 1974. Australian Devonian and Carboniferous conodont faunas. pp. 1–18 *in* J. Bouckaert & M. Streel (eds), *International Symposium on Belgian Micropalaeontological Limits from Emsian to Viséan. Namur, 1974, Publication* 5.

Druce, E.C., 1976. Conodont biostratigraphy of the Upper Devonian reef complexes of the Canning Basin, Western Australia. *Bureau of Mineral Resources, Geology & Geophysics, Bulletin* 158, 303 pp.

Druce, E.C., 1978*a*. *Clavohamulus primitus*—a key North American conodont found in the Georgina Basin. *BMR Journal of Australian Geology & Geophysics* 3, 351–55.

Druce, E.C., 1978*b*. Correlation of the Cambrian/Ordovician boundary in Australia. *Bureau of Mineral Resources, Geology & Geophysics, Bulletin* 192, 49–60.

Druce, E.C. & Jones, P.J., 1968. Stratigraphical significance of conodonts in the Upper Cambrian and Lower Ordovician sequence of the Boulia region, western Queensland. *Australian Journal of Science* 31 (2), 88.

Druce, E.C. & Jones, P.J., 1971. The Cambro–Ordovician conodonts from the Burke River Structural Belt, Queensland. *Bureau of Mineral Resources, Geology & Geophysics, Bulletin* 110, 158 pp.

Druce, E.C., Shergold, J.H. & Radke, B.M., 1982. A reassessment of the Cambrian–Ordovician boundary section at Black Mountain, western Queensland, Australia. pp. 193–209 *in* M.G. Bassett & W.T. Dean (eds), The Cambrian–Ordovician boundary: sections, fossil distributions, and correlations. *National Museum of Wales, Geological Series*, 3.

Dumoulin, J.A. & Bown, P.R., 1992. Depositional history, nannofossil biostratigraphy, and correlation of the Argo Abyssal Plain sites 765 and 261. *Proceedings of the Ocean Drilling Program, Scientific Results* 123, 3–56.

Dun, W.S., 1902. Notes on some Carboniferous brachiopods from Clarencetown. *Geological Survey of New South Wales, Record* 7(2), 72–88.

Dun, W.S. & Benson, W.N., 1920. The geology, palaeontology and petrography of the Currabubula District, with notes on adjacent regions. *Proceedings of the Linnean Society of New South Wales* 45, 337–74.

Dziewa, T.J., 1980. Early Triassic osteichthyans from the Knocklofty Formation of Tasmania. *Papers and Proceedings of the Royal Society of Tasmania* 114, 145–60.

Edgell, H.S., 1964. Triassic ammonite impressions from the type section of the Minchin Siltstone, Perth Basin. *Geological Survey of Western Australia, Annual Report for 1963*, 55–57.

Edwards, A.R., Hornibrook, N. de B., Raine, J.I., Scott, G.H., Stevens, G.R., Strong, C.P. & Wilson, G.J., 1989. A New Zealand Cretaceous–Cenozoic geological time scale. *New Zealand Geological Survey Record* 35, 135–49.

Edwards, L.E., 1982. Quantitative biostratigraphy: the methods should suit the data. pp. 45–60 *in* J.M. Cubitt & R.A. Reyment (eds), *Quantitative Stratigraphic Correlation*, Wiley & Sons.

Edwards, L.E., 1984. Insights on why graphic correlation (Shaw's method) works. *Journal of Geology* 92, 583–97.

Edwards, L.E., 1985. Insights on why graphic correlation (Shaw's method) works: a reply. *Journal of Geology* 93, 507–9.

Edwards, L.E., 1989. Supplemented graphic correlation: a powerful tool for paleontologists and non-paleontologists. *Palaios* 4, 127–143.

Egerton, P., 1864. On some ichthyolites from New South Wales forwarded by the Rev. W.B. Clarke. *Quarterly Journal of the Geological Society of London* 20, 1–20.

Eisenack, A. & Cookson, I.C., 1960. Microplankton from Australian Cretaceous sediments. *Proceedings of the Royal Society of Victoria* 72, 1–11.

Elphinstone, R. & Walter, M.R., 1991. Late Proterozoic and Early Cambrian trace fossils. *Bureau of Mineral Resources, Geology & Geophysics, Bulletin* 236, 97–111.

Embleton, B.J., 1972. The palaeomagnetism of some Palaeozoic sediments from central Australia. *Journal and Proceedings of the Royal Society of New South Wales* 105, 86–93.

Embleton, B.J. & McDonnell, K.L., 1980. Magnetostratigraphy in the Sydney Basin, south-eastern Australia. *Journal of Geomagnetism and Geoelectricity* 32, Suppl. III, 1–10.

Enay, R. & Geyssant, J.R., 1975. Faunes tithoniques des chaines bétiques (Espagne méridionale). *Mémoire du Bureau de Recherches Géologiques et Minières* 86, 39–55.

Engel, B.A., 1975. A new ?bryozoan from the Carboniferous of eastern Australia. *Palaeontology* 18, 571–605.

Engel, B.A., 1980. Carboniferous faunal zones in eastern Australia. *Bulletin of the Geological Survey of New South Wales* 26, 326–39.

Engel, B.A., 1989. Carboniferous bryozoans as stratigraphic indicators in eastern Australia. *Compte Rendu 11ème Congrès International de Stratigraphie et de Géologie du Carbonifère, Beijing, 1987*, 3, 33–40.

Engel, B.A. & Morris, N., 1975. *Linguaphillipsia* (Trilobita) in the Carboniferous of eastern Australia. *Senckenbergiana lethaea* 56, 147–85.

Engel, B.A. & Morris, N., 1980. New Cyrtosymbolinae (Trilobita) from the Lower Carboniferous of eastern Australia. *Senckenbergiana lethaea* 60, 265–89.

Engel, B.A. & Morris, N., 1983. *Phillipsia–Weberiphillipsia* in the Early Carboniferous of eastern New South Wales. *Alcheringa* 7, 223–51.

Engel, B.A. & Morris, N., 1984. *Conophillipsia* (Trilobita) in the Early Carboniferous of eastern Australia. *Alcheringa* 8, 23–64.

Engel, B.A. & Morris, L.N., 1985. The biostratigraphy of Carboniferous trilobites in eastern Australia. *Compte Rendu 10ème Congrès International de Stratigraphie et de Géologie du Carbonifère, Madrid, 1983*, 4, 491–499.

Engel, B.A. & Morris, N., 1989. Early Carboniferous trilobites (Weaniinae) of eastern Australia. *Alcheringa* 13, 305–46.

Engel, B.A. & Morris, N., 1991*a* [1990 imprint]. The biostratigraphic potential of Early Carboniferous trilobites from eastern Australia. *Courier Forschungsinstitut Senckenberg* 130, 189–98.

Engel, B.A. & Morris, N., 1991*b*. Aulacopleuridae and Brachymetopidae (Trilobita) from the Lower Carboniferous of eastern Australia (i) *Namuropyge* and *Brachymetopus* (*Spinimetopus*). *Geologica et Palaeontologica* 25, 123–35.

Englund, K.J., Henry, T.W., Gillespie, W.H., Pfefferkorn, H.W. & Mackenzie Gordon Jr, 1985. Boundary stratotype for the base of the Pennsylvanian System, east-central Appalachian Basin, USA. *Compte Rendu 10ème Congrès International de Stratigraphie et de Géologie du Carbonifère, Madrid, 1983*, 4, 371–82.

Erdtmann, B.D. & VandenBerg, A.H.M., 1985. *Araneograptus* gen. nov. and its two species from the Late Tremadocian (Lancefieldian La2) of Victoria. *Alcheringa* 9, 49–63.

Etheridge, R. Jr, 1874. Observations on a few graptolites from the Lower Silurian rocks of Victoria, Australia. *Annals and Magazine of Natural History* 14(4), 1–10.

Etheridge, R. Jr, 1883. A description of the remains of trilobites from the Lower Silurian rocks of the Mersey River district, Tasmania. *Papers and Proceedings of the Royal Society of Tasmania, 1882*, 150–62.

Etheridge, R. Jr, 1888. The invertebrate fauna of the Hawkesbury–Wianamatta Series (beds above the productive coal measures) of New South Wales. *Geological Survey of New South Wales Memoirs* (*Palaeontology*) 1, 1–21.

Etheridge, R. Jr, 1890*a*. On some Australian species of the Family Archaeocyathinae. *Transactions of the Royal Society of South Australia* 13, 10–22.

Etheridge, R. Jr, 1890*b*. Carboniferous mollusca in the Lower Carboniferous or *Rhacopteris* Series of the Port Stephens district, New South Wales. *Department of Mines, New South Wales, Annual Report for 1889*, 239.

Etheridge, R. Jr, 1890*c*. Additional Carboniferous mollusca in the Lower Carboniferous Series of the Port Stephens district, New South Wales. *Department of Mines, New South Wales, Annual Report for 1889*, 240.

Etheridge, R. Jr, 1891*a*. On the occurrence of an *Orthis*, allied to *O. actoniae*, J. de C. Sby., and *O. flabellulum*, J. de C. Sby., in the Lower Silurian rocks of central Australia. *South Australian Parliamentary Papers* 158, 13–14, pl. 1.

Etheridge, R. Jr, 1891*b*. Descriptions of some South Australian Silurian and Mesozoic fossils. pp. 9–12 *in* H.Y.L. Brown, Reports on coal-bearing area in neighbourhood of Leigh's Creek, etc. *South Australian Parliamentary Paper* 158.

Etheridge, R. Jr, 1892. On a species of *Asaphus* from the Lower Silurian rocks of Central Australia. pp. 8–9 *in* H.Y.L. Brown, Report on Leigh Creek and Hergott Springs. *South Australian Parliamentary Paper* 23.

Etheridge, R. Jr, 1893. Additional Silurian and Mesozoic fossils from Central Australia, being a supplement to Parliamentary Papers no. 158 of 1891 and no. 23 of 1892. *South Australian Parliamentary Paper* 52, 5–8.

Etheridge, R. Jr, 1894. Further additions to the Lower Silurian fauna of Central Australia. pp. 23–26 *in* H.Y.L. Brown, Report of government geologists for the year ended June 30th, 1894. *South Australian Parliamentary Paper* 25.

Etheridge, R. Jr, 1895. On the occurrence of a stromatoporoid, allied to *Labechia* and *Rosenella*, in Siluro–Devonian rocks of New South Wales. *Records of the Geological Survey of New South Wales* 4, pp. 134–40.

Etheridge, R. Jr, 1896*a*. Palaeontologia Novae Cambriae Meridionalis. Occasional descriptions of New South Wales fossils, no. 2. *Records of the Geological Survey of New South Wales* 5(1), 14–18.

Etheridge, R. Jr, 1896*b*. Evidence of the existence of a Cambrian fauna in Victoria. *Proceedings of the Royal Society of Victoria* 8, 52–64.

Etheridge, R. Jr, 1897. Official contributions to the palaeontology of South Australia. No. 9, On the occurrence of *Olenellus* in the Northern Territory. *South Australian Parliamentary Paper (1896)*, 13–16.

Etheridge, R. Jr, 1898. Palaeontologia Novae Cambriae Meridionalis. Occasional descriptions of New

South Wales fossils, no. 3. *Records of the Geological Survey of New South Wales* 5(4), 175–79.

Etheridge, R. Jr, 1901. *Ctenostreon pectiniformis* Schlotheim, an Australian fossil. *Records of the Australian Museum* 4, 13.

Etheridge, R. Jr, 1902. Official contributions to the palaeontology of South Australia. No. 12, Evidence of further Cambrian trilobites. *South Australian Parliamentary Paper (1902)*, 3–4.

Etheridge, R. Jr, 1904*a*. A monograph of the Silurian and Devonian corals of New South Wales, with illustrations from other parts of Australia. Part I—the genus *Halysites*. *Geological Survey of New South Wales, Palaeontology Memoir* 13, 39 pp.

Etheridge, R. Jr, 1904*b*. Trilobite remains collected in the Florentine Valley, west Tasmania, by Mr T. Stephens, M.A. *Records of the Australian Museum* 5, 98–101.

Etheridge, R. Jr, 1905. Additions to the Cambrian fauna of South Australia. *Transactions and Proceedings of the Royal Society of South Australia* 29, 246–51.

Etheridge, R. Jr, 1907. Palaeontologia Novae Cambriae Meridionalis. Occasional descriptions of New South Wales fossils, no. 5. *Records of the Geological Survey of New South Wales* 8(3), 192–96.

Etheridge, R. Jr, 1909. An organism allied to *Mitcheldeania* Wethered, of the Carboniferous Limestone, in the Upper Silurian of Malongulli. *Records of the Geological Survey of New South Wales* 8, 308–11.

Etheridge, R. Jr, 1910. Oolitic fossils from the Greenough River district, Western Australia. *Bulletin of the Geological Survey of Western Australia* 36, 29.

Etheridge, R. Jr, 1916. *Hyalostelia australia*, the anchoring spicules of an hexactinellid sponge from the Ordovician rocks of the MacDonnell Ranges, Central Australia. *Transactions of the Royal Society of South Australia* 40, 148–50.

Etheridge, R. Jr, 1919. The Cambrian trilobites of Australia and Tasmania. *Transactions and Proceedings of the Royal Society of South Australia* 43, 373–93.

Etheridge, R. Jr & Mitchell, J., 1890. On the identity of *Bronteus partschi*, de Koninck (non-Barrande), from the Upper Silurian rocks of New South Wales. *Proceedings of the Linnean Society of New South Wales* 5, 501–4.

Etheridge, R. Jr & Mitchell, J., 1892. The Silurian trilobites of New South Wales, with reference to those of other parts of Australia, Part I. The Proetidae. *Proceedings of the Linnean Society of New South Wales* 6, 311–20.

Etheridge, R. Jr & Mitchell, J., 1894. The Silurian trilobites of New South Wales, with reference to those of other parts of Australia, Part II. The genera *Proetus* and *Cyphaspis*. *Proceedings of the Linnean Society of New South Wales* 8, 169–78.

Etheridge, R. Jr & Mitchell, J., 1896. The Silurian trilobites of New South Wales, with reference to those of other parts of Australia, Part III. The Phacopidae. *Proceedings of the Linnean Society of New South Wales* 10, 486–511.

Etheridge, R. Jr & Mitchell, J., 1916. The Silurian trilobites of New South Wales, with reference to those of other parts of Australia, Part V. The Encrinuridae. *Proceedings of the Linnean Society of New South Wales* 40, 646–80.

Etheridge, R. Jr & Mitchell, J., 1917. The Silurian trilobites of New South Wales, with reference to those of other parts of Australia, Part VI. The Calymenidae, Cheiruridae, Harpeidae, Brontidae etc., with an appendix. *Proceedings of the Linnean Society of New South Wales* 42, 480–510.

Etheridge, R. Sr, 1872. Description of the Palaeozoic and Mesozoic fossils of Queensland. *Quarterly Journal of the Geological Society* 28, 317–50.

Ethington, R.L. & Clark, D.L., 1971. Lower Ordovician conodonts in North America. *Geological Society of America, Memoir* 127, 63–82.

Ethington, R.L. & Clark, D.L., 1981. Lower and Middle Ordovician conodonts from the Ibex area, western Millard County, Utah. *Brigham Young University Geology Studies* 28(2), 1–155.

Evans, J.W., 1956. Palaeozoic and Mesozoic Hemiptera (Insecta). *Australian Journal of Zoology* 4, 165–258.

Evans, J.W., 1961. Some Upper Triassic Hemiptera from Queensland. *Memoirs of the Queensland Museum* 14(1), 13–23.

Evans, J.W., 1963. The systematic position of the Ipsvidiidae (Upper Triassic Hemiptera) and some new Upper Permian and Middle Triassic Hemiptera from Australia. *Journal of the Entomological Society of Queensland* 2, 17–29.

Evans, J.W., 1971. Some Upper Triassic Hemiptera from Mount Crosby, Queensland. *Memoirs of the Queensland Museum* 16(1), 145–61.

Evans, P.R., 1966*a*. Mesozoic stratigraphic palynology in Australia. *Australasian Oil and Gas Journal* 12 (6), 58–63.

Evans, P.R., 1966*b*. Contributions to the palynology of northern Queensland and Papua. *Bureau of Mineral Resources, Geology & Geophysics, Record* 1966/19, 32 pp.

Evans, P.R., 1968. Upper Devonian and Lower Carboniferous miospores from the Mulga Downs Beds, NSW. *Australian Journal of Science* 31, 45–46.

Evitt, W.R., 1985. *Sporopollen in Dinoflagellate Cysts*, American Association of Stratigraphic Palynologists, 333 pp.

Exon, N.F. & Burger, D., 1981. Sedimentary cycles in the Surat Basin and global changes of sea level. *BMR Journal of Australian Geology & Geophysics* 6, 153–59.

Exon, N.F. & Colwell, J.B., 1994. Geological history of the outer North West Shelf of Australia: a synthesis. *AGSO Journal of Australian Geology and Geophysics* 15, 177–90.

Exon, N.F. & Williamson, P.E., 1988. Preliminary post-cruise report: *Rig Seismic* Research Cruises 7 & 8: sedimentary basin framework of the northern and western Exmouth Plateau. *Bureau of Mineral Resources, Geology & Geophysics, Record* 1988/30, 1–62.

Facer, R.A., 1981. Palaeomagnetic data for Permian and Triassic rocks from drill holes in the southern Sydney Basin, New South Wales. *Tectonophysics* 74, 305–21.

Fairfax, W., 1859. *Handbook to Australasia: Being a Brief Historical and Descriptive Account of Victoria, Tasmania, South Australia, New South Wales, Western Australia and New Zealand*, Melbourne, 244 pp., 1 map.

Faller, A.M., Briden, J.C. & Morris, W.A., 1977. Palaeomagnetic results from the Borrowdale Volcanic Group, English Lake District. *Geophysical Journal of the Royal Astronomical Society* 48, 111–21.

Fang, W.R., Van der Voo, R. & Liang, Q., 1990. Ordovician palaeomagnetism of eastern Yunnan, China. *Geophysical Research Letters* 17, 953–56.

Farmer, N., 1985. Geological Atlas 1:50 000 Series, Sheet 88, Kingsborough. *Geological Survey of Tasmania Explanatory Report*, 1–109.

Feistmantel, O., 1890. Geological and palaeontological relations of the coal and plant-bearing beds of Palaeozoic and Mesozoic age in eastern Australia and Tasmania. *Memoir of the Geological Survey of New South Wales, Palaeontology* 3, 11–183.

Fergusson, C.L., Cas, R.A., Collins, W.J., Craig, G.Y., Crook, K.A.W., Powell, C.McA., Scott, P.A. & Young, G.C., 1979. The Late Devonian Boyd Volcanic Complex, Eden, NSW. *Journal of the Geological Society of Australia* 26, 87–105.

Fielding, C.R. & McLoughlin, S., 1992. Sedimentology and palynostratigraphy of Permian rocks exposed at Fairbairn Dam, central Queensland. *Australian Journal of Earth Sciences* 39, 631–49.

Filatoff, J., 1975. Jurassic palynology of the Perth Basin, Western Australia. *Palaeontographica* B 154, 1–20.

Filatoff, J. & Price, P.L., 1988. A pteridacean spore lineage in the Australian Mesozoic. *Association of Australasian Palaeontologists Memoir* 5, 89–124.

Fisher, M.J., 1972. The Triassic palynofloral succession in England. *Geoscience and Man* 4, 101–9.

Fisher, M.J., 1979. The Triassic palynofloral succession in the Canadian Arctic. *American Association of Stratigraphic Palynologists, Contribution Series* 5B, 83–100.

Flajs, G. & Feist, R., 1988. Index conodonts, trilobites and environment of the Devonian–Carboniferous boundary beds at La Serre (Montagne Noire), France. *Courier Forschungsinstitut Senckenberg* 100, 53–107.

Flajs, G., Feist, R. & Ziegler, W. (eds), 1988. Devonian—Carboniferous boundary—results of recent studies. *Courier Forschungsinstitut Senckenberg* 100, 1–245.

Fleming, C.A., 1959. *Buchia plicata* (Zittel) and its allies, with a description of a new species, *Buchia hochstetteri*. *New Zealand Journal of Geology & Geophysics* 2, 889–904.

Fleming, P.J.G., 1966*a*. Eotriassic marine bivalves from the Maryborough Basin, south-east Queensland. *Geological Survey of Queensland, Publication* 333, *Palaeontological Paper* 20, 17–29.

Fleming, P.J.G., 1966*b*. Notes on Triassic fossils in the Parish of Chuwar. *Queensland Government Mining Journal* 67, 119–20.

Fleming, P.J.G., 1967. Names for Carboniferous and Permian formations of the Yarrol Basin in the Stanwell area, central Queensland. *Queensland Government Mining Journal* 68, 113–16.

Fleming, P.J.G., 1969. Fossils from the Neerkol Formation of central Queensland. pp. 264–75 *in* K.S.W.

Campbell (ed.), *Stratigraphy and Palaeontology*, ANU Press, Canberra.

Fleming, P.J.G., 1973. Bradoriids from the *Xystridura* Zone of the Georgina Basin, Queensland. *Geological Survey of Queensland, Publication* 356, *Palaeontological Papers* 31, 1–9.

Fletcher, H.O., 1964. New linguloid shells from Lower Ordovician and Middle Palaeozoic rocks of New South Wales. *Records of the Australian Museum* 26(10), 283–94, pls 31–32.

Fletcher, H.O., 1975. Silurian and Lower Devonian fossils from the Cobar area of New South Wales. *Records of the Australian Museum* 30, 63–85.

Flood, P.G. & Aitchison, J.C., 1992. Late Devonian accretion of the Gamilaroi terrane to Gondwana: provenance linkage provided by quartzite clasts in the overlap sequence. *Australian Journal of Earth Sciences* 39, 539–44.

Foord, A.H., 1890. Notes on the palaeontology of Western Australia. Descriptions of fossils from the Kimberley district, Western Australia. *Geological Magazine*, Decade 3, 7, 98–106, 145–55.

Fordham, B., 1992. Chronometric calibration of mid-Ordovician to Tournaisian conodont zones: a compilation from recent graphic-correlation and isotope studies. *Geological Magazine* 129, 709–21.

Forster, S.C. & Warrington, G., 1985 Geochronology of the Carboniferous, Permian and Triassic. pp. 99–113 *in* N.J. Snelling (ed.), q.v.

Fortey, R.A. & Owens, R.M., 1987. The Arenig in South Wales. *Bulletin of the British Museum of Natural History* 36, 69–307.

Fortey, R.A. & Shergold, J.H., 1984. Early Ordovician trilobites, Nora Formation, central Australia. *Palaeontology* 27, 315–66.

Fortey, R.A., Harper, D.A.T., Ingham, J.K., Owen, A.W. & Rushton, A.W.A., 1995. A revision of Ordovician series and stages from the historical type area. *Geological Magazine* 132, 15–30.

Foster, C.B., 1979. Permian plant microfossils of the Blair Athol Coal Measures, Baralaba Coal Measures, and basal Rewan Formation of Queensland. *Geological Survey of Queensland, Publication* 372, *Palaeontological Paper* 45, 244 pp.

Foster, C.B., 1982*a*. Biostratigraphic potential of Permian spore–pollen floras from GSQ Mundubbera 5 & 6, Taroom Trough. *Queensland Government Mining Journal* 83(964), 82–96.

Foster, C.B., 1982*b*. Spore–pollen assemblages of the Bowen Basin, Queensland, Australia: their relationship to the Permian/Triassic boundary. *Review of Palaeobotany and Palynology* 36, 165–83.

Foster, C.B., 1983. Illustrations of Early Tertiary (Eocene) plant microfossils from the Yaamba Basin, Queensland. *Geological Survey of Queensland, Publication* 381, 1–32.

Foster, C.B., 1989. Palynological report. Elf Aquitaine Barnett No. 2, Bonaparte Basin. (unpublished report, Elf Aquitaine) 30 pp.

Foster, C.B., 1992. A review of eastern Australian palynozones. *Bureau of Mineral Resources, Geology & Geophysics, Record* 1992/5, 1–11.

Foster, C.B. & Jones, P.J., 1994. Correlation between Australia and the type Tartarian, Russian Platform, evidence from palynology and Conchostraia: a discussion. *Permophiles* 24, 36–43.

Foster, C.B. & Waterhouse, J.B., 1988. The *Granulatisporites confluens* Oppel-zone and Early Permian marine faunas from the Grant Formation on the Barbwire Terrace, Canning Basin, Western Australia. *Australian Journal of Earth Sciences* 35, 135–57.

Foster, C.B., Cernovskis, A. & O'Brien, G.W., 1985. Organic-walled microfossils from the Early Cambrian of South Australia. *Alcheringa* 9, 259–68.

Frakes, L.A., Burger, D., Apthorpe, M., Wiseman, J., Dettmann, M.E., Alley, N.F., Flint, R., Gravestock, D., Ludbrook, N.H., Backhouse, J., Skwarko, S., Scheibnerová, V., McMinn, A., Moore, P.S., Bolton, B.R., Douglas, J.G., Christ, R., Wade, M., Molnar, R.E., McGowran, B., Balme, B.E. & Day, R.W., 1987. Australian Cretaceous shorelines, stage by stage. *Palaeogeography, Palaeoclimatology, Palaeoecology* 59, 31–48.

French, A.N. & Van der Voo, R., 1979. The magnetization of the Rose Hill Formation at the classical site of Graham's fold test. *Journal of Geophysical Research* 84, 7688–96.

French, R.B. & Van der Voo, R., 1977. Remagnetisation problems with the palaeomagnetism of the Middle Silurian Rose Hill Formation of the central Appalachians. *Journal of Geophysical Research* 82, 5803–06.

Freytag, I.B., 1964. Reptilian vertebrate remnants from Lower Cretaceous strata near Oodnadatta. *Geological Survey of South Australia, Quarterly Geological Notes* 10, 1–2.

Galbrun, B., 1985. Magnetostratigraphy of the Berriasian stratotype section (Berrias, France). *Earth and Planetary Science Letters* 74, 130–36.

Galbrun, B., 1992*a*. Triassic (Upper Carnian–Lower Rhaetian) magnetostratigraphy of Leg 122 sediments, Wombat Plateau, north-west Australia. *Proceedings of the Ocean Drilling Program, Scientific Results* 122, 685–93.

Galbrun, B., 1992*b*. Magnetostratigraphy of Upper Cretaceous and Lower Tertiary sediments, Site 761 and 762, Exmouth Plateau, north-west Australia. *Proceedings of the Ocean Drilling Program, Scientific Results* 122, 699–716.

Galbrun, B., Crasquin-Soleau, S. & Jaugey, J.M., 1992. Magnétostratigraphie des sédiments Triasiques des sites 759 et 760, ODP Leg 122, Plateau de Wombat, Nord-Ouest de l'Australie. *Marine Geology* 107, 293–98.

Galbrun, B., Gabilly, J. & Rasplus, L., 1988. Magnetostratigraphy of the Toarcian stratotype sections at Thouars and Airvault (Deux Sèvres, France). *Earth and Planetary Science Letters* 87, 453–62.

Galbrun, B., Baudin, F., Fourcade, E. & Rivas, P., 1990. Magnetostratigraphy of the Toarcian ammonitico rosso Limestone at Iznalloz, Spain. *Geophysical Research Letters* 17(13), 2441–44.

Gales, J.E., Van der Pluijm, B.A. & Van der Voo, R., 1989. Palaeomagnetism of the Lawrenceton Formation, volcanic rocks, Silurian Botwood Group, Change Islands, Newfoundland. *Canadian Journal of Earth Sciences* 26, 296–304.

Gallet, Y., Besse, J., Krystyn, L., Marcoux, J. & Théveniaut, H., 1992. Magnetostratigraphy of the Late Triassic Bolücektasi Tepe section (south-western Turkey): implications for changes in magnetic reversal frequency. *Physics of the Earth and Planetary Interiors* 93, 273–82.

Gallet, Y., Besse, J., Krystyn, L., Théveniaut, H. & Marcoux, J., 1993. Magnetostratigraphy of the Kavur Tepe section (south-western Turkey): A magnetic polarity time scale for the Norian. *Earth and Planetary Science Letters* 117, 443–56.

Gao, E.F., Huang, H.L., Zhu, Z.W., Liu, H.S., Fan, Y.Q. & Qing, Z.J., 1993. The study of paleomagnetism in northeastern Sino–Korean Massif during pre-Late Paleozoic (in Chinese). pp. 265–74 *in Contributions to Project of Plate Tectonics of Northern China*, 1, Geological Publishing House, Beijing.

Garratt, M.J., 1978. New evidence for a Silurian (Ludlow) age for the earliest *Baragwanathia* flora. *Alcheringa* 2, 217–24.

Garratt, M.J., 1980. Siluro–Devonian Notanopliidae (Brachiopoda). *National Museum of Victoria, Memoir* 41, 15–41.

Garratt, M.J., 1983*a*. Silurian and Devonian biostratigraphy of the Melbourne Trough, Victoria. *Proceedings of the Royal Society of Victoria* 95(2), 77–98.

Garratt, M.J., 1983*b*. Silurian to Early Devonian facies and biofacies patterns for the Melbourne Trough, central Victoria. *Journal of the Geological Society of Australia* 30, 121–47.

Garratt, M.J. & Rickards, R.B., 1984. Graptolite biostratigraphy of early land plants from Victoria, Australia. *Proceedings of the Yorkshire Geological Society* 44, 377–84.

Garratt, M.J. & Rickards, R.B., 1987. Přídolí (Silurian) graptolites in association with *Baragwanathia* (*Lycophytina*). *Geological Society of Denmark, Bulletin* 35, 135–39.

Garratt, M.J. & Wright, A.J., 1989. Late Silurian to Early Devonian biostratigraphy of southeastern Australia. pp. 647–62 *in* N.J. McMillan, A.F. Embry & D.J. Glass (eds), vol. 3, q.v.

Gatehouse, C.G., 1986. The geology of the Warburton Basin in South Australia. *Australian Journal of Earth Sciences* 33, 161–80.

Geiger, M.E. & Hopping, C.A., 1968. Triassic stratigraphy of the North Sea. *Philosophical Transactions of the Royal Society, Series B* 254, 1–36.

Geological Society of London, 1964. The Phanerozoic time scale. *Geological Society of London Quarterly Journal* 120 Supplement, 260–62.

George, T.N., Johnson, G.A.L., Mitchell, M., Prentice, J.E., Ramsbottom, W.H.C., Sevastopulo, G.D. & Wilson, R.B., 1976. A correlation of Dinantian rocks in the British Isles. *Special Report of the Geological Society of London* 7, 1–87.

George, T.N., Harland, W.B., Ager, D.V., Ball, H.W., Blow, W.H., Casey, R., Holland, C.H., Hughes, N.F., Kellaway, G.A., Kent, P.E., Ramsbottom, W.H.C., Stubblefield, J. & Woodland, A.W., 1969. Recommendations on stratigraphical usage. *Proceedings of the Geological Society of London* 1656, 139–66.

Gilbert-Tomlinson, J., 1973. The Lower Ordovician gastropod *Teiichispira* in northern Australia.

Bureau of Mineral Resources, Geology & Geophysics, Bulletin 126, 65–88.

Gill, E.D., 1940. A new trilobite from Cootamundra, NSW. *Proceedings of the Royal Society of Victoria* 52, 106–10.

Gill, E.D., 1945. Trilobita of the family Calymenidae from the Palaeozoic rocks of Victoria. *Proceedings of the Royal Society of Victoria* 56, 172–86.

Gill, E.D., 1948. Eldon Group fossils from the Lyell Highway, western Tasmania. *Records of the Queen Victoria Museum* 2, 57–74.

Gill, E.D., 1949. Palaeozoology and taxonomy of some Australian homalonotid trilobites. *Proceedings of the Royal Society of Victoria* 61, 61–73.

Ginter, M. & Ivanov, A., 1992. Devonian phoebodont shark teeth. *Acta Palaeontologica Polonica* 37, 55–75.

Glaessner, M.F., 1979. Lower Cambrian Crustacea and annelid worms from Kangaroo Island, South Australia. *Alcheringa* 3, 21–31.

Glass, B.P., Hall, C.M. & York, D., 1986. $^{40}Ar/^{39}Ar$ laser-probe dating of North American tektite fragments from Barbados and the age of the Eocene–Oligocene boundary. *Chemical Geology (Isotope Science Section)* 59, 181–86.

Glen, R.A., Macrae, G.P., Pogson, D.J., Scheibner, E., Agostini, A. & Sherwin, L., 1985. Summary of the geology and controls of mineralization in the Cobar region. *Department of Mineral Resources, New South Wales Geological Survey Report* GS 1985/203.

Glenister, B.F., 1958. Upper Devonian ammonoids from the *Manticoceras* Zone, Fitzroy Basin, Western Australia. *Journal of Paleontology* 32, 58–96.

Glenister, B.F., 1960. Carboniferous conodonts and ammonoids from Western Australia. *Compte Rendu 4ème Congrès International de Stratigraphie et du Géologie du Carbonifère, Heerlen, 1958* 1, 213–17.

Glenister, B.F. & Furnish, W.M., 1961. The Permian ammonoids of Australia. *Journal of Paleontology* 35, 673–736.

Glenister, B.F. & Glenister, A.T., 1958. Discovery of subsurface Ordovician strata, Broome area, Western Australia. *Australian Journal of Science* 20, 183–84.

Glenister, B.F., Windle, D.L. Jr & Furnish, W.M., 1973. Australian Metalegoceratidae (Lower Permian ammonoids). *Journal of Paleontology* 47, 1031–43.

Glenister, B.F., Baker, C., Furnish, W.M. & Dickins, J.M., 1990*a*. Late Permian ammonoid cephalopod *Cyclolobus* from Western Australia. *Journal of Paleontology* 64(3), 399–402.

Glenister, B.F., Baker, C., Furnish, W.M. & Thomas, G.A., 1990*b*. Additional Early Permian ammonoid cephalopods from Western Australia. *Journal of Paleontology* 64(3), 392–99.

Gomankov, A.V., 1988. Comment on correlation chart of the Upper Permian. *Permophiles* 13, 17–19.

Gonzalez, C.R., 1985. Esquema biostratigrafico del Paleozoico Superior marino de la Cuence, Uspallata-Iglesia, Re. Argentina. *Acta Geologica Lilloana* 16(2), 231–44.

Gonzalez, C.R., 1989. Relaciones biostratigraficas del Paleozoico Superior marino en el Gondwana Sudamericano. *Acta Geologica Lilloana* 17(1), 5–20.

Gonzalez, C.R., 1990. Development of the Late Paleozoic glaciations of the South American Gondwana in western Argentina. *Palaeogeography, Palaeoclimatology, Palaeoecology* 79, 275–87.

Gonzalez, C.R., 1993. Late Paleozoic faunal succession in Argentina. *Compte Rendus 12ème Congrès International de la Stratigraphie et Géologie du Carbonifère et Permien* 1, 537–50.

Gooday, A.J. & Becker, G., 1979. Ostracods in Devonian biostratigraphy. *Special Papers in Palaeontology* 23, 193–97.

Gorter, J.D., 1992. Ordovician petroleum in Australia in relation to eustasy. pp. 433–43 *in* B.D. Webby & J.R. Laurie (eds), q.v.

Gould, C., 1867. River Forth and north coast. *Tasmania House of Assembly, Paper* 74, 1–5.

Gould, R.E., 1975. The succession of Australian pre-Tertiary megafossil floras. *Botanical Review* 41(4), 453–81.

Gould, R.E., 1980. The coal-forming flora of the Walloon Coal Measures. *Coal Geology* 1, 83–105.

Gradstein, F.M. & Sheridan, R.E., 1983. On the Jurassic Atlantic Ocean and a synthesis of results of Deep Sea Drilling Project Leg 76. *Initial Reports of the Deep Sea Drilling Project* 76, 913–43.

Gradstein, F.M., Agterberg, F.P., Brower, J.C. & Schwarzacher, W.S., 1985. *Quantitative Stratigraphy*, D. Reidel, Dordrecht.

Gradstein, F.M., Agterberg, F.P., Ogg, J.G., Hardenbol, J. & Huang, Z., 1994. A Mesozoic time scale. *Journal of Geophysical Research*, 99, 24 051–74.

Graham, J.W., 1955. Evidence of polar shift since Triassic time. *Journal of Geophysical Research* 60, 329–424.

Gravestock, D.I., 1984. Archaeocyatha from lower parts of the Lower Cambrian carbonate sequence in South Australia. *Association of Australasian Palaeontologists, Memoir* 2, 139 pp.

Gravestock, D.I., in press. Early and Middle Palaeozoic. Chapter 7 *in* Geology of South Australia. *Geological Society of Australia.*

Gregory, J.W. Jr, 1903. The Heathcotian—a pre-Ordovician series—and its distribution in Victoria. *Proceedings of the Royal Society of Victoria* n.s. 15, 148–75.

Gregory, J.W. Sr, 1849. Notes on the geology of Western Australia. *Western Australian Almanac* (for 1849), 107–12.

Gregory, J.W. Sr, 1861. On the geology of a part of Western Australia. *Quarterly Journal of the Geological Society* 17, 475–83.

Grey, K., 1991. A mid-Givetian miospore age for the onset of reef development on the Lennard Shelf, Canning Basin, Western Australia. *Review of Palaeobotany and Palynology* 68, 37–48.

Grey, K., 1992. Miospore assemblages from the Devonian reef complexes, Canning Basin, Western Australia. *Geological Survey of Western Australia, Bulletin* 140, 1–139.

Griesbach, C.L., 1880. Palaeontological notes on the Lower Trias of the Himalayas. *Records of the Geological Survey of India* 13, 94–113.

Groos-Uffenorde, H. & Wang, S.-Q., 1989. The entomozoacean succession of South China and Germany (Ostracoda, Devonian). *Courier Forschungsinstitut Senckenberg* 110, 61–79.

Groves, J.R., 1988. Calcareous foraminifers from the Bashkirian stratotype (Middle Carboniferous, South Urals) and their significance for intercontinental correlations and the evolution of the Fusulinidae. *Journal of Paleontology* 62(3), 368–99.

Guex, J., 1991. *Biochronological Correlations*, Springer Verlag, Berlin, 252 pp.

Gümbel, K.W., 1861. *Geognostische beschreibung des bayerischen Alpengebirges und seines vorlandes*, T. Fischer, Kassel, 200 pp.

Guppy, D.J. & Öpik, A.A., 1950. Discovery of Ordovician rocks, Kimberley Division, WA. *Australian Journal of Science* 12, 205–6.

Haag, M. & Heller, F., 1991. Late Permian to Early Triassic magnetostratigraphy. *Earth and Planetary Science Letters* 107, 42–54.

Habib, D. & Drugg, W.S., 1983. Dinoflagellate age of Middle Jurassic–Early Cretaceous sediments in the Blake–Bahama Basin. *Initial Reports of the Deep Sea Drilling Project* 76, 623–38.

Hahn, G. & Hahn, R., 1988. The biostratigraphical distribution of Carboniferous Limestone trilobites in Belgium and adjacent areas. *Bulletin de la Société Belge de Géologie* 97 (1), 77–93.

Haig, D.W., 1979. Cretaceous foraminiferal biostratigraphy of Queensland. *Alcheringa* 3, 171–87.

Hailwood, E.A., 1989. Magnetostratigraphy. *Geological Society Special Report* 19, 84 pp.

Hall, R.L., 1975. Upper Ordovician coral faunas from north-eastern New South Wales. *Journal and Proceedings of the Royal Society of New South Wales* 108, 75–93.

Hall, T.S., 1888. On two new fossil sponges from Sandhurst. *Proceedings of the Royal Society of Victoria* 1, 60–61.

Hall, T.S., 1889. Two new Victorian Palaeozoic sponges. *Proceedings of the Royal Society of Victoria* 2, 152-55.

Hall, T.S., 1895. The geology of Castlemaine, with a subdivision of part of the Lower Silurian rocks of Victoria, and a list of minerals. *Proceedings of the Royal Society of Victoria* 7, 55–58.

Hall, T.S., 1897. Victorian graptolites. Part I. Ordovician from Matlock. (b) *Dictyonema macgillivrayi*, nom. mut. *Proceedings of the Royal Society of Victoria* 10, 13–16.

Hall, T.S., 1899. The graptolite-bearing rocks of Victoria, Australia. *Geological Magazine* 46, 439–51.

Hall, T.S. & Pritchard, G.B., 1902. A suggested nomenclature for the marine Tertiary deposits of southern Australia. *Proceedings of the Royal Society of Victoria* 14, 75–81.

Hallam, A., 1977. Eustatic cycles in the Jurassic. *Palaeogeography, Palaeoclimatology, Palaeoecology* 23, 1–32.

Hallam, A., 1984. Pre-Quaternary sea level changes. *Annual Review of Earth and Planetary Sciences* 12, 205–43.

Hallam, A., Hancock, J.M., Labrecque, J.L., Lowrie, W. & Channell, J.E.T., 1985. Jurassic to Paleogene: Part I, Jurassic and Cretaceous geochronology and Jurassic to Paleogene magnetostratigraphy. pp. 118–40 *in* N.J. Snelling (ed.), q.v.

d'Halloy, J.G.J.d'O., 1822. Observations sur un essai de carte géologique de la France, des Pays-Bas, et des contrées voisines. *Annales des Mines* 7, 353–76

Hamilton, G.B., 1979. Lower and Middle Jurassic calcareous nannofossils from Portugal. *Eclogae Geologicae Helvetiae* 72 (1), 1–17.

Hamilton, G.B., 1982. Triassic and Jurassic calcareous nannofossils. pp. 17–39 *in* A.R. Lord (ed.), *A Stratigraphic Index of Calcareous Nannofossils*, Ellis Horwood Ltd, Chichester.

Hancock, J.M., 1991. Ammonite scales for the Cretaceous System. *Cretaceous Research* 12, 259–91.

Handschumacher, D.V., Sager, W.W., Hilde, T.W.C. & Bracey, D.R., 1988. Pre-Cretaceous tectonic evolution of the Pacific Plate and extension of the geomagnetic polarity time scale with implications for the origin of the Jurassic 'Quiet Zone'. *Tectonophysics* 155, 365–80.

Haq, B.U. & Van Eysinga, F.W.B., 1987. *Geologic Time Table. Fourth revised, enlarged, and updated edition*, Elsevier, Amsterdam (wall chart).

Haq, B.U., Hardendbol, J. & Vail, P.R., 1987. Chronology of fluctuating sea-levels since the Triassic. *Science* 235, 1156–67.

Haq, B.U., Hardenbol, J. & Vail, P.R., 1988. Mesozoic and Cenozoic chronostratigraphy and cycles of sea level change. *SEPM Special Publication* 42, 71–108.

Hardenbol, J. & Berggren, W.A., 1978. A new Paleogene numerical time scale. pp. 213–34 *in* G.V. Cohee et al., q.v.

Hardman, E.T., 1884. Report on the geology of the Kimberley district, Western Australia. *Western Australia Parliamentary Paper* 31, 22 pp.

Hardman, E.T., 1885. Report on the geology of the Kimberley district, Western Australia. *Western Australia Parliamentary Paper* 34, 38 pp.

Harland, W.B., Armstrong, R.L., Cox, A.V., Craig, L.E., Smith, A.G. & Smith, D.G., 1990. *A Geological Time Scale 1989*, Cambridge University Press, Cambridge, pp. ix + 263.

Harland, W.B., Cox, A.B., Llewellyn, P.G., Pickton, C.A.G., Smith, A.G. & Walters, R., 1982. *A Geologic Time Scale*, Cambridge University Press, Cambridge, 131 pp.

Harper, C.W., 1984. Ranking algorithms in quantitative biostratigraphy. *Computer Geoscience* 10, 3–15.

Harper, C.W. & Crowley, K.D., 1985. Insights on why graphic correlation (Shaw's method) works: a discussion. *Journal of Geology* 93, 503–6.

Harris, W.J., 1916. The palaeontological sequence of the Lower Ordovician rocks of the Castlemaine district, Part I. *Proceedings of the Royal Society of Victoria* 29, 50–74.

Harris, W.J., 1933. *Isograptus caduceus* and its allies in Victoria. *Proceedings of the Royal Society of Victoria* 46, 79–114.

Harris, W.J., 1935. The graptolite succession of Bendigo East, with suggested zoning. *Proceedings of the Royal Society of Victoria* 47, 314–37.

Harris, W.J. & Keble, R.A., 1931. Victorian graptolite zones, with correlations and description of species. *Proceedings of the Royal Society of Victoria* 44, 25–48.

Harris, W.J. & Thomas, D.E., 1938. A revised classification and correlation of the Ordovician graptolite beds of Victoria. *Victorian Mining and Geological Journal* 1(3), 62–72.

Harris, W.K., 1971. Tertiary stratigraphic palynology, Otway Basin. pp. 67–87 *in* H. Wopfner & J.G. Douglas (eds), The Otway Basin of south-eastern Australia. *Geological Surveys of South Australia and Victoria, Special Bulletin.*

Harris, W.K., 1985. Middle to Late Eocene depositional cycles and dinoflagellate zones in southern Australia. pp. 133–44 *in* J.M. Lindsay (ed.), Stratigraphy, palaeontology, malacology. *Department of Mines and Energy, South Australia, Special Publication* 5.

Harris, W.K. & Daily, B., 1966. Palaeontology. Appendix 3, pp. 88–111 *in* Delhi-Santos Gidgealpa no. 1 Well, South Australia. *Bureau of Mineral Resources, Australia, Petroleum Search Subsidy Acts, Publication* 73.

Heath, R.S., 1979. Neogene planktonic Foraminifera: studies on Indo–Pacific oceanic sections. Unpublished PhD thesis, University of Adelaide.

Heath, R.S. & Apthorpe, M.C., 1984. Late Cretaceous and Tertiary stratigraphy, southern North West Shelf. Woodside Petroleum Pty Ltd, Perth, unpublished report.

Heath, R.S. & Apthorpe, M.C., 1986. Middle and (?) Early Triassic Foraminifera from the Northwest Shelf, Western Australia. *Journal of Foraminiferal Research* 16(4), 313–33.

Heath, R.S. & McGowran, B., 1984. Neogene datum planes: foraminiferal successions in Australia with reference sections from the Ninetyeast Ridge and the Ontong Java Plateau. pp. 1889–93 *in* N. Ikebe (ed.), *IGCP 114: Pacific Neogene Datum Planes; Contribution to Biostratigraphy and Chronology.*

Heirtzler, J.R., Dickson, G.O., Herron, E.M., Pitman, W.C. III & Le Pichon, X., 1968. Marine magnetic anomalies, geomagnetic field reversals, and motions of the ocean floor and continents. *Journal of Geophysical Research* 73, 2119–36.

Hekel, H., 1972. Pollen and spore assemblages from Queensland Tertiary sediments. *Geological Survey of Queensland, Publication* 355, 1–34.

Helby, R.J., 1967. Triassic plant microfossils from a shale within the Wollar Sandstone, NSW. *Journal and Proceedings of the Royal Society of New South Wales* 100(2), 61–73.

Helby, R.J., 1969. Plant microfossils in the Wianamatta Group. p. 423 *in* G.H. Packham (ed.), Geology of New South Wales, *Journal of the Geological Society of Australia* 16.

Helby, R.J., 1973. Review of Late Permian and Triassic palynology of New South Wales. *Geological Society of Australia, Special Publication* 4, 141–55.

Helby, R., 1987. A palynological study of the Cambridge Gulf Group (Triassic–Early Jurassic). *Association of Australasian Palaeontologists, Memoir* 4, microfiche 1, 1–47; fiche 2, figs 1–5; fiche 3, figs 6–12.

Helby, R.J. & Martin, A.R.H., 1965. *Cylostrobus* gen. nov., cones of lycopsidean plants from the Narrabeen Group (Triassic) of NSW, Australia. *Australian Journal of Botany* 13, 389–404.

Helby, R.J., Morgan, R. & Partridge, A.D., 1987. A palynological zonation of the Australian Mesozoic. *Association of Australasian Palaeontologists, Memoir* 4, 1–94.

Heller, F., 1977. Palaeomagnetism of Upper Jurassic limestones from southern Germany. *Journal of Geophysics* 42, 475–88.

Heller, F., 1978. Rock magnetic studies of Upper Jurassic limestones from southern Germany. *Journal of Geophysics* 44, 525–43.

Heller, F., Lowrie, W., Huamei, L. & Junda, W., 1988. Magnetostratigraphy of the Permo–Triassic boundary section at Shangsi (Guangyuan, Sichuan Province, China). *Earth and Planetary Science Letters* 88, 348–56.

Helsley, C.E., 1965. Paleomagnetic results from the Lower Permian Dunkard Series of West Virginia. *Journal of Geophysical Research* 70, 413–24.

Helsley, C.E., 1969. Magnetic reversal stratigraphy of the Lower Triassic Moenkopi Formation of Western Colorado. *Geological Society of America Bulletin* 80, 2431–50.

Helsley, C.E. & Steiner, M.B., 1969. Evidence for long intervals of normal polarity during the Cretaceous Period. *Earth and Planetary Science Letters* 5, 325–32.

Helsley, C.E. & Steiner, M.B., 1974. Paleomagnetism of the Lower Triassic Moenkopi Formation of western Colorado. *Geological Society of America Bulletin* 85, 457–64.

Henderson, R.A., 1973. Clarence and Raukumara Series (Albian–?Santonian) Ammonoidea from New Zealand. *Journal of the Royal Society of New Zealand* 3, 71–123.

Henderson, R.A., 1976. Idamean (early Upper Cambrian) trilobites from north-western Queensland, Australia. *Palaeontology* 19, 325–64.

Henderson, R.A., 1977. Stratigraphy of the Georgina Limestone and a revised zonation for the Upper Cambrian Idamean Stage. *Journal of the Geological Society of Australia* 23, 423–33.

Henderson, R.A., 1983. Early Ordovician faunas from the Mount Windsor Subprovince, north-eastern Queensland. *Association of Australasian Palaeontologists, Memoir* 1, 145–75.

Henderson, R.A., 1990. Late Albian ammonites from the Northern Territory, Australia. *Alcheringa* 14, 109–48.

Henderson, R.A. & MacKinnon, D.I., 1981. New Cambrian inarticulate Brachiopoda from Australasia and the age of the Tasman Formation. *Alcheringa* 5, 289–309.

Henderson, R.A. & McNamara, K.J., 1985. Maastrichtian nonheteromorph ammonites from the Miria Formation, Western Australia. *Palaeontology* 28, 35–88.

Henderson, R.A., Kennedy, W.J. & McNamara, K.J., 1992. Maastrichtian heteromorph ammonites from the Carnarvon Basin, Western Australia. *Alcheringa* 16, 133–70.

Hess, J.C. & Lippolt, H.J., 1986. $^{40}Ar/^{39}Ar$ ages of tonstein and tuff sanidines: new calibration points for the improvement of the Upper Carboniferous timescale. *Chemical Geology (Isotope Geoscience Section)* 59, 143–54.

Hess, J.C., Backfisch, S. & Lippolt, H.J., 1983. Konkordantes Sanidin- und diskordante Biotitalter eines Karbontuffs der Baden-Badener Senke, Nordschwarzwald. *Neues Jahrbuch für Geologie und Paläontologie Monatshefte* 1983(5), 277–92.

Heywood, P.B., 1978. Stratigraphic drilling report GSQ Eddystone 1. *Queensland Government Mining Journal* 79 (922), 407–17.

Higgins, A.C., 1975. Conodont zonation of the late Viséan–early Westphalian strata of the south and central Pennines of northern England. *Bulletin of the Geological Survey of Great Britain* 53, 1–90.

Higgins, A.C., 1981. The position and correlation of the boundary between the proposed Mississippian–Pennsylvanian subsystems. *Newsletters on Stratigraphy* 9 (3), 176–82.

Higgins, A.C., 1985. The Carboniferous System: Part 2 Conodonts of the Silesian Subsystem from Great Britain and Ireland. pp. 210–27 *in* A.C. Higgins & R.L. Austin (eds), q.v.

Higgins, A.C. & Austin, R.L. (eds), 1985. *A Stratigraphical Index of Conodont*, Ellis Horwood Ltd, Chichester (for the British Micropalaeontological Society), 263 pp.

Higgins, A.C., Richards, B.C. & Henderson, C.M., 1991. Conodont biostratigraphy and paleoecology of the uppermost Devonian and Carboniferous of the Western Canada Sedimentary Basin. *Geological Survey of Canada Bulletin* 417, 215–51.

Hill, D., 1937. The Permian corals of Western Australia. *Journal of the Royal Society of Western Australia* 23, 43–63.

Hill, D., 1940. The Silurian Rugosa of the Yass–Bowning district, NSW. *Proceedings of the Linnean Society of New South Wales* 65, 388–420.

Hill, D., 1942. Some Tasmanian Palaeozoic corals. *Papers and Proceedings of the Royal Society of Tasmania* (1941), 3–12.

Hill, D., 1955. Ordovician corals from Ida Bay, Queenstown and Zeehan, Tasmania. *Papers and Proceedings of the Royal Society of Tasmania* 89, 237–54.

Hill, D., 1957. Ordovician corals from New South Wales. *Journal and Proceedings of the Royal Society of New South Wales* 91, 97–107.

Hill, D., 1968. Devonian of eastern Australia. pp. 613–30 *in International Symposium on the Devonian System*, vol. 1, Alberta Society of Petroleum Geologists, Calgary.

Hill, D. & Jell, J.S., 1970. Devonian corals from the Canning Basin of Western Australia. *Geological Survey of Western Australia, Bulletin* 121, 1–157.

Hill, D., Playford, G. & Woods, J.T., 1965. *Triassic Fossils of Queensland*, Queensland Palaeontographical Society, Brisbane, 32 pp.

Hill, D., Playford, G. & Woods, J.T., 1969. *Ordovician and Silurian Fossils of Queensland*, Queensland Palaeontographical Society, Brisbane, 18 pp.

Hills, E.S., 1929. The geology and palaeontography of the Cathedral Range and Blue Hills in northwestern Gippsland. *Proceedings of the Royal Society of Victoria* 41, 176–201.

Hills, E.S., 1931. The Upper Devonian fishes of Victoria, Australia, and their bearing on the stratigraphy of the state. *Geological Magazine* 68, 206–31.

Hills, E.S., 1958. A brief review of Australian fossil vertebrates. pp. 86–107 *in* T.S. Westoll (ed.), *Studies on Fossil Vertebrates*, Athlone Press, London.

Hills, E.S., 1959. Record of *Bothriolepis* and *Phyllolepis* (Upper Devonian) from the Northern Territory of Australia. *Journal and Proceedings of the Royal Society of New South Wales* 92, 174–75.

Hills, S.J. & Thierstein, H.R., 1989. Plio–Pleistocene calcareous plankton biochronology. *Marine Micropaleontology* 14, 67–96.3

Hinde, G.L. 1899. On the Radiolaria in the Devonian rocks of New South Wales. *Quarterly Journal of the Geological Society of London* 55, 83–164.

Hintze, L.F., 1951. Lower Ordovician detailed stratigraphic sections for western Utah. *Utah Geological and Mineralogical Survey, Bulletin* 39, 99 pp.

Hintze, L.F., 1953 [imprint 1952]. Lower Ordovician trilobites from western Utah and eastern Nevada. *Utah Geological and Mineralogical Survey, Bulletin* 48, 249 pp.

Hintze, L.F., 1979. Preliminary zonations of Lower Ordovician of western Utah by various taxa. *Brigham Young University, Geology Series* 26(2), 13–19.

Hintze, L.F., Braithwaite, L.F., Clark, D.L., Ethington, R.L. & Flower, R.H., 1972 [imprint 1968]. A fossiliferous Lower Ordovician reference section from western United States. *Proceedings of the 23rd International Geological Congress*, 385–400.

Hinz, I.C.U., 1991*a*. On *Ulopsis ulula* gen. et sp. nov. *Stereo-Atlas of Ostracod Shells* 18(2), 69–72.

Hinz, I.C.U., 1991*b*. On *Capricambria cornucopiae* gen. et sp. nov. *Stereo-Atlas of Ostracod Shells* 18(16), 65–68.

Hinz, I.C.U. & Jones, P.J., 1992. On *Tubupestis tuber* Hinz & Jones gen. et sp. nov. *Stereo-Atlas of Ostracod Shells* 19(3), 9–12.

Hocking, R.M., Moors, H.T. & Van De Graaff, W.J.E., 1987. Geology of the Carnarvon Basin, Western Australia. *Geological Survey of Western Australia Bulletin* 133, 1–289.

Hölder, H., 1979. Jurassic. A390–A417 *in* R.A. Robison & C. Teichert (eds), *Treatise on Invertebrate Paleontology. Part A Introduction*, Geological Society of America, Boulder, Colorado, and University of Kansas, Lawrence, Kansas.

Holdsworth, B.K. & Jones, D.L., 1980. Preliminary radiolarian zonation for Late Devonian through Permian time. *Geology* 8, 281–85.

Holland, C.H. 1985. Series and stages of the Silurian System. *Episodes* 8(2), 101–3.

Holland, C.H. & Bassett, M.G. (eds), 1989. A global standard for the Silurian System. *National Museum of Wales, Geological Series* 9, Cardiff, 325 pp.

Holland, C.H. & Palmer, D.C., 1974. *Bohemograptus*, the youngest graptoloid known from the British Silurian sequence. *Special Papers in Palaeontology* 13, 215–36, 21.

Holloway, D.J. & Campbell, K.S.W., 1974. The Silurian trilobite *Onycopyge* Woodward. *Palaeontology* 17, 409–21.

Holloway, D.J. & Neil, J.V., 1982. Trilobites from the Mount Ida Formation (Late Silurian–Early Devonian), Victoria. *Proceedings of the Royal Society of Victoria* 94, 133–54.

Holloway, D.J. & Sandford, A., 1993. An Early Silurian trilobite fauna from Tasmania. *Association of Australasian Palaeontologists, Memoir* 15, 85–102.

Holmes, W.B.K., 1982. The Middle Triassic flora from Benolong, near Dubbo, central-western New South Wales. *Alcheringa* 6, 1–33.

Horn, M., 1960. Zur stratigraphischen Gliederung des tieferen Namur. V. Die Zone des *Eumorphoceras pseudobilingue* im Sauerland. *Fortschritte im der Geologie von Rheinland und Westfalen* 3 (1), 303–42.

Horner, F. & Heller, F., 1983. Lower Jurassic magnetostratigraphy at the Breggia Gorge (Ticino, Switzerland) and Alpe Turati (Como, Italy). *Geophysical Journal of the Royal Astronomical Society* 73, 705–18.

Hornibrook, N. de B., 1978. Correlation beyond New Zealand. pp. 428–36 *in* R.P. Suggate, G.R. Stevens & M.T. Te Punga (eds), *The Geology of New Zealand*, Government Printer, Wellington.

Hornibrook, N. de B., Brazier, R.C. & Strong, C.P., 1989. Manual of New Zealand Permian to Pleistocene foraminiferal biostratigraphy. *New Zealand Geological Survey Paleontological Bulletin* 56, 1–175.

House, M.R., 1985. Correlation of mid-Palaeozoic ammonoid evolutionary events with global sedimentary perturbations. *Nature* 313, 17–22.

House, M.R., 1987. Early Devonian goniatite faunas and their bearing on the definition of the base of the Emsian. *Submission to the Devonian Subcommission, IUGS. Calgary Symposium 1987*, 3 pp.

House, M.R., 1988. International definition of Devonian System boundaries. *Proceedings of the Ussher Society* 7, 41–46.

House, M.R., 1989. Analysis of mid-Palaeozoic extinctions. *Bulletin de la Société Géologique de Belgique* 98, 99–107.

House, M.R., Kirchgasser, W.T., Price, J.D. & Wade, G., 1985. Goniatites from Frasnian (Upper Devonian) and adjacent strata of the Montagne Noire. *Hercynica* 1, 1–21.

Hueber, F., 1983. A new species of *Baragwanathia* from the Sextant Formation (Emsian) northern Ontario, Canada. *Botanical Journal of the Linnean Society* 86, 57–79.

Hughes, N.F., 1989. *Fossils as Information*, Cambridge University Press, Cambridge, 136 pp.

Humboldt, F.W.H.A. von, 1799. *Über die Unterirdischen Gasarten und die Mittel ihren Nachtheil zu Vermindern: Ein Beitrag zur Physik der Praktischen Bergbaukunde*, 384 pp. Wiewag, Braunschweig.

Hurley, N.F. & Van der Voo, R., 1987. Paleomagnetism of Upper Devonian reefal limestones, Canning Basin, Western Australia. *Geological Society of America Bulletin* 98, 138–46.

Hurley, N.F. & Van der Voo, R., 1990. Magnetostratigraphy, Late Devonian iridium anomaly, and impact hypotheses. *Geology* 18, 291–94.

Hutchinson, P., 1973. A revision of the Redfieldiiform and Perleidiform fishes from the Triassic of Bekker's Kraal (South Africa) and Brookvale New South Wales. *British Museum (Natural History) Bulletin, Geology* 22, 233–354.

Idnurm, M. & Cook, P.J., 1980. Palaeomagnetism of beach ridges in South Australia and the

Milankovitch theory of ice ages. *Nature* 287, 699–702.

Imlay, R.W., 1974. Jurassic ammonite succession in the United States. *Mémoire du Bureau de Recherches Géologiques et Minières* 75, 709–24.

Irving, E., 1963. Palaeomagnetism of the Narrabeen Chocolate Shales and the Tasmanian Dolerite. *Journal of Geophysical Research* 68, 2283–87.

Irving, E., 1966. Paleomagnetism of some Carboniferous rocks from New South Wales and its relation to geological events. *Journal of Geophysical Research* 71, 6025–51.

Irving, E., 1971. Nomenclature in magnetic stratigraphy. *Geophysical Journal of the Royal Astronomical Society* 24, 529–31.

Irving, E. & Parry, L.G., 1963. The magnetism of some Permian rocks from New South Wales. *Geophysical Journal of the Royal Astronomical Society* 7, 395–411.

Irving, E. & Pullaiah, G., 1976. Reversals of the geomagnetic field, magnetostratigraphy, and relative magnitude of palaeosecular variation in the Phanerozoic. *Earth Science Reviews* 12, 35–64.

Ishiga, H., Leitch, E.C., Naka, T., Watanabe, T. & Iwasaki, M., 1987. Late Devonian Palaeoscenidiidae from the Hastings Block, New England Fold Belt, NSW, Australia. *Earth Science (Chikyu Kagaku)* 41, 297–302.

Ishiga, H., Leitch, E.C., Watanabe, T., Naka, T. & Iwasaki, M., 1988. Radiolarian and conodont biostratigraphy of siliceous rocks from the New England Fold Belt. *Australian Journal of Earth Sciences* 35(1), 73–80.

Jaanusson, V., 1960. On the series of the Ordovician System. *XXI International Geological Congress, Proceedings, Section* 7, 70–81.

Jaanusson, V., 1982. Introduction to the Ordovician of Sweden. pp. 1–10 *in* D.L. Bruton & S.H. Williams (eds), Field excursion guide, IV International Symposium on the Ordovician System. *Palaeontological Contributions from the University of Oslo* 279.

Jack, R.L. & Etheridge, R. Jr, 1892. The geology and palaeontology of Queensland and New Guinea. *Geological Survey of Queensland, Publication* 92, 1–768, 2 vols.

Jaeger, H., 1966. Two late *Monograptus* species from Victoria, and their significance for dating the *Baragwanathia* flora. *Proceedings of the Royal Society of Victoria* 79, 393–413.

Jaeger, H., 1967. Preliminary stratigraphic results from graptolite studies in the Upper Silurian and Early Devonian rocks of south-eastern Australia. *Journal of the Geological Society of Australia* 14, 281–86.

Jaeger, H., 1977. Graptolites. pp. 337–45 *in* A. Martinsson, q.v.

Jaeger, H., 1989. Devonian Graptoloidea. pp. 431–38 *in* N.J. McMillan, A.F. Embry & D.J. Glass (eds), vol. 3, q.v.

Jago, J.B., 1972*a*. Biostratigraphic and taxonomic studies of some Tasmanian Cambrian trilobites. Unpublished PhD thesis, University of Tasmania.

Jago, J.B., 1972*b*. Two new Cambrian trilobites from Tasmania. *Palaeontology* 15 (2), 226–37.

Jago, J.B., 1974*a*. *Glyptagnostus reticulatus* from the Huskisson River, Tasmania. *Papers and Proceedings of the Royal Society of Tasmania* 107, 117–26.

Jago, J.B., 1974*b*. A new Middle Cambrian polymerid trilobite from north-western Tasmania. *Papers and Proceedings of the Royal Society of Tasmania* 108, 141–49.

Jago, J.B., 1976*a*. Late Middle Cambrian agnostid trilobites from north-western Tasmania. *Palaeontology* 19(1), 133–72.

Jago, J.B., 1976*b*. Late Middle Cambrian agnostid trilobites from the Gunns Plains area, north-western Tasmania. *Papers and Proceedings of the Royal Society of Tasmania* 110, 1–18.

Jago, J.B., 1977. A late Middle Cambrian fauna from the Que River Beds, western Tasmania. *Papers and Proceedings of the Royal Society of Tasmania* 111, 41–57.

Jago, J.B., 1978. Late Cambrian fossils from the Climie Formation, western Tasmania. *Papers and Proceedings of the Royal Society of Tasmania* 112, 137–53.

Jago, J.B., 1979. Tasmanian Cambrian biostratigraphy—a preliminary report. *Journal of the Geological Society of Australia* 26, 223–30.

Jago, J.B., 1981. A late Middle Cambrian damesellid trilobite cranidium from Beaconsfield, Tasmania. *Papers and Proceedings of the Royal Society of Tasmania* 115, 19–20.

Jago, J.B., 1986. An early Late Cambrian fauna from Tom Creek, western Tasmania. *Papers and Proceedings of the Royal Society of Tasmania* 120, 97–98.

Jago, J.B., 1987. Idamean (Late Cambrian) trilobites from the Denison Range, south-west Tasmania. *Palaeontology* 30(2), 207–31.

Jago, J.B. & Brown, A.V., 1989. Middle to Upper Cambrian fossiliferous sedimentary rocks. pp. 74–82 *in* C.F. Burrett & E.L. Martin, q.v.

Jago, J.B. & Brown, A.V., 1992. Early Idamean (Late Cambrian) agnostoid trilobites from the Huskisson River, Tasmania. *Papers and Proceedings of the Royal Society of Tasmania* 126, 59–65.

Jago, J.B. & Corbett, K.D., 1990. Latest Cambrian trilobites from Misery Hill, western Tasmania. *Alcheringa* 14, 233–46.

Jago, J.B. & Daily, B., 1974. The trilobite *Clavagnostus* Howell from the Cambrian of Tasmania. *Palaeontology* 17(1), 95–109.

James, N.P. & Gravestock, D.I., 1990. Lower Cambrian shelf and shelf margin carbonate buildups, Flinders Ranges, South Australia. *Sedimentology* 37, 455–80.

Jeletzky, J.A., 1971. Marine Cretaceous biotic provinces and paleogeography of western and Arctic Canada: illustrated by a detailed study of ammonites. *Canadian Geological Survey Paper* 70–22, 92 pp.

Jeletzky, J.A., 1973. Biochronology of the marine Boreal latest Jurassic, Berriasian and Valanginian in Canada. pp. 41–80 *in* R. Casey & P.F. Rawson (eds), The Boreal Lower Cretaceous. *Geological Journal Special Issue* 5.

Jell, J.S., 1989. Lower and Middle Devonian of Queensland, Australia. pp. 755–72 *in* N.J. McMillan, A.F. Embry & D.J. Glass (eds), vol. 1, q.v.

Jell, J.S. & Talent, J.A., 1989. The Silurian of Australia: the most instructive sections. pp. 183–200 *in* C.H. Holland & M.G. Bassett (eds), A global standard for the Silurian System. *National Museum of Wales, Geological Series* 9, Cardiff.

Jell, P.A., 1977. *Penarosa netenta*, a new Middle Cambrian trilobite from northwestern Queensland. *Memoirs of the Queensland Museum* 18(1), 119–23.

Jell, P.A., 1978. *Asthenopsis* Whitehouse, 1939 (Trilobita, Middle Cambrian) in northern Australia. *Memoirs of the Queensland Museum* 18(2), 219–31.

Jell, P.A., 1985. Tremadoc trilobites of the Digger Island Formation, Waratah Bay, Victoria. *Memoirs of the Museum of Victoria* 46, 53–88, pls 19–33.

Jell, P.A., & Robison, R.A., 1978. Revision of a late Middle Cambrian faunule from northwestern Queensland. *University of Kansas Paleontological Contributions, Paper* 90, 21 pp.

Jell, P.A. & Stait, B.A., 1985*a*. Tremadoc trilobites from the Florentine Valley Formation, Tim Shea area, Tasmania. *Memoirs of the Museum of Victoria* 46, 1–34.

Jell, P.A. & Stait, B.A., 1985*b*. Revision of an early Arenig trilobite faunule from the Caroline Creek Sandstone, near Latrobe, Tasmania. *Memoirs of the Museum of Victoria* 46, 35–51.

Jell, P.A., Hughes, N.C. & Brown, A.V., 1991. Late Cambrian (post-Idamean) trilobites from the Higgins Creek area, western Tasmania. *Memoirs of the Queensland Museum* 30 (3), 455–85.

Jenkins, C.J., 1982. Late Pridolian graptolites from the Elmside Formation near Yass, New South Wales. *Proceedings of the Linnean Society of New South Wales* 106, 167–72.

Jenkins, C.J., Garratt, M.G., Strusz, D.L., Bischoff, G.O., Jell, J.A., Talent, J.A., Powell, C.McA., Sherwin, L., VandenBerg, A.H.M., Crook, K.A.W., Philip, G.M., Packham, G.H., Pickett, J., McLean, R.A., Jell, P., Holloway, D., Campbell, K.S.W, Link, A.G., Druce, E., McQueen, K., Rickards, R.B., Glen, R.A., Birkhead, P.K., Cas, R.A. & Creaser, P., 1986. *The Silurian of Mainland South-eastern Australia: a Field Guide. IUGS International Subcommission on Silurian Stratigraphy, Field Meeting, August 16th–27th, 1986*, University of Sydney.

Jenkins, D.G., 1958. Correspondence: Pelagic Foraminifera in the Tertiary of Victoria. *Geological Magazine* 95, 438–39.

Jenkins, D.G., 1960. Planktonic Foraminifera from the Lakes Entrance Oil Shaft, Victoria, Australia. *Micropaleontology* 6, 345–71.

Jenkins, D.G., 1965. Planktonic foraminiferal zones and new taxa from the Danian to Lower Miocene of New Zealand. *New Zealand Journal of Geology & Geophysics* 8, 1088–126.

Jenkins, D.G., 1967. Planktonic foraminiferal zones and new taxa from the Lower Miocene to the Pleistocene of New Zealand. *New Zealand Journal of Geology & Geophysics* 10, 1064–78.

Jenkins, D.G., 1985. Southern mid-latitude Paleocene to Holocene planktic Foraminifera. pp. 263–82 *in* H.M. Bolli, J.B. Saunders & K. Perch-Nielson (eds), *Plankton Stratigraphy*, Cambridge University Press, Cambridge.

Jenkins, T.B.H., 1966. The Upper Devonian index ammonoid *Cheiloceras* from New South Wales. *Palaeontology* 9, 458–63.

Jenkins, T.B.H., 1968. Famennian ammonoids from New South Wales. *Palaeontology* 11, 535–48.

Jenkins, T.B.H., 1974. Lower Carboniferous conodont biostratigraphy of New South Wales. *Palaeontology* 17, 909–24.

Jenkins, T.B.H., Crane, D.T. & Mory, A.J., 1993. Conodont biostratigraphy of the Viséan Series in eastern Australia. *Alcheringa* 17, 211–83.

Ji Qiang, 1987. Early Carboniferous conodonts from Jianghua County of Hunan Province and their stratigraphic value—with a discussion on the mid-Aikuanian Event. *Bulletin of the Institute of Geology, Chinese Academy of Geological Sciences* 16, 115–38.

Johnson, B.D. & Veevers, J.J., 1984. Oceanic palaeomagnetism. pp. 17–38 *in* J.J. Veevers (ed.), *Phanerozoic earth history of Australia*, Clarendon Press, Oxford.

Johnson, J.G., 1992. Belief and reality in biostratigraphic zonation. *Newsletters on Stratigraphy* 26, 41–48.

Johnson, J.G. & Sandberg, C.A., 1989. Devonian eustatic events in the western United States and their biostratigraphic responses. pp. 171–78 *in* N.J. McMillan, A.F. Embry & D.J. Glass (eds), vol. 3, q.v.

Johnson, J.G., Klapper, G. & Sandberg, C.A., 1985. Devonian eustatic fluctuations in Euramerica. *Geological Society of America, Bulletin* 96, 567–87.

Johnson, M.E., Kaljo, D. & Rong, J.-Y., 1991. Silurian eustasy. *Special Papers in Palaeontology* 44, 145–63.

Johnston, R.M., 1888. *Systematic Account of the Geology of Tasmania*, J. Walch & Sons, Hobart, 408 pp., 57 pls.

Jones, B.G., Carr, P.F. & Wright, A.J., 1981. Silurian and Early Devonian geochronology—a reappraisal, with new evidence from the Bungonia Limestone. *Alcheringa* 5, 197–207.

Jones, M.J. & Truswell, E.M., 1992. Palynology of the Late Carboniferous and Early Permian Joe Joe Group, southern Galilee Basin, Queensland. *BMR Journal of Australian Geology & Geophysics*. 13, 143–85.

Jones, O.A. & de Jersey, N.J., 1947. The flora of the Ipswich Coal Measures—morphology and floral succession. *Papers of the Department of Geology, University of Queensland* 3(3), 1–88.

Jones, P.J., 1958. Preliminary report on the micropalaeontology of samples from the Bonaparte Gulf Basin. *Bureau of Mineral Resources, Geology & Geophysics, Record* 1958/26.

Jones, P.J., 1959. Preliminary report on Ostracoda from Bore BMR No. 2, Laurel Downs, Fitzroy Basin, Western Australia. *Bureau of Mineral Resources, Geology & Geophysics, Report* 38, 37–52.

Jones, P.J., 1961*a*. Ostracod assemblages near the Upper Devonian–Lower Carboniferous boundary in the Fitzroy and Bonaparte Gulf basins. *Bureau of Mineral Resources, Geology & Geophysics, Bulletin* 60, 277–81.

Jones, P.J., 1961*b*. Discovery of conodonts in the Upper Cambrian of Queensland. *Australian Journal of Science* 24(3), 143–44.

Jones, P.J., 1962*a*. Micropalaeontological examination of the Lower Carboniferous–Upper Devonian sequence of Meda No. 1. *Bureau of Mineral Resources, Geology & Geophysics, Petroleum Search Subsidy Acts Publication* 7, 31–32.

Jones, P.J., 1962*b*. Preliminary notes on Upper Devonian Ostracoda from Frome Rocks No. 2 Well. *Bureau of Mineral Resources, Geology & Geophysics, Petroleum Search Subsidy Acts Publication* 8, 35–39.

Jones, P.J., 1962*c*. The ostracod genus *Cryptophyllus* in the Upper Devonian and Carboniferous of Western Australia. *Bureau of Mineral Resources, Geology & Geophysics, Bulletin* 62(3), 1–37.

Jones, P.J., 1968. Upper Devonian Ostracoda and Eridostraca from the Bonaparte Gulf Basin, northwestern Australia. *Bureau of Mineral Resources, Geology & Geophysics, Bulletin* 99, 1–109.

Jones, P.J., 1970. Marine Ostracoda (Palaeocopa, Podocopa) from the Lower Triassic of the Perth Basin, Western Australia. *Bureau of Mineral Resources, Geology & Geophysics, Bulletin* 108, 115–143.

Jones, P.J., 1971. Lower Ordovician conodonts from the Bonaparte Gulf Basin and the Daly River Basin, northwestern Australia. *Bureau of Mineral Resources, Geology & Geophysics, Bulletin* 117, 80 pp.

Jones, P.J., 1974. Australian Devonian and Carboniferous (Emsian–Viséan) ostracod faunas: a review. pp. 1–19 *in* J. Bouckaert & M. Streel (eds), *International Symposium on Belgian Micropaleontological Limits from Emsian to Viséan, Namur, 1974, Publication* 6.

Jones, P.J., 1985. Treposellidae (Beyrichiacea: Ostracoda) from the latest Devonian (Strunian) of the Bonaparte Basin, Western Australia. *BMR Journal of Australian Geology & Geophysics* 9, 149–62.

Jones, P.J., 1988. Comments on some Australian, British and German isotopic age data for the Carboniferous System. *Newsletter on Carboniferous Stratigraphy* 6, 26–29.

Jones, P.J., 1989. Lower Carboniferous Ostracoda (Beyrichicopida and Kirkbyocopa) from the Bonaparte Basin, north-western Australia. *Bureau of Mineral Resources, Geology & Geophysics, Bulletin* 228, 1–97.

Jones, P.J., 1991. Australian Phanerozoic Timescales 5. Carboniferous. Biostratigraphic chart and explanatory notes. *Bureau of Mineral Resources, Geology & Geophysics, Record* 1989/35, 1–43.

Jones, P.J., 1995. Australian Phanerozoic Timescales 5. Carboniferous. Biostratigraphic chart and explanatory notes, second series. *Australian Geological Survey Organisation, Record* 1995/34, 1–45.

Jones, P.J. & Druce, E.C., 1966. Intercontinental conodont correlations of the Palaeozoic sediments of the Bonaparte Gulf Basin, northwestern Australia. *Nature* 211, 357–59.

Jones, P.J., & McKenzie, K.G., 1980. Queensland Middle Cambrian Bradoriida (Crustacea): new taxa, palaeobiogeography and biological affinities. *Alcheringa* 4, 203–55.

Jones, P.J. & Nicoll, R.S., 1985. Late Triassic conodonts from Sahul Shoals No. 1, Ashmore Block, northwestern Australia. *Bureau of Mineral Resources Journal of Australian Geology & Geophysics* 9, 361–64.

Jones, P.J. & Roberts, J., 1976. Some aspects of Carboniferous biostratigraphy in eastern Australia: a review. *Bureau of Mineral Resources Journal of Australian Geology & Geophysics* 1, 141–51.

Jones, P.J. & Young, G.C., 1992. Biostratigraphic summary of Mimosa 1 well, Canning Basin, Western Australia. *Bureau of Mineral Resources, Geology & Geophysics Professional Opinion* 1992/2.

Jones, P.J., Campbell, K.S.W. & Roberts, J., 1973. Correlation chart for the Carboniferous System of Australia. *Bureau of Mineral Resources, Geology & Geophysics, Bulletin* 156A, 1–40.

Jones, P.J., Shergold, J.H. & Druce, E.C., 1971. Late Cambrian and Early Ordovician stages in western Queensland. *Journal of the Geological Society of Australia* 18 (1), 1–32.

Jupp, R. & Warren. A.A., 1986. The mandibles of the Triassic temnospondyl amphibians. *Alcheringa* 10, 99–124.

Kammer, T.W., Brenckle, P.L., Carter, J.L. & Ausich, W.I., 1991. Redefinition of the Osagean–Meramecian boundary in the Mississippian stratotype region. *Palaios* 5, 414–31.

Kapoor, H.M., 1992. Permo–Triassic boundary of the Indian subcontinent and its international correlation. pp. 21–36 *in* W.C. Sweet, Yang Zunyi, J.M. Dickins & Yin Hingfu (eds), *Permo–Triassic Events in the Eastern Tethys*, Cambridge University press, Cambridge.

Kauffman, E.G. & Hazel, J.E. (eds), 1977. *Concepts and Methods of Biostratigraphy*, Dowden, Hutchinson & Ross, Stroudsberg, 658 pp.

Kauffman, E.G. & Walliser, O.H., 1990. *Extinction Events in Earth History*, Springer-Verlag, Berlin, 432 pp.

Kemp, A., 1982. Australian Mesozoic and Cenozoic lungfish. pp. 133–43 *in* P.V. Rich & E.M. Thompson (eds), q.v.

Kemp, A., 1991. Australian Mesozoic and Cenozoic lungfish. pp. 465–96 *in* P.V. Rich et al. (eds), q.v.

Kemp, E.M., Balme, B.E., Helby, R.J., Kyle, R.A., Playford, G. & Price, P.L., 1977. Carboniferous and Permian palynostratigraphy in Australia and Antarctica: a review. *Bureau of Mineral Resources Journal of Australian Geology & Geophysics* 2, 177–208.

Kemp, N.R., 1991. Chondrichthyans in the Cretaceous and Tertiary of Australia. pp. 497–568 *in* P.V. Rich et al. (eds), q.v.

Kemper, E., 1973*a*. The Valanginian and Hauterivian stages in northwest Germany. pp. 327–44 *in* R. Casey & P.F. Rawson (eds), The Boreal Lower Cretaceous. *Geological Journal Special Issue* 5.

Kemper, E., 1973*b*. The Aptian and Albian stages in northwest Germany. pp. 345–60 *in* R. Casey & P.F. Rawson (eds), The Boreal Lower Cretaceous. *Geological Journal Special Issue* 5.

Kennedy, W.J., 1984*a*. Systematic palaeontology and stratigraphic distribution of the ammonite faunas of the French Coniacian. *Special Papers in Palaeontology* 31, 160 pp.

Kennedy, W.J., 1984*b*. Ammonite faunas and the 'standard zones' of the Cenomanian to Maastrichtian stages in their type areas, with some proposals for the definition of the stage boundaries by ammonites. *Geological Society of Denmark Bulletin* 33, 147–61.

Kennedy, W.J., 1986. Campanian and Maastrichtian ammonites from northern Aquitaine, France. *Special Papers in Palaeontology* 36, 145 pp.

Kennedy, W.J., 1987. Ammonites from the type Santonian and adjacent parts of northern Aquitaine, western France. *Palaeontology* 30, 765–82.

Kennedy, W.J. & Odin, G.S., 1982. The Jurassic and Cretaceous time scale in 1981. pp. 557–92 *in* G.S. Odin (ed.), q.v.

Kennett, J.P. & Srinivasan, M.S., 1983. *Neogene Planktonic Foraminifera. A Phylogenetic Atlas*, Hutchinson Ross, Stroudsburg, Pennsylvania, 265 pp.

Kent, D.V., 1988. Further palaeomagnetic evidence for oroclinal rotation in the central folded Appalachians from the Mauch Chunk Formation. *Tectonics* 7, 749–59.

Kent, D.V. & Gradstein, F.M., 1985. A Cretaceous and Jurassic geochronology. *Geological Society of America Bulletin* 96, 1419–27.

Kent D.V., Witte, W.K. & Olsen, P.E., 1993. A complete Triassic magnetostratigraphy from the Newark Basin. *Eos*, supplement 74, 109.

Khramov, A.N., 1987. *Paleomagnetology*, Springer Verlag, Berlin, 308 pp.

Khramov, A.N. & Rodionov, V.P., 1981. The geomagnetic field during Palaeozoic time. pp. 99–115 *in* M.W. McElhinny, A.N. Khramov, M. Ozima & D.A. Valencio (eds), Global reconstruction and the geomagnetic field during the Palaeozoic. *Advances in Earth and Planetary Sciences* 10.

Khramov, A.N., Rodionov, V.P. & Komissarova, R.A., 1965. *New Data on the Palaeozoic History of the Geomagnetic Field in the USSR*. Translated from *Nastoyashcheye i Proshloye Magnitonogo Polia Zemli*, 'Nauka' Press, Moscow, 206–13.

Kirchgasser, W.T. & Oliver, W.A., 1993. Correlation of stage boundaries in the Appalachian Devonian, eastern United States. *IUGS Subcommission on Devonian Stratigraphy, Newsletter* 10, 5–8.

Kirchgasser, W.T., Oliver, W.A. & Rickard, L.V., 1985. Devonian series boundaries in the eastern United States. *Courier Forschungsinstitut Senckenberg* 75, 233–59.

Kirk, R.B., 1985. A seismic stratigraphic case history in the eastern Barrow Subbasin, North West Shelf, Australia. *American Association of Petroleum Geologists Memoir* 39, 183–207.

Kirschvink, J.L., 1976. The magnetic stratigraphy of Late Proterozoic to Early Cambrian sediments of the Amadeus Basin, central Australia: a palaeomagnetic approach to the Precambrian–Cambrian boundary problem. *25th International Geological Congress, Sydney, Abstracts* 3, 858.

Kirschvink, J.L., 1978*a*. The Precambrian–Cambrian boundary problem: magnetostratigraphy of the Amadeus Basin, central Australia. *Geological Magazine* 115(2), 139–50.

Kirschvink, J.L., 1978*b*. The Precambrian–Cambrian boundary problem: palaeomagnetic directions from the Amadeus Basin, central Australia. *Earth and Planetary Science Letters* 40, 91–100.

Kirschvink, J.L., 1991. A palaeogeographic model for Vendian and Cambrian time. pp. 569–81 *in* J.W. Schopf, C. Klein & D. des Maris (eds), *The Proterozoic Biosphere: A Multidisciplinary Study*, Oxford University Press, Oxford.

Kirschvink, J.L. & Rozanov, A. Yu., 1984. Magnetostratigraphy of Lower Cambrian strata from the Siberian Platform: a palaeomagnetic pole and a preliminary polarity time-scale. *Geological Magazine* 121, 189–203.

Kirschvink, J.L., Magaritz, M., Ripperdan, R.L., Zhuravlev, A. Yu., & Rozanov, A. Yu., 1991. The Precambrian–Cambrian boundary: magnetostratigraphy and carbon isotopes resolve correlation problems between Siberia, Morocco and South China. *GSA Today* 1, 69–91.

Klapper, G., 1989. The Montagne Noire Frasnian (Upper Devonian) conodont succession. pp. 449–68 *in* N.J. McMillan, A.F. Embry & D.J. Glass (eds), vol. 3, q.v.

Klapper, G. & Foster, C.T., 1993. Shape analysis of Frasnian species of the Late Devonian conodont genus *Palmatolepis*. *Palaeontological Society Memoir* 32, 1–35.

Klapper, G. & Johnson, J.G., 1990. Revisions of Middle Devonian conodont zones. *Journal of Paleontology* 64, 934–36

Klapper, G. & Ziegler, W., 1979. Devonian conodont biostratigraphy. *Special Papers in Palaeontology* 23, 199–244.

Klapper, G., Feist, R. & House, M.R., 1987. Decision on the boundary stratotype for the Middle/

Upper Devonian Series Boundary. *Episodes* 10, 97–101.

Kleffner, M.A., 1989. A conodont-based Silurian chronostratigraphy. *Geological Society of America Bulletin* 101, 904–12.

Klootwijk, C.T., 1980. Early Palaeozoic magnetism in Australia. *Tectonophysics* 64, 249–332.

Klootwijk, C.T., Idnurm, M., Théveniaut, H. & Trench, A., 1994. Phanerozoic magnetostratigraphy: a contribution to the Timescales Project. *Australian Geological Survey Organisation, Record* 1994/45, 52 pp.

Kobayashi, T., 1940*a*. Lower Ordovician fossils from Junee, Tasmania. *Papers and Proceedings of the Royal Society of Tasmania 1939*, 61–66.

Kobayashi, T., 1940*b*. Lower Ordovician fossils from Caroline Creek, near Latrobe, Mersey River district, Tasmania. *Papers and Proceedings of the Royal Society of Tasmania 1939* 67–76.

Kobayashi, T., 1949. The *Glyptagnostus* hemera, the oldest world-instant. *Journal of Japanese Geology and Geography* 21, 1–6.

Koren, T.N. & Karpinsky, A.P., 1984. Graptolite zones and standard stratigraphic scale of Silurian. *Proceedings of the 27th International Geological Congress* 1, 47–76.

Koren, T.N., & Modzalevskaya, T.L., 1991. Siluriyskaya Sistema. pp. 34–48 *in* T.N. Koren et al. (eds), Zonalnaya Stratigrafiya Fanerozoya SSSR. *Vsesoyuznyy Nauchno-Issledovatelskiy Geologicheskiy Institut (VSEGEI)*, 'Nedra', Moscow.

Kotlyar, G.V. & Stepanov, D.L. (eds), 1984. Main features of stratigraphy of the Permian System in the USSR. *Vsesoyuznyy Nauchno-Issledovatelskiy Geologicheskiy Institut, Trudy* 286, 1–280.

Kotlyar, G.V., Leven, E.Y., Bogoslovskaya, M.F. & Dmitriev, V.U., 1987. Stages of the Permian deposits of the Tethyan Region. *Soviet Geology* 1987/2, 53–62.

Kotlyar, G.V., Zhakharov, Yu. D., Kropatcheva, G.S., Pronina, G.P., Che Dija, I.O. & Burago, V.I., 1989. *Evolution of the latest Permian Biota, Midian Regional Stage in the USSR*, 'Nauka', Leningrad, 1–185.

Kozur, H., 1992. Boundaries and stage subdivision of the mid-Permian (Guadalupian Series) in the light of new micropaleontological data. *International Geological Review* 34, 907–32.

Kozur, H., Leve, E.Y., Losovskiy, V.R. & Pyatakova, M.V., 1978. Subdivision of Permian–Triassic boundary beds in Transcaucasia on the basis of conodonts. *International Geology Review* 22(3), 361–68.

Krishna, J., 1984. Current status of the Jurassic stratigraphy of Kachchh, western India. pp. 732–42 *in* O. Michelsen & A. Zeiss (eds), *IUGS International Symposium on Jurassic Stratigraphy*, vol. III, Geological Survey of Denmark, Copenhagen.

Kristan-Tollman, E., 1986. Beobachtungen zur Trias am Südostende der Tethys Papua Neuguinea, Australien, Neuseeland. *Neues Jahrbuch für Geologie und Paläontologie, Monatschefte* 1986 (4), 201–22.

Kruse, P.D., 1978. New Archaeocyatha from the Early Cambrian of the Mt Wright area, New South Wales. *Alcheringa* 2, 27–47.

Kruse, P.D., 1982. Archaeocyathan biostratigraphy of the Gnalta Group at Mt Wright, New South Wales. *Palaeontographica* [A], 177, 1–212.

Kruse, P.D., 1991. Cambrian of the Top Springs Limestone, Georgina Basin. *The Beagle, Records of the Northern Territory Museum of Arts and Sciences* 8 (1), 169–88.

Kruse, P.D. & West, P.W., 1980. Archaeocyatha of the Amadeus and Georgina basins. *BMR Journal of Australian Geology & Geophysics* 5, 165–81.

Krystyn, L., 1973. Zur Ammoniten- und Conodonten-Stratigraphie der Hallstätter Obertrias (Salzkammergut, Österreich). *Verhandlungen Geologisches Bundesanstalt, Austria* 1973/1, 113–53.

Krystyn, L., 1980. Triassic conodont localities of the Salzkammergut region (northern Calcareous Alps). pp. 61–98 *in* H.P. Schönlaub (ed.), *Second European Conodont Symposium, Guidebook and Abstracts.* Abhandlungen des Geologischen Bundesanstalt, Austria.

Krystyn, L., 1988. Zur Rhät-Stratigraphie in den Zlambach-Schichten (vorläufiger Bericht). *Sitzungsberichten der Österreichische Akademie der Wissenschaften in Wien, Mathematisch-Naturwissenschaftliche Klasse* 196, 21–33.

Kullmann, J., 1993. Paleozoic ammonoids of Mexico and South America. *Compte Rendu 12ème Congrès International de la Stratigraphie et Géologie du Carbonifère et Permien* 1, 557–62.

Kullmann, J., Korn, D. & Weyer, D., 1991 [imprint 1990]. Ammonoid zonation of the Lower Carboniferous Subsystem. *Courier Forschungsinstitut Senckenberg* 130, 127–31.

Kummel, B., 1972. The Lower Triassic (Scythian) ammonoid *Otoceras*. *Harvard University, Museum of Comparative Zoology, Bulletin* 143, 365–417.

Kutek, J., Matyja, B.A. & Wierzbowski, A., 1984. Late Jurassic biogeography in Poland and its stratigraphical implications. pp. 743–54 *in* O. Michelsen & A. Zeiss (eds), *IUGS International Symposium on Jurassic Stratigraphy*, vol. III, Geological Survey of Denmark, Copenhagen.

Kuznetsova, K.I., 1974. Distribution of benthonic Foraminifera in Upper Jurassic and Lower Cretaceous deposits at site 261, DSPD Leg 27, in the eastern Indian Ocean. *Initial Reports of the Deep Sea Drilling Project* 27, 673–81.

Laird, M.G., 1981. The Late Mesozoic fragmentation of the New Zealand segment of Gondwana. pp. 311–18 *in* M.M. Cresswell & P. Vella (eds), *Gondwana Five*, Balkema, Rotterdam.

Landing, E., 1992*a*. Precambrian–Cambrian GSSP, SE Newfoundland: biostratigraphy and geochronology. *Phanerozoic Time Scale, Bulletin de Liaison et Informations* 11, 6–8.

Landing, E., 1992*b*. Lower Cambrian of southeastern Newfoundland. Epeirogeny and Lazarus faunas, lithofacies–biofacies linkages, and the myth of global chronology. pp. 283–309 *in* J.H. Lipps & P.W. Signor (eds), *Origin and Early Evolution of the Metazoa*, Plenum Press, New York.

Landing, E., 1994. Precambrian–Cambrian boundary global stratotype ratified and a new perspective of Cambrian time. *Geology* 22, 179–82.

Lane, H.R. & Straka, J.J., 1974. Late Mississippian and Early Pennsylvanian conodonts, Arkansas and Oklahoma. *Geological Society of America Special Paper* 152, 1–144.

Lane H.R. & Ziegler, W., 1983. Taxonomy and phylogeny of *Scaliognathus* Branson & Mehl 1941 (Conodonta, Lower Carboniferous). *Senckenbergiana lethaea* 64(2/4), 199–225.

Lane, H.R. & Ziegler, W. (eds), 1985. Toward a boundary in the Carboniferous: Stratigraphy and Palaeontology. *Courier Forschungsinstitut Senckenberg* 74, 1–195.

Lane, H.R., Baesemann, J.F. & Groves, J.R., 1985*a*. Is the base of the *Reticuloceras* Zone a reliably recognizable biostratigraphic level? pp. 137–48 *in* H.R. Lane & W. Ziegler (eds), q.v.

Lane, H.R., Sandberg, C.A. & Ziegler, W., 1980. Taxonomy and phylogeny of some Lower Carboniferous conodonts and preliminary standard post-*Siphonodella* zonation. *Geologica et Palaeontologica* 14, 117–64.

Lane, H.R., Merrill, G.K., Straka, J.J. & Webster, G.D., 1971. North American Pennsylvanian conodont biostratigraphy. *Geological Society of America Memoir* 127, 395–414.

Lane, H.R., Bouckaert, J., Brenckle, P., Einor, O.L., Havlena, V., Higgins, A.C., Yang, Jing-Zhi, Manger, W.L., Nassichuk, N., Nemirovskaya, T., Owens, B., Ramsbottom, W.H.C., Reitlinger, E.A. & Weyant, M., 1985*b*. Proposal for an international Mid-Carboniferous Boundary. *Compte Rendu 10ème Congrès International de Stratigraphie et de Géologie du Carbonifère, Madrid, 1983*, 4, 323–39.

Lane, P.D. & Thomas, A.T., 1978. Silurian trilobites from north-east Queensland and the classification of effaced trilobites. *Geological Magazine* 115, 351–58.

Lanphere, M.A. & Jones, D.L., 1978. Cretaceous time scale from North America. pp. 259–68 *in* G.V. Cohee et al., q.v.

Lapworth, C., 1879. On the tripartite classification of the Lower Palaeozoic rocks. *Geological Magazine* 16, 1–15.

Larson, E.E., Walker, T.R., Patterson, P.E., Hoblitt, R.P. & Rosenbaum, J.G., 1982. Paleomagnetism of the Moenkopi Formation, Colorado Plateau: Basis for a long term model of acquisition of chemical remanent magnetism in red beds. *Journal of Geophysical Research* 87, 1081–106.

Larson, M.L. & Jackson, D.E., 1966. Biostratigraphy of the Glenogle Formation (Ordovician) near Glenogle, British Columbia. *Bulletin of Canadian Petroleum Geology* 14, 486–503.

Larson, R.L. & Hilde, T.W.C., 1975. A revised time scale of magnetic reversals for the Early Cretaceous and Late Jurassic. *Journal of Geophysical Research* 80, 2586–94.

Laurie, J.R., 1980. Early Ordovician orthide brachiopods from southern Tasmania. *Alcheringa* 4, 11–23.

Laurie, J.R., 1987*a*. A re-assessment of the brachiopod genus *Spanodonta* Prendergast from the Lower Ordovician of Western Australia. *Alcheringa* 11, 43–49.

Laurie, J.R., 1987*b*. Early Ordovician orthide brachiopods from the Digger Island Formation, Waratah Bay, Victoria. *Memoirs of the Museum of Victoria* 48 (2), 101–6.

Laurie, J.R., 1991*a*. Ordovician brachiopod biostratigraphy of Tasmania. pp. 303–10 *in* D.I. MacKinnon, D.E. Lee & J.D. Campbell (eds), *Brachiopods through Time*, Balkema, Rotterdam.

Laurie, J.R., 1991*b*. Articulate brachiopods from the Ordovician and Lower Silurian of Tasmania. *Association of Australasian Palaeontologists Memoir* 11, 1–106.

Laurie, J.R., in prep. Early Ordovician trilobite faunas of the Horn Valley Siltstone, Amadeus Basin, central Australia.

Laurie, J.R. & Shergold, J.H., 1985. Phosphatic organisms and correlation of the Early Cambrian carbonate formations in central Australia. *BMR Journal of Australian Geology & Geophysics* 9, 83–89.

Laurie, J.R. & Shergold, J.H., in press. Early Ordovician trilobite fauna of the Emanuel Formation, Canning Basin, Western Australia.

Lavering, I.H., 1993. Gradational benthic marine communities in the Early Carboniferous *Rhipidomella fortimuscula* Zone (late Viséan), New South Wales. *AGSO Journal of Australian Geology & Geophysics* 14(4), 361–70.

Lees, T., 1986*a*. Catalogue of type, figured and mentioned fossil fish, amphibians and reptiles held by the Queensland Museum. *Memoirs of the Queensland Museum* 22, 265–88.

Lees, T., 1986*b*. A new chimaeroid *Ptyktoptychion tayyo* gen. et sp. nov. (Pisces: Holocephali) from the marine Cretaceous of Queensland. *Alcheringa* 10, 187–93.

Legg, D.P., 1976. Ordovician trilobites and graptolites from the Canning Basin, Western Australia. *Geologica et Palaeontologica* 10, 1–37.

Legg, D.P., 1978. Ordovician biostratigraphy of the Canning Basin, Western Australia. *Alcheringa* 2, 321–34.

Lehman, T.M., 1987. Late Maastrichtian palaeoenvironments and dinosaur biogeography in the Western Interior of North America. *Palaeogeography, Palaeoclimatology, Palaeoecology* 60, 189–217.

Leichhardt, L., 1847. *Journal of an Overland Expedition in Australia from Moreton Bay to Port Essington, a Distance of Upwards of 3000 Miles, during the Years 1844–1845*, T. & W. Boone, London.

Lenz, A.C., 1988. Ordovician series/stages, graptolite basinal facies, northern Yukon, Canada. *New York State Museum Bulletin* 462, 59–64.

Lenz, A.C. & Chen, X., 1985. Middle to Upper Ordovician graptolite biostratigraphy of Peel River and other areas of the northern Canadian Cordillera. *Canadian Journal of Earth Sciences* 22, 227–39.

Lenz, A.C. & Jackson, D.E., 1986. Arenig and Llanvirn graptolite biostratigraphy, Canadian Cordillera. pp. 27–45 *in* C.P. Hughes & R.B. Rickards (eds), Palaeoecology and biostratigraphy of graptolites. *Geological Society Special Publication* 20.

Leonova, T.B. & Bogoslovskaya, M.F., 1990. Filogeneticheskie svyazi v nadsemeistve Adrianitaceae. *Trudy Paleontologicheskogo Instituta* 243, 87–97.

Li Qianyu & McGowran, B., 1994. Miocene upwelling events: neritic foraminiferal evidence from southern Australia. *Australian Journal of Earth Sciences* 41, 593–603.

Lin, J.R., Fuller, M. & Zhang, W.Y., 1985. Paleogeography of the North and South China Blocks during the Cambrian. *Journal of Geodynamics* 2, 91–114.

Lindsay, J.M., 1967. Foraminifera and stratigraphy of the type section of the Port Willunga Beds, Aldinga Bay, South Australia. *Transactions of the Royal Society of South Australia* 91, 93–110.

Lindsay, J.M., 1969. Cainozoic Foraminifera and stratigraphy of the Adelaide Plains Sub-Basin, South Australia. *Geological Survey of South Australia, Bulletin* 42, 62 pp.

Lindsay, J.M., 1985. Aspects of South Australian Tertiary foraminiferal biostratigraphy, with emphasis on studies of *Massilina* and *Subbotina*. *Geological Survey of South Australia, Special Publication* 5, 187–231.

Link, A.G. & Druce, E.C., 1972. Ludlovian and Gedinnian conodont stratigraphy of the Yass Basin, New South Wales. *Bureau of Mineral Resources, Geology & Geophysics, Bulletin* 134, 136 pp.

Lipina, O.A. & Reitlinger, E.A., 1970. Stratigraphie zonale et Paléozoogéographie du Carbonifère Inférieur d'après les Foraminifères. *Compte Rendu 6ème Congrès International de Stratigraphie et de Géologie du Carbonifère, Sheffield, 1967*, 3, 1101–12.

Lipina, O.A. & Tschigova, V.A., 1979. The Tournaisian–Viséan boundary on the Russian Platform and Urals according to foraminifers and ostracods. *Compte Rendu 8ème Congrès International de Stratigraphie et de Géologie du Carbonifère, Moscow, 1975*, 3, 283–87.

Lippolt, H.J. & Hess, J.C., 1983. Isotopic evidence for the stratigraphic position of the Saar–Nahe

Rotliegend volcanism I. $^{40}Ar/^{40}K$ and $^{40}Ar/^{39}Ar$ investigations. *Neues Jahrbuch für Geologie und Paläontologie Monatshefte* 1983(12), 713–30.

Lippolt, H.J. & Hess, J.C., 1985. Ar^{40}/Ar^{39} dating of sanidines from Upper Carboniferous tonsteins. *Compte Rendu 10ème Congrès International de Stratigraphie et de Géologie du Carbonifère, Madrid, 1983*, 4, 175–81.

Loboziak, S., Coquel, R. & Owens, B., 1984. Les miospores des Formations Hale et Bloyd du Nord Arkansas. pp. 385–90 *in* P.K. Sutherland & W.L. Manger (eds), 1984*b*, q.v.

Lock, J. & McElhinny, M.W., 1991. The global palaeomagnetic database. *Surveys in Geophysics* 12, 317–491.

Long, J., 1982. The history of fishes on the Australian continent. pp. 53–85 *in* P.V. Rich & E.M. Thompson (eds), q.v.

Long, J.A., 1983*a*. New bothriolepid fish from the Late Devonian of Victoria, Australia. *Palaeontology* 26, 295–320.

Long, J.A., 1983*b*. A new diplacanthoid acanthodian from the Late Devonian of Victoria. *Association of Australasian Palaeontologist Memoir* 1, 51–65.

Long, J.A., 1984*a*. New placoderm fishes from the Early Devonian Buchan Group, eastern Victoria. *Proceedings of the Royal Society of Victoria* 96, 173–86.

Long J.A., 1984*b*. New phyllolepids from Victoria, and the relationships of the group. *Proceedings of the Linnean Society of New South Wales* 107, 263–308.

Long, J.A., 1985. The structure and relationships of a new osteolepiform fish from the Late Devonian of Victoria, Australia. *Alcheringa* 9, 1–22.

Long, J.A., 1986. New ischnacanthid acanthodians from the Early Devonian of Australia, with comments on acanthodian interrelationships. *Zoological Journal of the Linnean Society of London* 87, 321–39.

Long, J.A., 1987. A redescription of the lungfish *Eoctenodus microsoma* Hills 1929, with reassessment of other Australian records of the genus *Dipterus* Sedgwick & Murchison 1828. *Records of the Western Australian Museum* 13, 297–314.

Long, J.A., 1988*a*. New palaeoniscoid fishes (Osteichthyes, Actinopterygii) from the Late Devonian and Early Carboniferous of Victoria, Australia. *Association of Australasian Palaeontologists Memoir* 7, 1–64.

Long, J.A., 1988*b*. Late Devonian fishes from Gogo, Western Australia. *National Geographic Research* 4, 436–50.

Long, J.A., 1989. A new rhizodontiform fish from the Early Carboniferous of Victoria, Australia, with remarks on the phylogenetic position of the group. *Journal of Vertebrate Paleontology* 9, 1–17.

Long, J.A., 1990. Late Devonian chondrichthyans and other microvertebrate remains from northern Thailand. *Journal of Vertebrate Paleontology* 10, 59–71.

Long, J., 1991. The long history of Australian fossil fishes. pp. 337–428 *in* P.V. Rich et al. (eds), q.v.

Long, J.A., 1992. Cranial anatomy of two new Late Devonian lungfishes (Pisces: Dipnoi) from Mount Howitt, Victoria. *Records of the Australian Museum* 44, 299–318.

Long, J. & Turner, S., 1984. A checklist and bibliography of Australian fossil fish. pp. 235–54 *in* M. Archer & G. Clayton (eds), q.v.

Long, J.A. & Werdelin, L., 1986. A new Late Devonian bothriolepid (Placodermi, Antiarcha) from Victoria, with descriptions of other species from the state. *Alcheringa* 10, 355–99.

Lopez-Gamundi, O.R. & Espejo, I.S., 1993. Correlation of a paleoclimatic mega-event: the Carboniferous glaciation in Argentina. *Compte Rendu 12ème Congrès International de la Stratigraphie et Géologie du Carbonifère et Permien* 1, 313–24.

Lopez-Gamundi, O.R. & Rossello, E.A., 1993. Devonian–Carboniferous unconformity in Argentina and its relation to the Eo-Hercynian orogeny in southern South America. *Geologische Rundschau* 82, 136–47.

Lovering, J.F., 1953. A microfossil assemblage from the Minchinbury Sandstone. *Australian Journal of Science* 15, 171–73.

Lowrie, W. & Alvarez, W., 1981. One hundred million years of geomagnetic polar history. *Geology* 9, 392–97.

Lowrie, W. & Alvarez, W., 1984. Lower Cretaceous magnetic stratigraphy in Umbrian pelagic limestone sections. *Earth and Planetary Science Letters* 71, 315–28.

Lowrie, W. & Ogg, J.G., 1986. A magnetic polarity time scale for the Early Cretaceous and Late Jurassic. *Earth and Planetary Science Letters* 76, 341–49.

Lowrie, W., Alvarez, W., Premoli-Silva, I. & Monechi, S., 1980. Lower Cretaceous magnetic stratigraphy in Umbrian pelagic carbonate rocks. *Royal Astronomical Society Geophysical Journal* 60, 263–81.

Luck, G.R., 1973. Palaeomagnetic results from Palaeozoic rocks of south-east Australia. *Geophysical Journal of the Royal Astronomical Society* 32, 35–52.

Ludbrook, N.H., 1961. Mesozoic non-marine Mollusca (Pelecypoda: Unionidae) from the north of South Australia. *Transactions of the Royal Society of South Australia* 84, 137–47.

Ludbrook, N.H., 1963. Correlation of the Tertiary rocks of South Australia. *Transactions of the Royal Society of South Australia* 87, 5–15.

Ludbrook, N.H., 1966. Cretaceous biostratigraphy of the Great Artesian Basin in South Australia. *Geological Survey of South Australia Bulletin* 40, 223 pp.

Ludbrook, N.H., 1971. Stratigraphy and correlation of marine sediments in the western part of the Gambier Embayment. pp. 47–66 *in* H. Wopfner & J.G. Douglas (eds), The Otway Basin of South-eastern Australia. *Geological Surveys of South Australia and Victoria, Special Bulletin.*

Ludbrook, N.H., 1973. Distribution and stratigraphic utility of Cenozoic molluscan faunas in southern Australia. *Scientific Reports of the Tohoku University, Series 2 (Geology), Special Volume, Hatai Memorial Volume* 6, 241–61.

Ludbrook, N.H. & Lindsay, J.M., 1966. The Aldingan Stage. *Geological Survey of South Australia, Quarterly Geological Notes* 19, 1–2.

Ludbrook, N.H. & Lindsay, J.M., 1969. Tertiary foraminiferal zones in South Australia. pp. 366–74 *in* P. Brönnimann & H.H. Renz (eds), *Proceedings of the 1st International Conference on Planktonic Microfossils, Geneva, 1967*, II, Brill, Leiden.

Lutke, F., 1979. Biostratigraphical significance of the Devonian Dacryoconarida. *Special Papers in Palaeontology* 23, 281–89.

Ma, X., McElhinny, M.W., Embleton, B.J.J. & Zhang, Z., 1993. Permo–Triassic palaeomagnetism in the Emei Mountain region, south-west China. *Geophysical Journal International* 114, 293–303.

MacFadden, B.J., Whitelaw, M.J., McFadden, P. & Rich, T.H.V., 1987. Magnetic polarity stratigraphy of the Pleistocene section at Portland (Victoria), Australia. *Quaternary Research* 28, pp. 364–73.

MacPhail, M.K. & Hill, R.S., in press. Potassium argon dated palynofloras 1: the Early–Middle Oligocene Wilmot Dam site. *Papers and Proceedings of the Royal Society of Tasmania.*

MacPhail, M.K. & Kellett, J.R., 1993. Palynostratigraphy of the Bookpurnong Beds and related Late Miocene–Early Pliocene facies in the central west Murray Basin, part 1: dinoflagellates. *AGSO Journal of Australian Geology & Geophysics* 14, 371–82.

MacPhail, M.K. & Truswell, E.M., 1989. Palynostratigraphy of the central west Murray Basin. *BMR Journal of Australian Geology & Geophysics* 11, 301–31.

MacPhail, M.K. & Truswell, E.M., 1993. Palynostratigraphy of the Bookpurnong Beds and related Late Miocene–Early Pliocene facies in the central west Murray Basin, part 2: spores and pollen. *AGSO Journal of Australian Geology & Geophysics* 14, 383–409.

MacPhail, M.K., Kellett, J.R., Rexilius, J.P. & O'Rorke, M.E., 1993. The 'Geera Clay equivalent': a regressive marine unit that sheds new light on the age of the Mologa weathering surface in the Murray Basin. *AGSO Journal of Australian Geology & Geophysics* 14, 47–63.

Magnus, G. & Opdyke, N.D., 1991. A paleomagnetic investigation of the Minturn Formation, Colorado: a study in establishing the timing of remanence acquisition. *Tectonophysics* 187, 181–89.

Maitland, A.G., 1919. A summary of the geology of Western Australia. *Memoir of the Geological Survey of Western Australia* 1, 1–55.

Mallett, C.W., 1977. Studies in Victorian Tertiary Foraminifera: Neogene planktonic faunas. Unpublished PhD Thesis, University of Melbourne.

Mallett, C.W., 1978. Sea level changes in the Neogene of southern Australia. *APEA Journal* 18, 64–69.

Mallett, C.W., 1982. Late Pliocene planktonic Foraminifera from subsurface shell beds, Jandakot, near Perth, western Australia. *Search* 13, 35–36.

Malone, E.J., Jensen, A.R., Gregory, C.M. & Forbes, V.R., 1966. Geology of the southern half of the Bowen 1:250,000 sheet area, Queensland. *Bureau of Mineral Resources, Geology & Geophysics, Report* 100, 1–87.

Mamet, B., 1974. Une zonation par Foraminifères due Carbonifère Inférieur de la Téthys Occidentale. *Compte Rendu 7ème Congrès International de Stratigraphie et de Géologie du Carbonifère, Krefeld, 1971*, 3, 391–408.

Mamet, B.L. & Belford, D.J., 1968. Carboniferous Foraminifera, Bonaparte Gulf Basin, northwestern Australia. *Micropaleontology* 14, 339–347.

Mamet, B.L. & Playford, P.E., 1968. Sur la présence de Quaisiendothyrinae (Foraminifères), en Australie occidentale (Canning Basin). *Compte Rendu Sommaire des Séances de la Société Géologique de France 1968, fascicule F., Séance du 28 Octobre 1968*, 229.

Mamet, B.L. & Roux, A., 1983. Algues Dévono-carbonifères de l'Australie. *Revue de Micropaléontologie* 26(2), 63–131.

Manger, W.L. & Saunders, W.B., 1980. Lower Pennsylvanian (Morrowan) ammonoids from the North American midcontinent. *The Paleontological Society Memoir* 10, 1–56.

Manger, W.L. & Sutherland, P.K., 1991 [imprint 1990]. Comparative ammonoid/conodont-based and foraminifer-based Middle Carboniferous correlations. *Courier Forschungsinstitut Senckenberg* 130, 345–50.

Maples, C.G. & Waters, J.A., 1987. Redefinition of the Meramecian–Chesterian boundary (Mississippian). *Geology* 15 (7), 647–51.

Marsden, M.A.H., (ed.) 1988. Upper Devonian–Carboniferous. pp. 147–94 *in* J.G. Douglas & J.A. Ferguson (eds), q.v.

Marshall, C.R., 1990. Confidence intervals on stratigraphic ranges. *Paleobiology* 16, 1–10.

Martin, H.A., 1973. Upper Tertiary palynology in southern New South Wales. *Geological Society of Australia, Special Publication* 4, 35–54.

Martin, H.A., 1977. The Tertiary stratigraphic palynology of the Murray Basin in New South Wales. I, The Hay–Balranald–Wakool districts. *Journal and Proceedings of the Royal Society of New South Wales* 110, 41–47.

Martin, H.A., 1984. The use of qualitative relationships and palaeoecology in stratigraphic palynology of the Murray Basin in New South Wales. *Alcheringa* 8, 253–72.

Martin, H.A., 1991. Dinoflagellate and spore–pollen biostratigraphy of the S.A.D.M.E. MC63 bore, western Murray Basin. *Alcheringa* 15, 107–44.

Martini, E., 1971. Standard Tertiary and Quaternary calcareous nannoplankton zonation. pp. 739–85 *in* A. Farinacci (ed.), *Proceedings of the Second Planktonic Conference, Roma 1970*, Edizioni Tecnoscienza, Roma.

Martinsson, A. (ed.), 1977. *The Silurian–Devonian Boundary*. Final report of the Committe of the Siluro–Devonian Boundary within IUGS Commission on Stratigraphy and a state of the art report for Project Ecostratigraphy, Schweizerbartsche, Stuttgart.

Martinsson, A., Bassett, M.G. & Holland, C.H., 1981. Ratification of standard chronostratigraphical divisions and stratotypes for the Silurian System. *Lethaia* 14(2), 168.

Marton, E., Marton, P. & Heller, F., 1980. Remanent magnetization of a Pliensbachian limestone sequence at Bakonycsernye (Hungary). *Earth and Planetary Science Letters* 57, 182–90.

Matheson, R.S. & Teichert, C., 1948. Geological reconnaissance in the eastern portion of the Kimberley Division, Western Australia. *Annual Report of the Department of Mines, Western Australia for 1945*, 73–87.

Matsumoto, T., 1980. Inter-regional correlation of transgressions and regressions in the Cretaceous Period. *Cretaceous Research* 1, 359–73.

Matthews, S.C., 1970. Comments on palaeontological standards for the Dinantian. *Compte Rendu 6ème Congrès International de Stratigraphie et Géologie du Carbonifère, Sheffield, 1967*, 3, 1159–64.

Mauritsch, H.J. & Rother, K., 1983. Paleomagnetic investigations in the Thüringer Forest (GDR). *Tectonophysics* 99, 63–72.

Mawson, R., 1987. Documentation of conodont assemblages across the Early Devonian–Middle Devonian boundary, Broken River Formation, north Queensland, Australia. *Courier Forschungsinstitut Senckenberg* 92, 251–73.

Mawson, R. & Talent, J.A., 1989. Late Emsian–Givetian stratigraphy and conodont biofacies-carbonate slope and offshore shoal to sheltered lagoon and nearshore carbonate ramp—Broken River, north Queensland. *Courier Forschungsinstitut Senkenberg* 117, 205–209.

Mawson, R. & Talent, J.A., 1994. The Tamworth Group (mid-Devonian) at Attunga, New South Wales: conodont data and inferred ages. *Courier Forschungsinstitut Senckenberg* 168, 37–59.

Mawson, R., Jell, J.S., & Talent, J.A., 1985. Stage boundaries within the Devonian: implications for application to Australian sequences. *Courier Forschungsinstitut Senckenberg* 75, 1–16.

Mawson, R., Talent, J.A., Brock, G.A. & Engelbretsen, M.J., 1992. Conodont data in relation to sequences about the Pragian–Emsian boundary (Early Devonian) in south-eastern Australia. *Proceedings of the Royal Society of Victoria* 104, 23–56.

Mawson, R., Talent, J.A., Bear, V.C., Benson, D.S., Brock, G.A., Farrell, J.R., Hyland, K.A., Pyemont, B.D., Sloan, T.R., Sorentino, L., Stewart, M.I., Trotter, J.A., Wilson, G.A. & Simpson, A.G., 1989. Conodont data in relation to resolution of stage and zonal boundaries for the Devonian of Australia. pp. 485–527 *in* N.J. McMillan, A.F. Embry & D.J. Glass (eds), vol. 3, q.v.

Maxwell, W.G.H., 1954. Upper Palaeozoic formations in the Mt Morgan district—faunas. *Papers, Department of Geology, University of Queensland* 4 (5), 1–69.

Maxwell, W.G.H., 1961*a*. Lower Carboniferous brachiopod faunas from Old Cannindah, Queensland. *Journal of Paleontology* 35, 82–103.

Maxwell, W.G.H., 1961*b*. Lower Carboniferous gastropod faunas from Old Cannindah, Queensland. *Palaeontology* 4(1), 59–70.

Maxwell, W.G.H., 1964. The geology of the Yarrol Region. Part 1. Biostratigraphy. *Papers, Department of Geology, University of Queensland* 5(9), 1–79.

Mayne, S.J., Nicholas, E., Bigg-Withwer, A.L., Rasidi, J.S. & Raine, M.J., 1974. Geology of the Sydney Basin—a review. *Bureau of Mineral Resources, Geology & Geophysics, Bulletin* 149, 1–229.

McCabe, C. & Channell, J.E.T., 1990. Palaeomagnetic results from volcanic rocks of the Shelve Inlier, Wales: Evidence for a wide Late Ordovician Iapetus Ocean in Britain. *Earth and Planetary Science Letters* 96, 458–68.

McCabe, C., Van der Voo, R., Wilkinson, B.H. & Devaney, K., 1985. A Middle/Late Silurian palaeomagnetic pole from limestone reefs of the Wabash Formation (Indiana, USA). *Journal of Geophysical Research* 90, 2959–65.

McClung, G., 1978. Morphology, palaeoecology and biostratigraphy of *Ingelarella* (Brachiopoda: Spiriferacea) in the Bowen and Sydney basins of eastern Australia. *Geological Survey of Queensland Publications* 365, *Palaeontological Paper* 40, 18–60.

McClung, G., 1981. Review of the stratigraphy of the Permian Back Creek Group in the Bowen Basin, Queensland. *Geological Survey of Queensland Publication* 371, *Palaeontological Paper* 44, 1–31.

M'Coy, F. 1847. On the fossil botany and zoology of the rocks associated with the coal of Australia. *Annals and Magazine of Natural History* 20, 145–57.

M'Coy, F., 1874. *Prodromus of the Palaeontology of Victoria; or Figures and Descriptions of Victorian Organic Remains. Decade I.* Geological Survey of Victoria, Melbourne, 1–41.

M'Coy, F., 1875. *Prodromus of the Palaeontology of Victoria; or Figures and Descriptions of Victorian Organic Remains. Decade II.* Geological Survey of Victoria, Melbourne, 1–37.

McDougall, I., 1974. Potassium–argon ages on basaltic rocks recovered from DSDP Leg 22, Indian Ocean, 1974. *Initial Reports of the Deep Sea Drilling Project* XXII, 377–79.

McElhinny, M.W., 1969. The palaeomagnetism of the Permian of south-east Australia and its significance regarding the problem of intercontinental correlation. *Geological Society of Australia, Special Publication* 2, 61–67.

McElhinny, M.W. & Burek, P.J., 1971. Mesozoic palaeomagnetic stratigraphy. *Nature* 232, 98–102.

McElhinny, M. W. & Lock, J., 1990. Global Palaeomagnetic Database Project, *Physics of the Earth and Planetary Interiors* 63, 1–6.

McElhinny, M.W. & Lock, J., 1993. Global paleomagnetic database supplement number one: update to 1992. *Surveys in Geophysics* 14, 303–29.

McElroy, C.T., 1963. The geology of the Clarence–Moreton Basin. *Geological Survey of New South Wales, Memoir* 9, 172 pp.

McEwan Mason, J.R.C., 1991. The late Cainozoic magnetostratigraphy and preliminary palynology of Lake George, New South Wales. pp. 195–209 *in* M.A.J. Williams, P. De Deckker & A.P. Kershaw. The Cainozoic in Australia; a re-appraisal of the evidence. *Geological Society of Australia, Special Publication* 18.

McFadden, P.L., Ma, X.H., McElhinny, M.W. & Zhang, Z.K., 1988. Permo–Triassic magnetostratigraphy in China: northern Tarim. *Earth and Planetary Science Letters* 87, 152–60.

McGowran, B., 1973. Observation Bore No. 2, Gambier Embayment of the Otway Basin: Tertiary micropalaeontology, and stratigraphy. *South Australian Department of Mines, Mineral Resources Review* 135, 43–55.

McGowran, B., 1978. Early Tertiary foraminiferal biostratigraphy in southern Australia: a progress report. *Bureau of Mineral Resources, Geology & Geophysics, Bulletin* 192, 83–95.

McGowran, B., 1979. The Tertiary of Australia: Foraminiferal overview. *Marine Micropaleontology* 4, 235–64.

McGowran, B., 1986. Cainozoic oceanic and climatic events: the Indo–Pacific foraminiferal biostratigraphic record. *Palaeogeography, Palaeoclimatology, Palaeoecology* 55, 247–65.

McGowran, B., 1989. The later Eocene transgressions in southern Australia. *Alcheringa* 13, 45–68.

McGowran, B., 1991. Maastrichtian and Early Cainozoic, southern Australia: planktonic foraminiferal biostratigraphy. pp. 79–98 *in* M.A.J. Williams, P. De Deckker & A.P. Kershaw (eds), The Cainozoic in Australia: a re-appraisal of the evidence. *Geological Society of Australia, Special Publication* 18.

McGowran, B. & Beecroft, A., 1985. *Guembelitria* in the Early Tertiary of southern Australia and its palaeoceanographic significance. *South Australian Department of Mines and Energy, Special Publication* 5, 247–61.

McGowran, B. & Li Qianyu, 1993. Miocene planktonic Foraminifera from Lakes Entrance in Gippsland: midlatitude neritic signals from a transforming ocean. *Association of Australasian Palaeontologists Memoir* 15, 395–405.

McGowran, B., Lindsay, J.M. & Harris, W.K., 1971. Attempted reconciliation of Tertiary biostratigraphic systems. pp. 273–81 *in* H. Wopfner & J.G. Douglas (eds), The Otway Basin of southeastern Australia. *Geological Surveys of South Australia and Victoria, Special Bulletin.*

McGowran, B., Moss, G.D., & Beecroft, A., 1992. Late Eocene and Early Oligocene in southern Australia: local neritic signals of global oceanic changes. pp. 178–201 *in* D.R. Prothero & W.A. Berggren (eds), *Eocene–Oligocene Climatic and Biotic Evolution*, Princeton University Press, Princeton.

McGregor, D.C. & Playford, G., 1992. Canadian and Australian Devonian spores: zonation and correlation. *Geological Survey of Canada, Bulletin* 438, 1–83.

McIntosh, W.C., Hargraves, R.B. & West, C.L., 1985. Paleomagnetism and oxide mineralogy of Upper Triassic to Lower Jurassic red beds and basalts in the Newark Basin *Geological Society of America Bulletin* 96, 463–80.

McKellar, J.L., 1974. Jurassic miospores from the upper Evergreen Formation, Hutton Sandstone, and basal Injune Creek Group, north-eastern Surat Basin. *Geological Survey of Queensland Publication* 361, *Palaeontological Paper* 35, 1–89.

McKellar, J.L., 1977. Palynostratigraphy of core samples from the Hughenden 1: 250 000 sheet area, northern Galilee and Eromanga basins. *Queensland Government Mining Journal* 80, 295–302.

McKellar, R.G., 1967. The geology of the Cannindah Creek area, Monto district, Queensland. *Geological Survey of Queensland Publication* 331, 1–38.

McKellar, R.G., 1969. Brachiopods and trilobites from Siluro–Devonian strata in the Rockhampton district, Queensland. *Geological Survey of Queensland Publication* 337, *Palaeontological Paper* 11, 1–12.

McKellar, R.G., 1970. The Devonian productoid brachiopod faunas of Queensland. *Geological Survey of Queensland Publication* 342, *Palaeontological Paper* 18, 1–40.

McKeown, K.C., 1937. New fossil insect wings (Protohemiptera, family Mesotitanidae). *Records of the Australian Museum* 20, 31–37.

McKerrow, W.S., Lambert, R. St J. & Cocks, L.R.M., 1985. The Ordovician, Silurian and Devonian periods. pp. 73–80 *in* N.J. Snelling (ed.), q.v.

McKerrow, W.S., Scotese, C.R. & Brasier, M.D., 1992. Early Cambrian continental reconstructions. *Journal of the Geological Society* 149, 599–606.

McKinney, M.L., 1986*a*. Biostratigraphic gap analysis. *Geology* 14, 36–38.

McKinney, M.L., 1986*b*. How biostratigraphic gaps form. *Journal of Geology* 94, 875–84.

McLaren, D.J., 1973. The Siluro–Devonian Boundary. *Geological Magazine* 110, 302–303.

McLaurin, A.N., 1976. The Yapeenian (upper Lower Ordovician) succession in central Victoria, Australia. *Proceedings of the Royal Society of Victoria* 88, 23–30.

McLean, R.A., 1974*a*. Chonophyllinid corals from the Silurian of New South Wales. *Palaeontology* 17(3), 655–68.

McLean, R.A., 1974*b*. Cystiphyllidae and Goniophyllidae (Rugosa) from the Lower Silurian of New South Wales. *Palaeontographica* A 147(1/3), 1–38, pls1–6.

McLean, R.A., 1974*c*. The rugose coral genera *Streptelasma* Hall, *Grewingkia* Dybowski and *Calostylis* Lindström from the Lower Silurian of New South Wales. *Proceedings of the Linnean Society of New South Wales* 99(1), 36–53.

McLean, R.A., 1975*a*. Silurian rugose corals from the Mumbil area, central New South Wales. *Proceedings of the Linnean Society of New South Wales* 99(4), 181–96.

McLean, R.A., 1975*b*. Lower Silurian rugose corals from central New South Wales. *Journal and Proceedings of the Royal Society of New South Wales* 108(1/2), 54–69.

McLean, R.A., 1976. Aspects of the Silurian rugose coral fauna of the Yass region, New South Wales. *Proceedings of the Linnean Society of New South Wales* 100(3), 179–94.

McLean, R.A., 1977. Biostratigraphy and zoogeographic affinities of the Lower Silurian rugose corals of New South Wales, Australia. *Bureau des Recherches Géologiques et Minières, Mémoire* 89, 102–107.

McLean, R.A., 1985. New Early Silurian rugose corals from the Panuara area, central New South Wales. *Alcheringa* 9, 23–34.

McMillan, N.J., Embry, A.F. & Glass, D.J. (eds), 1989 [imprint 1988]. *Devonian of the World, vols 1–3*, Canadian Society of Petroleum Geologists, Calgary.

McMinn, A., 1985. Palynostratigraphy of the Middle Permian coal sequences of the Sydney Basin. *Australian Journal of Earth Sciences* 32, 301–309.

McMinn, A., 1987. Palynostratigraphy of the Stroud–Gloucester Trough, NSW. *Alcheringa* 11(2), 151–64.

McMinn, A., 1988. Outline of a Late Cretaceous dinoflagellate zonation of north-western Australia. *Alcheringa* 12, 137–56.

McMinn, A., 1992. Pliocene through Holocene dinoflagellate cyst biostratigraphy of the Gippsland Basin, Australia. pp. 147–61 *in* M.J. Head & J.H. Wrenn (eds), *Neogene and Quaternary Dinoflagellate Cysts and Acritarchs*, American Association of Stratigraphic Palynologists Foundation, Dallas.

McNamara, K.J., 1978. *Myloceras* (Ammonoidea) from the Albian of central Queensland. *Alcheringa* 2, 231–42.

McNamara, K.J., 1980. Heteromorph ammonites from the Albian of South Australia. *Royal Society of South Australia, Transactions* 104, 145–59.

McNamara, K.J., 1985. A new micromorph ammonite genus from the Albian of South Australia. *South Australian Department of Mines and Energy Special Publication* 5, 263–68.

McTavish, R.A., 1970. Triassic conodonts in Western Australia. *Search* 1(4), 159–60.

McTavish, R.A., 1973. Triassic conodont faunas from Western Australia. *Neues Jahrbuch für Geologie und Paläontologie, Abhandlung* 143(3), 275–303.

McTavish, R.A. & Legg, D.P., 1972. Middle Ordovician correlation—conodont and graptolite evidence from Western Australia. *Neues Jahrbuch für Geologie und Paläontologie Monatshefte* 8, 465–74.

McTavish, R.A. & Legg, D.P., 1976. The Ordovician of the Canning Basin, Western Australia. pp. 447–78 *in* M.G. Bassett (ed.). *The Ordovician System: Proceedings of a Palaeontological Association Symposium*, University of Wales Press and National Museum of Wales, Cardiff.

McWhae, J.R.H., Playford, P.E., Lindner, A.W., Glenister, B.F. & Balme, B.E., 1958. The stratigraphy of Western Australia. *Journal of the Geological Society of Australia* 4, 1–161.

McWilliams, M., 1993. On integrating palaeomagnetism and geochronology. p. 9 *in* C. Klootwijk (ed.). Palaeomagnetism in Australasia: dating, tectonic and environmental applications. *Australian Geological Survey Organisation Record* 1993/20.

Medd, A.W., 1966. The fine structure of some Lower Triassic acritarchs. *Palaeontology* 9 (2), 351–54.

Medd, A.W., 1982. Nannofossil zonation of the English Middle and Upper Jurassic. *Marine Micropaleontology* 7, 73–95.

Mégnien, C. & Mégnien, F. (eds), 1980. Synthèse géologique du Bassin de Paris. vol. 1: Stratigraphie et paléogéographie. *Mémoire du Bureau des Recherches Géologiques et Minières* 101, 466 pp.

Meischner, D., 1970. Conodonten-chronologie des Deutschen Karbons. *Compte Rendu 6ème Congrès International de Stratigraphie et de Géologie du Carbonifère, Sheffield, 1967*, 3, 1169–80.

Memmi, L. & Salaj, J., 1975. Le Berriasien de Tunisie, succession de faunes d'ammonites, de foraminifères et de tintinnides. *Mémoire du*

Bureau des Recherches Géologiques et Minières 86, 58–67.

Menning, M., 1986. Zur Dauer des Zechsteins aus magnetostratigraphischer Sicht. *Zeitschrift für Geologischen Wissenschaften* 14, 395–404.

Menning, M., Katzung, G. & Lützner, H., 1988. Magnetostratigraphic investigations in the Rotliegendes (300–252 Ma) of central Europe. *Zeitschrift für Geologischen Wissenschaften* 16, 1045–63.

Merrill, G.K., 1975. Pennsylvanian conodont biostratigraphy and paleoecology of northwestern Illinois. *Geological Society America Microform Publication* 3, 1–127.

Mesezhnikov, M.S., 1988. Tithonian (Volgian). pp. 50–62 *in* G.Y. Krymholts, M.S. Mesezhnikov & G.E.G. Westermann (eds). The Jurassic ammonite zones of the Soviet Union. *Geological Society of America Special Paper* 223.

Metcalfe, I., 1981. Conodont zonation and correlation of the Dinantian and early Namurian strata of the Craven Lowlands of northern England. *Institute of Geological Sciences Report* 80/10, 1–70.

Meyen, S.V., 1987. *Fundamentals of Palaeobotany*, Chapman and Hall, London, New York, 432 pp.

Michelsen, O. & Zeiss, A. (eds), 1984. *IUGS International Symposium on Jurassic Stratigraphy (Erlangen, 1984), vols I–III.* Geological Survey of Denmark, Copenhagen.

Miller, J.D. & Kent, D.V., 1989. Palaeomagnetism of the Upper Ordovician Juniata Formation of the central Appalachians revisited again. *Journal of Geophysical Research* 94, 1843–49.

Miller, J.D. & Opdyke, N.D., 1985. Magnetostratigraphy of the Red Sandstone Creek section—Vail, Colorado. *Geophysical Research Letters* 12, 133–36.

Miller, J.F., 1969. Conodont fauna of the Notch Peak Limestone (Cambro–Ordovician) House Range, Utah. *Journal of Paleontology* 43, 413–39.

Miller, J.F., 1980. Taxonomic revisions of some Upper Cambrian and Lower Ordovician conodonts with comments on their evolution. *University of Kansas Paleontological Contributions, Paper* 99, 39 pp.

Miller, J.F., 1984. Cambrian and earliest Ordovician conodont evolution, biofacies, and provincialism. *Geological Society of America Special Paper* 196, 43–68.

Miller, J.F., 1988. Conodonts as biostratigraphic tools for redefinition and correlation of the Cambrian–Ordovician Boundary. *Geological Magazine* 125, 349–62.

Miller, K.G., Aubry, M.-P., Khan, M.J., Melillo, A.J., Kent, D.V. & Berggren, W.A., 1985. Oligocene–Miocene biostratigraphy, magnetostratigraphy, and isotopic statigraphy of the western North Atlantic. *Geology* 13, 257–61.

Mitchell, J., 1887. On some new trilobites from Bowning, NSW. *Proceedings of the Linnean Society of New South Wales* 2, 435–40.

Mitchell, J., 1888. On a new trilobite from Bowning. *Proceedings of the Linnean Society of New South Wales* 3, 397–99.

Mitchell, J., 1919. On two new trilobites from Bowning. *Proceedings of the Linnean Society of New South Wales* 44, 441–49.

Mitchell, J., 1923. The Strophomenidae from the fossiliferous beds of Bowning, New South Wales, Part I: *Stropheodonta. Proceedings of the Linnean Society of New South Wales* 48, 465–74.

Mitchell, J., 1927. The fossil Estheriae of Australia. Part 1. *Proceedings of the Linnean Society of New South Wales* 52(2), 105–12.

Mitchell, T.L., 1838. *Three Expeditions into the Interior of Eastern Australia, with Descriptions of the Recently Explored Regions of Australia Felix and of the Present Colony of New South Wales.* 2 vols, Boone, London.

Mojsisovics, E. von, 1869. Über die Gliederung der oberen Triasbildungen der östlichen Alpen. *Jahrbuch Geolgischen Reichsanstalt* 19, 91–150.

Mojsisovics, E. von, Waagen, W. & Diener, C., 1895. Entwurf einer Gliederung der pelagischen Sedimente des Trias-Systems. *Sitzungberichte Akademie Wissenschaft en Wien* 104, 1271–302.

Molina-Garza, R.S., Geissman, J.W. & Van der Voo, R., 1989. Paleomagnetism of the Dewey Lake Formation (Late Permian), north-west Texas: end of the Kiaman Superchron in North America. *Journal of Geophysical Research* 94, 17881–88.

Mollan, R.G., Dickins, J.M., Exon, N.F. & Kirkegaard, A.G., 1969. Geology of the Springsure 1:250,000 sheet area, Queensland. *Bureau of Mineral Resources, Geology & Geophysics, Report* 123, 1–119.

Mollan, R.G., Forbes, V.R., Jensen, A.R., Exon, N.F. & Gregory, C.M., 1971. Geology of the Eddystone, Taroom and western part of the Mundubbera sheet areas, Queensland. *Bureau of Mineral Resources, Geology & Geophysics, Report* 142, 1–137.

Molnar, R.E., 1980. Australian Late Mesozoic terrestrial tetrapods: some implications. *Mémoire du Société Géologique de France* 139, 131–43.

Molnar, R.E., 1982*a*. Australian Mesozoic reptiles. pp. 169–225 *in* P.V. Rich & E.M. Thompson (eds), q.v.

Molnar, R.E., 1982*b*. A catalogue of fossil amphibians and reptiles in Queensland. *Memoirs of the Queensland Museum* 20, 613–33.

Molnar, R., 1984*a*. A checklist of Australian fossil amphibians. pp. 309–10 *in* M. Archer & G. Clayton (eds), q.v.

Molnar, R., 1984*b*. A checklist of Australian fossil reptiles. pp. 405–406 *in* M. Archer & G. Clayton (eds), q.v.

Molnar, R.E., 1984*c*. Ornithischian dinosaurs in Australia. pp. 151–56 *in* W.E. Reif & F. Westphal (eds), *Third Symposium on Mesozoic Terrestrial Ecosystems, Short Papers*, Attempto Verlag, Tübingen.

Molnar, R.E., 1986. An enantiornithine bird from the Lower Cretaceous of Queensland, Australia. *Nature* 322, 736–38.

Molnar, R.E., 1991. Fossil reptiles in Australia. pp. 605–702 *in* P.V. Rich et al. (eds), q.v.

Molostovskiy, E.A., 1992. Paleomagnetic stratigraphy of the Permian System. *International Geological Review* 34, 1001–1007.

Moore, C., 1870. Australian Mesozoic geology and palaeontology. *Quarterly Journal of the Geological Society* 26, 226–61.

Moores, E.M., 1991. South-west US—East Antarctic (SWEAT) connection: a hypothesis. *Geology* 19, 425–28.

Morante, R., 1993. Determining the Permian–Triassic boundary in Australia through C-isotope chemostratigraphy. pp. 293–98 *in* P.G. Flood & J.A. Aitchison (eds), *New England Orogen, Eastern Australia*, Department of Geology & Geophysics, University of New England.

Morgan, R., 1980*a*. Palynostratigraphy of the Australian Early and Middle Cretaceous. *Geological Survey of New South Wales Palaeontology Memoir*, 18, 153 pp.

Morgan, R., 1980*b*. Eustasy in the Australian Early and Middle Cretaceous. *Geological Survey of New South Wales Bulletin* 27, 105 pp.

Morris, L.N., 1975. The *Rhacopteris* flora in New South Wales. pp. 99–108 *in* K.S.W. Campbell (ed.), *Gondwana Geology*, ANU Press, Canberra.

Morris L.N., 1985. The floral succession in eastern Australia. pp. 118–23 *in* C. Martinez Diaz (ed.), The Carboniferous of the World. II Australia, Indian subcontinent, South Africa, South America and North Africa. *IUGS Publication* 20.

Morris, W.A., Briden, J.C., Piper, J.D.A. & Sallomy, J.T., 1973. Palaeomagnetic studies in the British Caledonides—V, Miscellaneous new data. *Geophysical Journal of the Royal Astronomical Society* 34, 69–105.

Morrison, J., 1983. Palaeomagnetism of the Silurian and Ordovician Red Mountain, Catheys and Sequatchie formations from the Valley and Ridge Province, north-west Georgia. Unpublished MS thesis, University of Georgia, Athens.

Morrison, J. & Ellwood, B.B., 1986. Palaeomagnetism of Silurian–Ordovician sediments from the Valley and Ridge Province, north–west Georgia. *Geophysical Research Letters* 13, 189–92.

Morton, N., 1974. The definition of standard Jurassic stages. *Mémoire du Bureau des Recherches Géologiques et Minières* 75, 83–93.

Mory, A.J., 1981. A review of Early Carboniferous stratigraphy and correlations in the northern Tamworth Belt, New South Wales. *Proceedings of the Linnean Society of New South Wales* 105(3), 213–36.

Mory, A.J. & Beere, G.M., 1988. Geology of the onshore Bonaparte and Ord basins in Western Australia. *Western Australian Geological Survey Bulletin* 134, 183 pp.

Mory, A.J. & Crane, D.T., 1982. Early Carboniferous *Siphonodella* (Conodonta) faunas from eastern Australia. *Alcheringa* 6, 275–303.

Müller, K.J. & Hinz, I., 1991. Upper Cambrian conodonts from Sweden. *Fossils & Strata* 28, 1–153.

Muller, S.M. & Schrenck, H.G., 1943. Standard of the Cretaceous System. *American Association of Petroleum Geologists Bulletin* 27, 262–78.

Murchison, R.I., 1835. On the Silurian System of rocks. *London and Edinburgh Philosophical Magazine and Journal of Science, Series* 3, 7, 46–52.

Murchison, R.I., 1839. *The Silurian System*, John Murray, London, 768 pp.

Murphy, M.A., 1977. On time-stratigraphic units. *Journal of Paleontology* 51, 213–19.

Murphy, M.A., 1987. The possibility of a Lower Devonian equal-increment time scale based on lineages in Lower Devonian conodonts. pp. 284–93 *in* R.L. Austin (ed.) *Conodonts: Investiga-*

tive Techniques and Applications, Ellis Horwood, Chichester.

Murphy, M.A. & Berry, W.B.N., 1983. Early Devonian conodont–graptolite collation and correlations with brachiopod and coral zones, central Nevada. *American Association of Petroleum Geologists Bulletin* 67, 371–79.

Murray, C.G., 1983. Permian geology of Queensland. pp. 1–32 *in Permian Geology of Queensland*, Geological Society of Australia, Queensland Division.

Mutterlose, J., 1992. Lower Cretaceous nannofossil biostratigraphy off northwestern Australia (Leg 123). *Proceedings of the Ocean Drilling Program, Scientific Results* 123, 343–68.

Nazarov, B.B. & Ormiston, A.R., 1983. Upper Devonian (Frasnian) radiolarian fauna from the Gogo Formation, Western Australia. *Micropaleontology* 29, 454–66.

Nazarov, B.B. & Ormiston, A.R., 1985. Evolution of Radiolaria in the Paleozoic and its correlation with the development of other marine fossil groups. *Senckenbergiana Lethaea* 66(3/5), 203–15.

Nazarov, B.B., Cockbain, A.E. & Playford, P.E., 1982. Upper Devonian Radiolaria from the Gogo Formation, Canning Basin, Western Australia. *Alcheringa* 6, 161–74.

Nicoll, R.S., 1984. Conodont studies in the Canning Basin—a review and update. pp. 439–44 *in* P.G. Purcell (ed.), q.v.

Nicoll, R.S., 1990. The genus *Cordylodus* and a latest Cambrian–earliest Ordovician conodont biostratigraphy. *BMR Journal of Australian Geology & Geophysics* 11, 529–58.

Nicoll, R.S., 1991. Differentiation of Late Cambrian–Early Ordovician species of *Cordylodus* (Conodonta) with biapical basal cavities. *BMR Journal of Australian Geology & Geophysics* 12, 223–44.

Nicoll, R.S., 1992. Evolution of the conodont genus *Cordylodus* and the Cambrian–Ordovician Boundary. pp. 105–13 *in* B.D. Webby & J.R. Laurie (eds), q.v.

Nicoll, R.S., 1993. Ordovician conodont distribution in selected petroleum wells, Canning Basin, Western Australia. *Australian Geological Survey Organisation Record* 1993/17, 136 pp.

Nicoll, R.S. & Druce, E.C., 1979. Conodonts from the Fairfield Group, Canning Basin, Western Australia. *Bureau of Mineral Resources, Geology & Geophysics, Bulletin* 190, 134 pp.

Nicoll, R.S. & Foster, C.B., 1994. Late Triassic conodont and palynomorph biostratigraphy and conodont thermal maturation, North West Shelf, Australia. *AGSO Journal of Australian Geology and Geophysics* 15, 101–18.

Nicoll, R.S. & Jenkins, T.B.H., 1985. Conodont biostratigraphy. pp. 114–17 *in* J. Roberts 1985*a*, q.v.

Nicoll, R.S. & Jones, P.J., 1981. Lower Carboniferous conodont and ostracod biostratigraphy and the lithofacies of the Bonaparte Gulf Basin, northwestern Australia. *Geological Society of Australia Abstracts* 3, 50.

Nicoll, R.S. & Playford, P.E., 1993. Upper Devonian iridium anomalies, conodont zonation and the Frasnian–Famennian boundary in the Canning Basin, Western Australia. *Palaeogeography, Palaeoclimatology, Palaeoecology* 104, 105–13.

Nicoll, R.S. & Shergold, J.H., 1991. Revised Late Cambrian (pre-Payntonian–Datsonian) conodont stratigraphy at Black Mountain, Georgina Basin, western Queensland, Australia. *BMR Journal of Australian Geology & Geophysics* 12, 93–18.

Nicoll, R.S., Laurie, J.R. & Roche, M., 1993. Revised stratigraphy for the Lower Ordovician Prices Creek Group, Canning Basin, Western Australia. *AGSO Journal of Australian Geology & Geophysics* 14 (1), 65–76.

Nicoll, R.S., Nielsen, A.T., Laurie, J.R. & Shergold, J.H., 1992. Preliminary correlation of latest Cambrian to Early Ordovician sea level events in Australia and Scandinavia. pp. 381–94 *in* B.D. Webby & J.R. Laurie (eds), q.v.

Nicolls, D.R., 1968. Studies in Victorian Foraminifera above the Orbulina universa datum. Unpublished MSc thesis, University of Melbourne.

Nielsen, A.T., 1992. Ecostratigraphy and the recognition of Arenigian (Early Ordovician) sea-level changes. pp. 355–66 *in* B.D.Webby & J.R. Laurie (eds), q.v.

Noltimier, N.C. & Ellwood, B.B., 1977. The coal pole: Palaeomagnetic results from the Westphalian B, C and D coals, Wales. *Transactions of the American Geophysical Union* 58, 375.

Norford, B.S., 1988. Introduction to papers on the Cambrian–Ordovician boundary. *Geological Magazine* 125, 323–26.

Nott, J., 1992. Long-term drainage evolution in the Shoalhaven catchment, south-east highlands, Australia. *Earth Surface Processes and Landforms* 17, 361–74.

Obradovich, J.D. & Cobban, W.A., 1975. A time-scale for the Late Cretaceous of the western interior of North America. *Geological Association of Canada Special Paper* 13, 31–54.

Odin, G.S. (ed.), 1982. *Numerical dating in Stratigraphy. Parts I and II*, John Wiley & Sons Ltd, Chichester, 1040 pp.

Odin, G.S., 1985*a*. Remarks on the numerical scale of Ordovician to Devonian times. pp. 93–98 *in* N.J. Snelling (ed.), q.v.

Odin, G.S., 1985*b*. Comments on the geochronology of the Carboniferous to Triassic times. pp. 114–17 *in* N.J. Snelling (ed.), q.v.

Odin, G.S. & Gale, N.H., 1982. Numerical dating of Hercynian times (Devonian to Permian). pp. 487–500 *in* G.S. Odin (ed.), q.v.

Odin, G.S. & Létolle, R., 1982. The Triassic time scale in 1981. pp. 523–33 *in* G.S. Odin (ed.), q.v.

Odin, G.S. & Odin, C., 1990. Échelle numérique des temps géologiques mise à jour 1990. *Géochronique* 35, 12–21.

Odin, G.S., Montanari, A., Deino, A., Guise, P.G., Kreuzer, H. & Rex, D.C., 1991. Reliability of volcano–sedimentary biotite ages across the Eocene–Oligocene boundary (Apennines, Italy). *Chemical Geology (Isotope Science Section)* 86, 203–24.

Odin, G.S., Gale, N.H., Auvray, B., Bielski, M., Doré, F., Lancelot, J.-R., & Pasteels, P., 1983. Numerical dating of the Precambrian–Cambrian boundary. *Nature* 301, 21–23.

Ogg, G., 1994. Dinoflagellate cysts of the Early Cretaceous North Atlantic Ocean. *Marine Micropaleontology* 23, 241–63.

Ogg, J.G., 1983. Magnetostratigraphy of Upper Jurassic and lowest Cretaceous sediments, Deep Sea Project Site 534, western North Atlantic. *Initial Reports of the Deep Sea Drilling Project* 76, 685–97.

Ogg, J.G. & Lowrie, W., 1986. Magnetostratigraphy of the Jurassic–Cretaceous boundary. *Geology* 14, 547–50.

Ogg, J.G. & Steiner, M.B., 1991. Early Triassic magnetic polarity time scale—integration of magnetostratigraphy, ammonite zonation and sequence stratigraphy from stratotype sections (Canadian Arctic Archipelago). *Earth and Planetary Science Letters* 107, 69–89.

Ogg, J.G., Kodama, K. & Wallick, B.P., 1992. Lower Cretaceous magnetostratigraphy and paleolatitudes off north-west Australia, ODP Site 765 and DSDP Site 261, Argo Abyssal Plain, and ODP Site 766, Gascoyne Abyssal Plain. *Proceedings of the Ocean Drilling Program, Scientific Results* 123, 523–48.

Ogg, J.G., Steiner, M.B., Company, M. & Tavera, J.M., 1988. Magnetostratigraphy across the Berriasian–Valanginian stage boundary (Early Cretaceous), at Cehegin (Murcia Province, southern Spain). *Earth and Planetary Science Letters* 87, 205–15.

Ogg, J.G., Steiner, M.B., Oloriz, F. & Tavera, J.M., 1984. Jurassic magnetostratigraphy, 1. Kimmeridgian–Tithonian of Sierra Gorda and Carcabuey, southern Spain. *Earth and Planetary Science Letters* 71, 147–62.

Ogg, J.G., Steiner, M.B., Wieczorek, J. & Hoffmann, M., 1991. Jurassic magnetostratigraphy, 4. Early Callovian through Middle Oxfordian of the Krakow Uplands (Poland). *Earth and Planetary Science Letters* 104, 488–504.

Okada, H. & Bukry, D., 1980. Supplementary modification and introduction of code numbers to the low-latitude coccolith biostratigraphic zonation (Bukry 1973, 1975). *Marine Micropaleontology* 5, 321–25.

Oliver, W.A. & Chlupáč, I., 1991. Defining the Devonian: 1979–89. *Lethaia* 24, 119–22.

Opdyke, N.D., 1986. The Kiaman Superchron in North America. *Eos* 67, 269.

Opdyke, N.D., Huang, K., Xu, G., Zhang, W.Y. & Kent, D.V., 1987. Palaeomagnetic results from the Silurian of the Yangtze paraplatform. *Tectonophysics* 139, 123–32.

Opdyke, N.D., Khramov, A.N., Gurevitch, E., Iosifidi, A.G. & Makarov, I.A., 1993. A paleomagnetic study of the Middle Carboniferous of the Donets Basin, Ukraine. *Eos* 74 Spring Meeting Supplement, 118.

Öpik, A.A., 1953. Lower Silurian fossils from the '*Illaenus* Band', Heathcote, Victoria. *Geological Survey of Victoria, Memoir* 19, 42 pp.

Öpik, A.A., 1956. Cambrian geology of Queensland. pp. 1–24 *in* J. Rodgers (ed.), El Sistema Cambrico, su paleogeografia y el problema de su base 2. *20th Session International Geological Congress, Mexico.*

Öpik, A.A., 1958. The Cambrian trilobite *Redlichia*: organisation and generic concept. *Bureau of*

Mineral Resources, Geology & Geophysics, Bulletin 42, 50 pp.

Öpik, A.A., 1961. The geology and palaeontology of the headwaters of the Burke River, Queensland. *Bureau of Mineral Resources, Geology & Geophysics, Bulletin* 53, 249 pp.

Öpik, A.A., 1963. Early Upper Cambrian fossils from Queensland. *Bureau of Mineral Resources, Geology & Geophysics, Bulletin* 64, 133 pp.

Öpik, A.A., 1966. The Early Upper Cambrian crisis and its correlation. *Journal and Proceedings of the Royal Society of New South Wales* 100, 9–14.

Öpik, A.A., 1967. The Mindyallan fauna of north-western Queensland. *Bureau of Mineral Resources, Geology & Geophysics, Bulletin* 74: vol. 1, 404 pp.; vol. 2, 167 pp.

Öpik, A.A., 1968*a*. The Ordian Stage of the Cambrian and its Australian Metadoxididae. *Bureau of Mineral Resources, Geology & Geophysics, Bulletin* 92, 133–70.

Öpik, A.A., 1968*b*. Ordian (Cambrian) Crustacea Bradoriida of Australia. *Bureau of Mineral Resources, Geology & Geophysics, Bulletin* 103, 44 pp.

Öpik, A.A., 1969. Appendix 3. The Cambrian and Ordovician sequence, Cambridge Gulf area. *Bureau of Mineral Resources, Geology & Geophysics, Report* 109, 74–77.

Öpik, A.A., 1970*a*. Nepeiid trilobites of the Middle Cambrian of northern Australia. *Bureau of Mineral Resources, Geology & Geophysics, Bulletin* 113, 48 pp.

Öpik, A.A., 1970*b*. *Redlichia* of the Ordian (Cambrian) of northern Australia and New South Wales. *Bureau of Mineral Resources, Geology & Geophysics, Bulletin* 114, 66 pp.

Öpik, A.A., 1975*a*. Templetonian and Ordian xystridurid trilobites of Australia. *Bureau of Mineral Resources, Geology & Geophysics, Bulletin* 121, 84 pp.

Öpik, A.A., 1975*b*. Cymbric Vale fauna of New South Wales and Early Cambrian biostratigraphy. *Bureau of Mineral Resources, Geology & Geophysics, Bulletin* 159, 78 pp.

Öpik, A.A., 1979. Middle Cambrian agnostids: systematics and biostratigraphy. *Bureau of Mineral Resources, Geology & Geophysics, Bulletin* 172: vol. 1, 188 pp.; vol. 2.

Öpik, A.A., 1982. Dolichometopid trilobites of Queensland, Northern Territory and New South Wales. *Bureau of Mineral Resources, Geology & Geophysics, Bulletin* 175, 85 pp.

Öpik, A.A., Casey, J.N., Gilbert-Tomlinson, J., Traves, D.M., Daily, B., Thomas, D.E., Singleton, O.P., Banks, M.R. & Noakes, L.C., 1957. The Cambrian geology of Australia. *Bureau of Mineral Resources, Geology & Geophysics, Bulletin* 49, 284 pp.

d'Orbigny, A., 1840–42. *Paléontologie Française, Terrains Crétacés. 1: Céphalopodes*, Masson, Paris, 662 pp.

Orchard, M.J., 1983. *Epigondolella* populations and their phylogeny and zonation in the Upper Triassic. *Fossils & Strata* 15, 177–92.

Orchard, M.J., 1991*a*. Late Triassic conodont biochronology and biostratigraphy of the Kunga Group, Queen Charlotte Islands, British Columbia. *Geological Survey of Canada, Paper* 90–10, 173–93.

Orchard, M.J., 1991*b*. Upper Triassic conodont biochronology and new index species from the Canadian Cordillera. *Geological Survey of Canada, Bulletin* 417, 299–335.

Osmolska, H., 1970. Revision of non-cyrtosymbolinid trilobites from the Tournaisian–Namurian of Eurasia. *Palaeontologica Polonica* 23, 1–165.

Owen, H.G., 1973. Ammonite faunal provinces in the middle and upper Albian and their palaeogeographical significance. pp. 145–54 *in* R. Casey & P.F. Rawson (eds), The Boreal Lower Cretaceous. *Geological Journal Special Issue* 5.

Owen, J.A., 1975. Palynology of some Tertiary deposits from New South Wales. Unpublished PhD thesis, Australian National University.

Owen, J.A., 1988. Miocene palynomorph assemblages from Kiandra, New South Wales. *Alcheringa* 12, 269–97.

Owen, M. & Wyborn, D., 1979. Geology and geochemistry of the Tantangara and Brindabella area. *Bureau of Mineral Resources, Geology & Geophysics, Bulletin* 204, 52 pp.

Owens, B., 1984. Miospore zonation of the Carboniferous. pp. 90–102 *in* P.K. Sutherland & W.L. Manger (eds), 1984*b*, q.v.

Owens, B., Loboziak, S. & Coquel, R., 1984. Late Mississippian–Early Pennsylvanian Miospore assemblages from northern Arkansas. pp. 377–84 *in* P.K. Sutherland & W.L. Manger (eds), 1984*b*, q.v.

Owens, B., Riley, N.J. & Calver, M.A., 1985. Boundary stratotypes and new stage names for the Lower and Middle Westphalian sequences in Britain. *Compte Rendu 10ème Congres International de Stratigraphie et de Géologie du Carbonifère, Madrid, 1983*, 4, 461–72.

Owens, R.M., 1986. The Carboniferous trilobites of Britain. Part 1. *Monographs of the Palaeontographical Society* 138, 1–26.

Pakistani–Japanese Research Group, 1985. Permian and Triassic systems in the Salt Range and Surghar Range, Pakistan. pp. 221–312 *in* K. Nakazawa & J.M. Dickins (eds), *The Tethys, her Paleogeography and Paleobiogeography from Paleozoic to Mesozoic*, Tokai University Press.

Palfreyman, W.D., 1984. Guide to the geology of Australia. *Bureau of Mineral Resources, Geology & Geophysics, Bulletin* 181, 1–111.

Palmer, J.A., Perry, S.P. & Tarling, D.H., 1985. Carboniferous magnetostratigraphy. *Journal of the Geological Society* 142, 945–55.

Palmieri, V., 1969. Upper Carboniferous conodonts from limestones near Murgon, south-east Queensland. *Geological Survey of Queensland Publication* 341, *Palaeontological Paper* 17, 1–13.

Palmieri, V., 1971. Tertiary subsurface biostratigraphy of the Capricorn Basin. *Geological Survey of Queensland, Report* 52, 1–18.

Palmieri, V., 1974. Correlation and environmental trends of the subsurface Tertiary Capricorn Basin. *Geological Survey of Queensland, Report* 86, 1–14.

Palmieri, V., 1983. Biostratigraphic appraisal of Permian Foraminifera from the Denison Trough–Bowen Basin (central Queensland). pp. 141–54 *in Permian Geology of Queensland*, Geological Society of Australia, Queensland Division.

Palmieri, V., 1984. Neogene Foraminiferida from GSQ Sandy Cape 1–3R bore, Queensland: a biostratigraphic appraisal. *Palaeogeography, Palaeoclimatology, Palaeoecology* 46, 165–183.

Paproth, E., 1980. The Devonian–Carboniferous boundary. *Lethaia* 13(4), 287.

Paproth, E. & Streel, M., 1979. The Devonian–Carboniferous boundary in western Europe. *XIV Pacific Science Congress USSR. Field Excursion Guidebook Supplement* 8, 155–64.

Paproth, E., Feist, R. & Flajs, G., 1991. Decision on the Devonian–Carboniferous boundary stratotype. *Episodes* 14(4), 331–36.

Paproth, E., Conil, R., Bless, M.J.M., Boonen, P., Carpentier, N., Coen, M., Delcambre, B., Deprijck, Ch., Deuzon, S., Dreesen, R., Groessens, E., Hance, L., Hennebert, M., Hibo, D., Hahn, G. & R., Hislaire, O., Kasig, W., Laloux, M., Lauwers, A., Lees, A., Lys, M., Op de Beek, K., Overlau, P., Pirlet, H., Poty, E., Ramsbottom, W., Streel, M., Swennen, R., Thorez, J., Vanguestaine, M., Van Steenwinkel, M. & Vieslet, J.L., 1983. Bio- and lithostratigraphic subdivisions of the Dinantian in Belgium, a review. *Annales de la Société Géologique de Belgique*. 106, 185–239.

Parfrey, S.M., 1988. Biostratigraphy of the Barfield Formation, south-eastern Bowen Basin, with a review of the fauna of the Ingelara and lower Peawaddy Formation, south-western Bowen Basin. *Queensland Department of Mines Report* 1, 1–53.

Paris, F., 1981. Les chitinozoaires dans le Paléozoique du sud-ouest de l'Europe. *Mémoires de la Société Géologique et Mineralogique de Bretagne* 26, 1–412.

Partridge, A.D., 1976. The geological expression of eustacy in the Early Tertiary of the Gippsland Basin. *APEA Journal* 16, 73–79.

Pedder, A.E.H., 1985. Lower Devonian rugose corals of Lochkovian age from Yukon Territory. *Geological Survey of Canada Paper* 85–1A, 587–602.

Peerdeman, F.M., Davies, P.J. & Chivas, A.R., 1993. The stable isotope signal in shallow-water, upper-slope sediments off the Great Barrier Reef (Hole 820A). *Proceedings of the Ocean Drilling Program, Scientific Results* 133, 163–72.

Peppers, R.A., 1984. Comparison of miospore assemblages in the Pennsylvanian System of the Illinois Basin with those in the Upper Carboniferous of western Europe. pp. 483–502 *in* P.K. Sutherland & W.L. Manger (eds), 1984*b*, q.v.

Perch-Nielsen, K., 1977. Albian to Pleistocene calcareous nannofossils from the western South Atlantic. *Initial Reports of the Deep Sea Drilling Project* 39, 699–825.

Perch-Nielsen, K. 1979. Calcareous nannofossils from the Cretaceous between the North Sea and the Mediterranean. pp. 223–72 *in* J. Wiedmann (ed.), Aspekte der Kreide Europas. *International Union of Geological Sciences, Series A* 6, Schweizerbart'sche Verlagsbuchhandlung, Stuttgart.

Perch-Nielsen, K., 1985*a*. Mesozoic calcareous nannofossils. pp. 329–426 *in* H.M. Bolli, J.B. Saunders & K. Perch-Nielsen (eds), *Plankton Stratigraphy*, Cambridge University Press, Cambridge.

Perch-Nielsen, K., 1985*b*. Cenozoic calcareous nannofossils. pp. 427–554 *in* H.M. Bolli, J.B. Saunders & K. Perch-Nielsen (eds), *Plankton Stratigraphy*, Cambridge University Press, Cambridge.

Percival, I.G., 1976. The geology of the Licking Hole Creek area, near Walli, central western New South Wales. *Journal and Proceedings of the Royal Society of New South Wales* 109, 7–23.

Percival, I.G., 1978. Inarticulate brachiopods from the Late Ordovician of New South Wales, and their palaeoecological significance. *Alcheringa* 2, 117–41.

Percival, I.G., 1979*a*. Ordovician plectambonitacean brachiopods from New South Wales. *Alcheringa* 3, 91–116.

Percival, I.G., 1979*b*. Late Ordovician articulate brachiopods from Gunningbland, central western New South Wales. *Proceedings of the Linnean Society of New South Wales* 103, 175–87.

Percival, I.G., 1991. Late Ordovician brachiopods from central New South Wales. *Association of Australasian Palaeontologists, Memoir* 11, 107–77.

Percival, I.G., 1992. Ordovician brachiopod biostratigraphy of central-western New South Wales. pp. 215–29 *in* B.D. Webby & J.R. Laurie (eds), q.v.

Perkins, C. & Walshe, J.L., 1993. Geochronology of the Mount Read Volcanics, Tasmania, Australia. *Economic Geology* 88, 1176–97.

Perroud, H., Bonhommet, N. & Thebault, J.P., 1985. Palaeomagnetism of the Ordovician Moulin de Chateaupanne Formation, Vendée, western France. *Geophysical Journal of the Royal Astronomical Society* 85, 573–82.

Peterson, D.N. & Nairn, A.E.M., 1971. Palaeomagnetism of Permian redbeds from the south-western United States. *Geophysical Journal of the Royal Astronomical Society* 23, 191–205.

Petersen, M.S., 1975. Upper Devonian (Famennian) ammonoids from the Canning Basin, Western Australia. *Paleontological Society Memoir* 8, 1–55.

Pflaumann, U. & Cepek, P., 1982. Cretaceous foraminiferal and nannoplankton biostratigraphy and paleoecology along the West African continental margin. pp. 309–53 *in* U. von Rad, K. Hinz, M. Sarnthein & E. Seibold (eds), *Geology of the North-west African Continental Margin*, Springer-Verlag, Berlin.

Philip, G.M. & Pedder, A.E.H., 1967. A correlation of some Devonian limestones of New South Wales and Victoria. *Geological Magazine* 104, 232–39.

Pickett, J.W., 1960. A clymenid from the *Wocklumeria* Zone of New South Wales. *Palaeontology* 3, 237–41.

Pickett, J., 1972. Correlation of the Middle Devonian formations of Australia. *Journal of the Geological Society of Australia* 18, 457–66.

Pickett, J., 1980. Conodont assemblages from the Cobar Supergroup (Early Devonian), New South Wales. *Alcheringa* 4, 67–88.

Pickett, J.W, 1981. Late Devonian and Early Carboniferous conodonts from the Myrtlevale Beds and Hardwick Formation, Star Basin. *Queensland Government Mining Journal* 82, 24–27.

Pickett, J.W., 1982. The Silurian System in New South Wales. *Geological Survey of New South Wales, Bulletin* 29, 1–164.

Pickett, J.W., 1983. An annotated bibliography and review of Australian fossil sponges. *Association of Australasian Palaeontologists, Memoir* 1, 93–120.

Pickett, J.W., 1985. Correlation of a mid to late Ordovician section near Parkes, New South Wales. *Records of the Geological Survey of New Zealand* 9, 7.

Pickett, J.W., 1993. The Late Devonian–Early Carboniferous succession at Slaughterhouse Creek, Gravesend. *Geological Survey of New South Wales, Quarterly Notes* 91, 1–10.

Pickett, J.W. & McClatchie, L., 1991. Age and relations of stratigraphic units in the Murda Syncline area. *New South Wales Geological Survey, Quarterly Notes* 85, 9–32.

Pickett, J.W. & Wu Wangshi, 1990. The succession of Early Carboniferous coral faunas in eastern Australia and China. *Alcheringa* 14, 89–108.

Piper, J.D.A., 1975. Palaeomagnetism of Silurian lavas of Somerset and Gloucestershire, England. *Earth and Planetary Science Letters* 25, 355–60.

Piper, J.D.A., 1987. *Palaeomagnetism and the Continental Crust*, Open University Press, Milton Keynes.

Pitman, W.C., 1978. Relationship between eustacy and stratigraphic sequences of passive margins. *Geological Society of America Bulletin* 89, 1389–403.

Playford, G., 1971. Lower Carboniferous spores from the Bonaparte Gulf Basin, Western Australia and Northern Territory. *Bureau of Mineral Resources, Geology & Geophysics, Bulletin* 115, 104 pp.

Playford, G., 1972. Trilete spores of *Umbonatisporites* in the Lower Carboniferous of north-western Australia. *Neues Jahrbuch für Geologie und Paläontologie, Abhandlungen* 141, 301–15.

Playford, G., 1976. Plant microfossils from the Upper Devonian and Lower Carboniferous of the Canning Basin, Western Australia. *Palaeontographica B* 158, 1–71.

Playford, G., 1977. A Lower Carboniferous palynoflora from the Drummond Basin, east-central Queensland. *Proceedings of the Royal Society of Queensland* 88, 75–81.

Playford, G., 1978. Lower Carboniferous spores from the Ducabrook Formation, Drummond Basin, Queensland. *Palaeontographica* B 167, 105–60.

Playford, G., 1982. A latest Devonian palynoflora from the Buttons Beds, Bonaparte Gulf Basin, Western Australia. *BMR Journal of Australian Geology* & *Geophysics* 7, 149–57.

Playford, G., 1983. The Devonian miospore genus *Geminospora* Balme 1962: a reappraisal based on topotypic *G. lemurata* (type species). *Association of Australasian Palaeontologists, Memoir* 1, 311–25.

Playford, G., 1985. Palynology of the Australian Lower Carboniferous: a review. *Compte Rendu 10ème Congrès International de Stratigraphie et de Géologie du Carbonifère, Madrid, 1983*, 4, 247–65.

Playford, G., 1991 [imprint 1990]. Australian Lower Carboniferous miospores relevant to extra-Gondwanic correlations: an evaluation. *Courier Forschungsinstitut Senckenberg* 130, 85–125.

Playford, G. & Dettmann, M.E., 1965. Rhaeto–Liassic plant microfossils from the Leigh Creek Coal Measures, South Australia. *Senckenbergiana Lethaea* 46, 127–81.

Playford, G. & Helby, R., 1968. Spores from a Carboniferous section in the Hunter Valley, New South Wales. *Journal of the Geological Society of Australia* 15, 103–19.

Playford, G. & Martin F., 1984. Ordovician acritarchs from the Canning Basin, Western Australia. *Alcheringa* 8, 187–23.

Playford, G. & Miller, M.A., 1988. Chitinozoa from Lower Ordovician strata of the Georgina Basin, Queensland (Australia). *Geobios* 21, 17–39.

Playford, G. & Satterthwait, D.F., 1985. Lower Carboniferous (Viséan) spores of the Bonaparte Gulf Basin, north-western Australia, Part One. *Palaeontographica Abt B* 195, 129–52.

Playford, G. & Satterthwait, D.F., 1986. Lower Carboniferous (Viséan) spores of the Bonaparte Gulf Basin, north-western Australia, Part Two. *Palaeontographica Abt B* 200, 1–32.

Playford, G. & Satterthwait, D.F., 1988. Lower Carboniferous (Viséan) spores of the Bonaparte Gulf Basin, north-western Australia: Part Three. *Palaeontographica Abt B* 208, 1–26.

Playford, G. & Wicander, R., 1988. Acritarch palynoflora of the Coolibah Formation (Lower Ordovician), Georgina Basin, Queensland. *Association of Australasian Palaeontologists, Memoir* 5, 5–40.

Playford, G., Haig, D.W. & Dettmann, M.E., 1975. A mid-Cretaceous microfossil assemblage from the Great Artesian Basin, northwestern Queensland. *Neues Jahrbuch für Geologie und Paläontologie, Abhandlungen* 149, 333–62.

Playford, G., Jones, B.G. & Kemp, E.M., 1976. Palynological evidence for the age of the synorogenic Brewer Conglomerate, Amadeus Basin, Central Australia. *Alcheringa* 1, 235–43.

Playford, P.E., 1980. Devonian 'great barrier reef' of the Canning Basin, Western Australia. *American Association of Petroleum Geologists Bulletin* 64, 814–40.

Playford, P.E. & Lowry, D.C., 1966. Devonian reef complexes of the Canning Basin, Western Australia. *Western Australian Geological Survey Bulletin* 118, 150 pp.

Pocock, S.A.J., 1976. A preliminary dinoflagellate zonation of the uppermost Jurassic and lower part of the Cretaceous, Canadian Arctic, and possible correlation in the Western Canada Basin. *Geoscience and Man* 15, 101–14.

Pocock, S.A.J., 1980. Palynology at the Jurassic–Cretaceous boundary in North America. *Fourth International Palynological Conference (Lucknow, 1976–1977), Proceedings* 2, 377–85.

Pojeta, J. & Gilbert-Tomlinson J., 1977. Australian Ordovician pelecypod molluscs. *Bureau of Mineral Resources, Geology* & *Geophysics, Bulletin* 174, 64 pp.

Pojeta, J., & Runnegar, B., 1976. The palaeontology of rostroconch molluscs and the early history of the Phylum Mollusca. *US Geological Survey Professional Paper* 968, 88 pp.

Pojeta, J., Gilbert-Tomlinson J. & Shergold, J.H., 1977. Cambrian and Ordovician rostroconch molluscs from northern Australia. *Bureau of Mineral*

Resources, Geology & Geophysics, Bulletin 171, 54 pp.

Pole, M.S., Hill, R.S., Green, N. & Macphail, M.K., 1993. The Oligocene Berwick Quarry flora—rainforest in a drying environment. *Australian Systematic Botany* 6, 399–427.

Potts, S.S., Van der Pluijm, B.A. & Van der Voo, R., 1993. Palaeomagnetism of the Ordovician Bluffer Pond Formation: Palaeogeographic implications for the Munsungun Terrane of northern Maine. *Journal of Geophysical Research* 98, 7987–96.

Powell, C. McA., Neef, G., Crane, D., Jell, P.A. & Percival, J.G., 1982. Significance of Late Cambrian (Idamean) fossils in the Cupala Creek Formation, north-western New South Wales. *Proceedings of the Linnean Society of New South Wales* 106(2), 127–50.

Powis, G.D., 1979. Palynology of the Late Palaeozoic glacial sequence, Canning Basin, Western Australia. Unpublished PhD thesis, University of Western Australia.

Powis, G.D., 1984. Palynostratigraphy of the Late Carboniferous sequence, Canning Basin, WA. pp. 429–38 *in* P.G. Purcell (ed.), q.v.

Price, P.L., 1980. Biostratigraphy of the Devonian section from selected wells in A. to P. 232PO, Adavale Basin, Queensland. Mines Administration Pty Ltd Palynology Facility, Report 208/1 (unpublished).

Price, P.L., 1983. A Permian palynostratigraphy for Queensland. pp. 155–78 *in Permian Geology of Queensland*, Geological Society of Australia, Queensland Division.

Price, P.L., Filatoff, J., Williams, A.J., Pickering, S.A. & Wood, G.R., 1985. Late Palaeozoic and Mesozoic palynostratigraphical units. CSR Oil and Gas Division, Palynology Facility, Report 274/25 (unpublished).

Proto Decima, F., 1974. Leg 27 calcareous nannoplankton. *Initial Reports of the Deep Sea Drilling Project* 27, 589–621.

Purcell, P.E. (ed.), 1984. The Canning Basin, West Australia. *Proceedings of the Geological Society of Australia/Petroleum Exploration Society of Australia Symposium, Perth, 1984*, 582 pp.

Purcell, P.G. & Purcell, R.R., 1988. The North West Shelf, Australia. *Proceedings of the Petroleum Exploration Society of Australia Symposium Perth, 1988*, 651 pp.

Qu Lifan, Yang Jiduan, Bai Yunhong & Zhang Zhenlai, 1983. A preliminary discussion on the characteristics and stratigraphic divisions of Triassic spores and pollen of China. *Chinese Academy of Geological Sciences* 5, 81–94.

Quilty, P.G., 1972. The biostratigraphy of the Tasmanian marine Tertiary. *Papers and Proceedings of the Royal Society of Tasmania* 105, 25–44.

Quilty, P.G., 1975. An annotated bibliography of the palaeontology of Western Australia. *Geological Survey of Western Australia Report* 3, 264 pp.

Quilty, P.G., 1978. The Late Cretaceous–Tertiary section in Challenger No. 1 (Perth Basin)—details and implications. *Bureau of Mineral Resources, Geology & Geophysics, Bulletin* 192, 109–35.

Quilty, P.G., 1980. Sedimentation cycles in the Cretaceous and Cenozoic of Western Australia. *Tectonophysics* 63, 349–66.

Quilty, P.G., 1984. Cretaceous foraminiferids from Exmouth Plateau and Kerguelen Ridge, Indian Ocean. *Alcheringa* 8, 225-41.

Quilty, P.G., 1990. Triassic and Jurassic foraminiferid faunas, northern Exmouth Plateau, eastern Indian Ocean. *Journal of Foraminiferal Research* 20(4), 349–67.

Quilty, P.G., 1993. Tasmantid and Lord Howe seamounts: biostratigraphy and palaeoceanographic significance. *Alcheringa* 17, 27–53.

Raggatt, H.G. & Fletcher, H.O., 1937. A contribution to the Permian–Upper Carboniferous problem and an analysis of the fauna of the Upper Palaeozoic (Permian) of Northwest Basin, Western Australia. *Australian Museum Records* 20, 150–84.

Raine, J.I., 1984. Outline of a palynological zonation of Cretaceous to Paleogene terrestrial sediments in West Coast region, South Island, New Zealand. *New Zealand Geological Survey Report* 109, 82 pp.

Ramsbottom, W.H.C., 1973. Transgressions and regressions in the Dinantian: a new synthesis of British Dinantian stratigraphy. *Proceedings of the Yorkshire Geological Society* 39, 567–607.

Ramsbottom, W.H.C., 1977. Major cycles of transgression and regression (mesothems) in the Namurian. *Proceedings of the Yorkshire Geological Society* 41, 261–91.

Ramsbottom, W.H.C., 1979. Rates of transgression and regression in the Carboniferous of NW Europe. *Journal of the Geological Society* 136, 147–53.

Ramsbottom, W.H.C. & Saunders, W.B., 1984. Carboniferous ammonoid zonation. pp. 52–64 *in* P.K. Sutherland & W.L. Manger (eds), 1984*b*, q.v.

Ramsbottom, W.H.C., Calver, M.A., Eager, R.M.C., Hodson, F., Holliday, D.W., Stubblefield, C.J. & Wilson, R.B., 1978. A correlation of Silesian rocks in the British Isles. *Special Report of the Geological Society of London* 10, 1–82.

Rawson, P.F., 1973. Lower Cretaceous (Ryazanian–Barremian) marine connections and cephalopod migrations between the Tethyan and Boreal realms. pp. 131–44 *in* R. Casey & P.F. Rawson (eds), The Boreal Lower Cretaceous. *Geological Journal Special Issue* 5.

Rawson, P.F., Curry, D., Dilley, F.C., Hancock, J.M., Kennedy, W.J., Neale, J.W., Wood, C.J. & Worssam, B.C., 1978. A correlation of Cretaceous rocks in the British Isles. *Geological Society of London Special Report* 9, 70 pp.

Reale, V., Baldanza, A., Monechi, S. & Mattioli, E., 1992. Calcareous nannofossil biostratigraphic events from the Early–Middle Jurassic sequences of the Umbria–Marche area, (central Italy). *Memorie di Scienze Geologiche Universita di Padova* 43, 41–75.

Reeve, S.C. & Helsley, C.E., 1972. Magnetic reversal sequence in the upper portion of the Chinle Formation, Montoya, New Mexico, *Geological Society of America Bulletin* 83, 3795–812.

Reiser, R.F. & Williams, A.J., 1969. Palynology of the Lower Jurassic sediments of the northern Surat Basin, Queensland. *Geological Survey of Queensland Publication* 339, 24 pp.

Remane, J., 1978. Calpionellids. pp. 161–70 *in* B.U. Haq & A. Boersma (eds), *Introduction to Marine Micropalaeontology*, Elsevier, New York.

Retallack, G.J., 1977. Reconstructing Triassic vegetation of eastern Australasia: a new approach for the biostratigraphy of Gondwanaland. *Alcheringa* 1, 247–78, microfiche A7-B6.

Retallack, G.J., 1980. Late Carboniferous to Middle Triassic megafossil floras from the Sydney Basin. *Geological Survey of New South Wales, Bulletin* 26, 384–430.

Riccardi, A.C., 1988. The Cretaceous System of southern South America. *Geological Society of America Memoir* 168, 161 pp.

Rich, P.V., 1991. The Mesozoic and Tertiary history of birds on the Australian Plate. pp. 721–808 *in* P.V. Rich et al. (eds), q.v.

Rich, P.V. & Thompson, E.W. (eds), 1982. *The Fossil Vertebrate Record of Australasia*, Monash University, Clayton, Victoria

Rich, P.V. & Van Tets, J., 1982. Fossil birds of Australia and New Guinea: their biogeographic, phylogenetic and biostratigraphic input. pp. 235–384 *in* P.V. Rich & E.W. Thompson (eds), q.v.

Rich, P.V., Monaghan, J.M., Baird, R.F. & Rich, T.H. (eds), 1991. *Vertebrate Palaeontology of Australasia*, Pioneer Design Studio, Lilydale, Victoria.

Rich, T.H., 1991. Monotremes, placentals and marsupials: their record in Australia and its biases. pp. 893–1004 *in* P.V. Rich et al. (eds), q.v.

Rich, T.H. & Rich, P.V., 1988. Tetrapod (terrestrial vertebrate) assemblages. pp. 240–43 *in* J.G. Douglas & J.A. Ferguson (eds), q.v.

Rich, T.H. & Rich, P.V., 1989. Polar dinosaurs and biotas of the Early Cretaceous of southeastern Australia. *National Geographic Research* 5, 15–53.

Rich, T.H., Flannery, T.F. & Archer, M., 1989. A second Cretaceous mammalian specimen from Lightning Ridge, NSW, Australia. *Alcheringa* 13, 85–88.

Rich, T.H., Archer, M., Hand, S.J., Godthelp, H., Muirhead, J., Pledge, N.S., Flannery, T.F., Woodburne, M.O., Case, J.A., Tedford, R.H., Turnbull, W.D., Lundelius, E.L., Rich, L.S.V., Whitelaw, M.J., Kemp, A. & Rich, P.V., 1991. Appendix I. Australian and Mesozoic and Tertiary terrestrial mammal localities. pp. 1005–58 *in* P.V. Rich et al. (eds), q.v.

Richards, J.R. & Singleton, O.P., 1981. Palaeozoic Victoria, Australia: igneous rocks, ages and their interpretation. *Journal of the Geological Society of Australia* 28, 395–421.

Richardson, J.B. & Edwards, D., 1989. Sporomorphs and plant megafossils. pp. 216–26 *in* C.H. Holland & M.G. Bassett (eds), A global standard for the Silurian System. *National Museum of Wales Geological Series* 9, Cardiff.

Richardson, J.B. & McGregor, D.C., 1986. Silurian and Devonian spore zones of the Old Red Sandstone Continent and adjacent regions. *Geological Survey of Canada, Bulletin* 364, 79 pp.

Rickards, R.B., 1976. The sequence of Silurian graptolite zones in the British Isles. *Geological Journal* 11(2), 153–88.

Rickards, R.B. & Riva, J., 1981. *Glyptograptus? persculptus* (Salter), its tectonic deformation, and its stratigraphic significance for the Carys Mills

Formation of NE Maine, USA. *Geological Journal* 16, 219–35.

Rickards, R.B. & Stait, B.A., 1984. *Psigraptus*, its classification, evolution and zooid. *Alcheringa* 8, 101–11.

Rickards, R.B., Baillie, P.W. & Jago, J.B., 1990. An Upper Cambrian (Idamean) dendroid assemblage from near Smithton, north-western Tasmania. *Alcheringa* 14, 207–32.

Rickards, R.B., Hutt, J.E. & Berry, W.B.N., 1977. Evolution of the Silurian and Devonian graptoloids. *British Museum (Natural History) (Geology), Bulletin* 28(1), 1–120.

Rickards, R.B., Packham, G.H., Wright, A.J. & Williamson, P.L., 1994. Silurian graptolite faunas from the Quarry Creek district, NSW. *Geological Society of Australia, Abstracts* 37, 373.

Riek, E.F., 1950. A fossil mecopteran from the Triassic beds at Brookvale, New South Wales. *Records of the Australian Museum* 22, 254–56.

Riek, E.F., 1954. Further Triassic insects from Brookvale, New South Wales. *Records of the Australian Museum* 23 (4), 161–68.

Riek, E.F., 1955. Fossil insects from the Triassic beds at Mt Crosby, Queensland. *Australian Journal of Zoology* 3 (4), 654–91.

Riek, E.F., 1962. Fossil insects from the Triassic at Hobart, Tasmania. *Papers and Proceedings of the Royal Society of Tasmania* 96, 39–40.

Riek, E.F., 1964. Merostomoidea (Arthropoda, Trilobitomorpha) from the Australian Middle Triassic. *Records of the Australian Museum* 26 (13), 327–32.

Riek, E.F., 1967. A fossil cockroach (Blattoidea: Poroblattinidae) from the Mount Nicholas Coal Measures, Tasmania. *Journal of the Australian Entomological Society* 6, 73.

Riek, E.F., 1968*a*. On the occurrence of fossil insects in the Mesozoic rocks of Western Australia. *Records of the Australian Museum* 27 (16), 311–12.

Riek, E.F., 1968*b*. Re-examination of two arthropod species from the Triassic of Brookvale, New South Wales. *Records of the Australian Museum* 27 (17), 313–21.

Riek, E.F., 1973. A Carboniferous insect [from Tasmania]. *Nature* 244, 455–56.

Riek, E.F., 1976. Neosecoptera, a new insect suborder based on a specimen discovered in the Late Carboniferous of Tasmania. *Alcheringa* 1(2), 227–34.

Rigby, J.F., 1973. *Gondwanidium* and other similar Upper Palaeozoic genera, and their stratigraphic significance. *Geological Survey of Queensland Publication* 350, *Palaeontological Paper* 24, 1–10.

Rigby, J.F., 1977. New collections of fossil plants from the Esk Formation, SE Queensland. *Queensland Government Mining Journal* 78, 320–25.

Rigby, J.F., 1985. Aspects of Carboniferous palaeobotany in eastern Australia. *Compte Rendu 10ème Congrès International de Stratigraphie et de Géologie du Carbonifère, Madrid, 1983*, 4, 307–12.

Rigby, J.K. & Webby B.D., 1988. Late Ordovician sponges from the Malongulli Formation of central New South Wales, Australia. *Palaeontographica Americana* 56, 1–147, pls 1-44.

Riley, N.J., 1990*a*. Revision of the *Beyrichoceras* Ammonoid Biozone (Dinantian), NW Europe. *Newsletters on Stratigraphy* 21 (3), 149–56.

Riley, N.J., 1990*b*. Stratigraphy of the Worston Shale Group (Dinantian), Craven Basin, north-west England. *Proceedings of the Yorkshire Geological Society* 48 (2), 163–87.

Riley, N.J., 1991 [imprint 1990]. A global review of mid-Dinantian ammonoid stratigraphy. *Courier Forschungsinstitut Senckenberg* 130, 133–43.

Riley, N.J., 1993. Dinantian (Lower Carboniferous) biostratigraphy and chronostratigraphy in the British Isles. *Journal of the Geological Society* 150 (3), 427–46.

Riley, N.J., Razzo, M.J. & Owens, B., 1985. A new boundary stratotype section for the Bolsovian (Westphalian C) in northern England. *Compte Rendu 10ème Congrès International de Stratigraphie et de Géologie du Carbonifère, Madrid, 1983*, 1, 35–44.

Riley, N.J., Claoué-Long, J., Higgings, A.C., Owens, B., Spears, A., Taylor, L. & Varker, W.J., 1993. Geochronometry and geochemistry of the European Mid-Carboniferous Boundary stratotype proposal, Stonehead Beck, North Yorkshire, England. *Newsletter on Carboniferous Stratigraphy* 11, 13.

Ripperdan, R.L., 1990. Magnetostratigraphic investigations of the Lower Palaeozoic System boundaries, and associated palaeogeographic implications. Unpublished PhD thesis, California Institute of Technology, 195 pp.

Ripperdan, R.L., & Kirschvink, J.L., 1992. Paleomagnetic results from the Cambrian–Ordovician boundary section at Black Mountain, Georgina

Basin, western Queensland, Australia. pp. 93–103 *in* B.D. Webby & J.R. Laurie (eds), q.v.

Ripperdan, R.L., Magaritz, M. & Kirschvink, J.L., 1993. Carbon isotope and magnetic polarity evidence for non-depositional events within the Cambrian–Ordovician boundary section near Dayangcha, Jilin Province, China. *Geological Magazine* 130, 443–52.

Ripperdan, R.L., Magaritz, M., Nicoll, R.S. & Shergold, J.H., 1992. Simultaneous changes in carbon isotopes, sea level, and conodont biozones within the Cambrian–Ordovician boundary interval at Black Mountain, Australia. *Geology* 20, 1039–42.

Ritchie, A., 1981. First complete specimen of the dipnoan *Gosfordia truncata* Woodward from the Triassic of New South Wales. *Records of the Australian Museum* 33(11), 606–16.

Ritchie, A. & Gilbert-Tomlinson, J., 1977. First Ordovician vertebrates from the southern hemisphere. *Alcheringa* 1, 351–68.

Riva, J., 1974. A revision of some Ordovician graptolites of eastern North America. *Palaeontology* 17, 1–40.

Robaszynski, F., 1984. The Albian, Cenomanian, and Turonian stages in their type regions. *Geological Society of Denmark, Bulletin* 33, 191–98.

Robaszynski, F. & Amedro, F., 1986. The Cretaceous of the Boulonnais (France) and a comparison with the Cretaceous of Kent (United Kingdom). *Proceedings of the Geologists Association* 97, 171–208.

Roberts, J., 1963. A Lower Carboniferous fauna from Lewinsbrook, New South Wales. *Journal and Proceedings of the Royal Society of New South Wales* 97, 1–29.

Roberts, J., 1965. Lower Carboniferous zones and correlation based on faunas from the Gresford–Dungog district, New South Wales. *Journal of the Geological Society of Australia* 12, 105–22.

Roberts J., 1971. Devonian and Carboniferous brachiopods from Bonaparte Gulf Basin, northwestern Australia. *Bureau of Mineral Resources, Geology & Geophysics, Bulletin* 122, vol. I, 319 pp; vol II, 59 pls.

Roberts, J., 1975. Early Carboniferous brachiopod zones of eastern Australia. *Journal of the Geological Society of Australia* 22, 1–32.

Roberts, J., 1976. Carboniferous chonetacean and productacean brachiopods from eastern Australia. *Palaeontology* 19 (1), 17–77.

Roberts, J., 1981. Control mechanisms of Carboniferous brachiopod zones in eastern Australia. *Lethaia* 14, 123–34.

Roberts, J., 1985*a*. Australia. pp. 9–145 *in* C. Martinez Diaz (ed.), The Carboniferous of the world: 2. Australia, Indian subcontinent, South Africa, South America and North Africa *International Union of Geological Sciences Publication* 20.

Roberts, J., 1985*b*. Carboniferous sea level changes derived from depositional patterns in Australia. *Compte Rendu 10ème Congrès International de Stratigraphie et de Géologie du Carbonifère, Madrid, 1983*, 4, 43–64.

Roberts, J., Claoué-Long, J.C. & Jones, P.J., 1991*a*. Calibration of the Carboniferous and Early Permian of the Southern New England Orogen by SHRIMP ion microprobe zircon analyses. *Advances in the Study of the Sydney Basin, 25th Newcastle Symposium*, 38–43.

Roberts, J., Claoué-Long, J.C. & Jones, P.J., 1991*b*. Calibration of the Carboniferous and Early Permian of the Southern New England Orogen by SHRIMP ion microprobe zircon analyses. *Newsletter on Carboniferous Stratigraphy, IUGS Subcommission on Carboniferous Stratigraphy* 9, 15–17.

Roberts, J., Claoué-Long, J.C. & Jones, P.J., 1993*a*. Revised correlation of Carboniferous and Early Permian units of the Southern New England Orogen, Australia. *Newsletter on Carboniferous Stratigraphy, IUGS Subcommission on Carboniferous Stratigraphy* 11, 23–26.

Roberts, J., Claoué-Long, J.C. & Jones, P.J., 1993*b*. SHRIMP zircon dating and Australian Carboniferous time. *Compte Rendu 12ème Congrès International de la Stratigraphie et Géologie du Carbonifère et Permien, Buenos Aires*, 2, 319-38.

Roberts, J., Claoué-Long, J.C. & Jones, P.J., in press. Australian Early Carboniferous time. *In* W.A. Berggren & D.V. Kent (eds), Geochronology, time scales and global stratigraphic correlations. *SEPM Special Publication*.

Roberts, J., Engel, B. & Chapman, J. (eds), 1991. *Geology of the Camberwell, Dungog and Buladelah 1:100,000 sheets*, New South Wales Geological Survey, Sydney, 382 pp.

Roberts, J., Hunt, J.W. & Thompson, D.M., 1976. Late Carboniferous marine invertebrate zones of eastern Australia. *Alcheringa* 1, 197–225.

Roberts, J. Jones, P.J. & Druce, E.C., 1967. Palaeontology and correlations of the Upper Devonian of the Bonaparte Gulf Basin, Western Australia and Northern Territory. pp. 565–77 *in International Symposium on the Devonian System*, vol. 2, Alberta Society of Petroleum Geologists, Calgary.

Roberts, J., Jones, P.J. & Jenkins, T.B.H., 1993*c*. Revised correlations for Carboniferous marine invertebrate zones of eastern Australia. *Alcheringa* 17(3), 353–376.

Roberts, J., Claoué-Long, J.C., Jones, P.J. & Foster, C.B., 1995. SHRIMP zircon age control of Gondwanan sequences in Late Carboniferous Australia. *In* R. Dunay & E. Hailwood (eds), Dating and correlating biostratigraphically barren strata. *Geological Society of London Special Publication* 89, 145–74.

Roberts, J., Jones, P.J., Jell, J.S. Jenkins, T.B.H., Marsden, M.A.H., McKellar, R.G., McKelvey, B.C. & Seddon, G., 1972. Correlation of the Upper Devonian rocks of Australia. *Journal of the Geological Society of Australia* 18, 467–90.

Robison, R.A., Rozova, A.V., Rowell, A.J. & Fletcher, T.P., 1977. Cambrian boundaries and divisions. *Lethaia* 10, 257–62.

Roden, M.K., Parrish, R.R., & Miller, D.S., 1990. The absolute age of the Eifelian Tioga ash bed, Pennsylvania. *Journal of Geology* 98, 282–85.

Rodionov, V.P., 1966. Dipole character of the geomagnetic field in the Late Cambrian and Ordovician in the south of the Siberian Platform. *Geologiya i Geofizika* 1, 94–101.

Rolfe, W.D.I. & Edwards, V.A., 1979. Devonian Arthropoda (Trilobita and Ostracoda excluded). *Special Papers in Palaeontology* 23, 325–29.

Ross, C.A., 1963. Standard Wolfcampian Series (Permian), Glass Mountains, Texas. *Geological Society of America Memoir* 88, 1–205.

Ross, C.A. & Ross, J.R.P., 1985. Late Paleozoic depositional sequences are synchronous and worldwide. *Geology* 13, 194–97.

Ross, C.A. & Ross, J.R.P., 1987. Late Paleozoic sea levels and depositional sequences. *Cushman Foundation for Foraminiferal Research, Special Publication* 24, 137–49.

Ross, J.R.P., 1961. Ordovician, Silurian and Devonian Bryozoa of Australia. *Bulletin of the Bureau of Mineral Resources, Geology & Geophysics* 50, 1–172.

Ross, R.J., 1951. Stratigraphy of the Garden City Formation in north-eastern Utah, and its trilobite faunas. *Peabody Museum of Natural History, Bulletin* 6, 161 pp.

Ross R.J. & Naeser C.W., 1984. The Ordovician time scale—new refinements. pp. 5–10 *in* D.L. Bruton (ed.), Aspects of the Ordovician System. *Palaeontological Contributions from the University of Oslo* 295.

Ross, R.J., Hintze, L.F., Ethington, R.L., Miller, J.F., Taylor, M.E. & Repetski, J.E., 1993. The Ibexian Series (Lower Ordovician), a replacement for 'Canadian Series' in North American chronostratigraphy. *United States Geological Survey, Open File Report* 93–598, 1–75.

Rotai, A.P., 1979. Carboniferous stratigraphy of the USSR: proposal for an international classification. pp. 225–47 *in* R.H. Wagner, A.C. Higgins & S.V. Meyen (eds), The Carboniferous of the USSR. *Yorkshire Geological Society Occasional Publication* 4, 225–47.

Roth, P., 1983. Cretaceous nannoplankton biostratigraphy and oceanography of the north-western Atlantic Ocean. *Initial Reports of the Deep Sea Drilling Project* 44, 587–621.

Rowell, A.J. & Henderson, R.A., 1978. New genera of acrotretids from the Cambrian of Australia and the United States. *University of Kansas Paleontological Contributions, Paper* 93, 12 pp.

Roy, J.L., 1977. La position stratigraphique déterminée paléomagnétiquement de sédiments Carbonifères de Minudie Point, Nouvelle Écosse: à prôpos de l'horizon repère magnétique du Carbonifère. *Canadian Journal of Earth Sciences* 14, 1116–27.

Roy, J.L. & Morris, W.A., 1983. A review of paleomagnetic results from the Carboniferous of North America: the concept of Carboniferous geomagnetic field horizon markers. *Earth and Planetary Science Letters* 65, 167–81.

Rozefelds, A.C.,1985. A fossil zygopteran nymph (Insecta: Odonata) from the Late Triassic Aberdare Conglomerate, south-east Queensland. *Proceedings of the Royal Society of Queensland* 96, 25–32.

Rui, L. & Zhang, L., 1987. A new chronostratigraphic unit at the base of the Upper Carboniferous, with reference to the mid-Carboniferous boundary in South China. pp. 107–21 *in* C.Y. Wang (ed.), q.v.

Rui, L. & Zhang, L., 1991 [imprint 1990]. Chronostratigraphic subdivision of the Upper Carboniferous of South China. *Courier Forschungsinstitut Senckenberg* 130, 339–44.

Runnegar, B., 1967. Preliminary faunal zonation of the eastern Australian Permian. *Queensland Government Mining Journal* 68, 552–56.

Runnegar, B.N., 1969*a*. The Permian faunal succession in eastern Australia. *Geological Society of Australia Special Publication* 2, 73–98.

Runnegar, B., 1969*b*. A Lower Triassic ammonoid fauna from south-east Queensland. *Journal of Paleontology* 43 (3), 818–28.

Runnegar, B.N., 1984. The Permian of Gondwanaland. *Proceedings of the 27th International Geological Congress* 1, 305–39.

Runnegar, B. & Campbell, K.S.W., 1976. Late Palaeozoic faunas of Australia. *Earth Science Review* 12, 235–57.

Runnegar, B. & Ferguson, J.A., 1969. Stratigraphy of the Permian and Lower Triassic marine sediments of the Gympie district. *Papers of the Department of Geology, University of Queensland* 6(9), 247–81.

Runnegar, B. & McClung, G., 1975. A Permian time scale for Gondwanaland. pp. 425–41 *in* K.S.W. Campbell (ed.), *Gondwana Geology*, ANU Press, Canberra.

Russell, D.A., 1975. Reptilian diversity and the Cretaceous–Tertiary transition in North America. *Geological Association of Canada Special Paper* 13, 119–36.

Ryan, R.J. & Boehner, R.C., 1994. Geology of the Cumberland Basin, Cumberland, Colchester and Pictou Counties, Nova Scotia. *Memoir, Nova Scotia Mines and Energy Branch*, 222 pp.

Sachs, V.N., Basov, V.A., Zakharov, V.A., Nalnjaeva, T.I. & Shulgina, N.I., 1975. Jurassic–Cretaceous boundary, position of Berriasian in the Boreal Realm and correlation with Tethys. *Mémoire du Bureau de Recherches Géologiques et Minières* 86, 135–41.

Sadler, P.M., 1981. Sediment accumulation rates and the completeness of stratigraphic sections. *Journal of Geology* 89, 569–84.

Sadler, P.M. & Dingus, L.W., 1982. Expected completeness of sedimentary sections: estimating a time-scale dependent limiting factor in the resolution of the fossil record. *Proceedings of the Third North American Palaeontological Convention* 2, 461–64.

Saint-Smith, E.C., 1924. Notes on the occurrence of Cambrian strata near Mt Isa, north-west Queensland. *Queensland Government Mining Journal* 25, p. 411.

Salvador, A. (ed.), 1994. *International Stratigraphic Guide: A guide to stratigraphic classification, terminology and procedure*, 2nd edn. The International Union of Geological Sciences and the Geological Society of America, Inc. Boulder, Colorado, 214 pp.

Sandberg, C.A. & Dreesen, R., 1984. Late Devonian icriodontid biofacies models and alternate shallow-water conodont zonation. *Geological Society of America, Special Paper* 196, 143–78.

Sandberg, C.A. & Poole, F.G., 1977. Conodont biostratigraphy and depositional complexes of Upper Devonian cratonic-platform and continental-shelf rocks in the western United States. pp. 144–82 *in* M.A. Murphy, W.B.N. Berry & C.A. Sandberg (eds), Western North America; Devonian. *California University, Riverside Campus Museum Contributions* 4.

Sandberg, C.A., Poole, F.G. & Johnson, J.G. 1989*a*. Upper Devonian of western United States. pp. 183–220 *in* N.J. McMillan, A.F. Embry & D.J. Glass (eds), vol. 1, q.v.

Sandberg, C.A., Ziegler, W. & Bultynck, P., 1989*b*. New standard conodont zones and early *Ancyrodella* phylogeny across the Middle–Upper Devonian boundary. *Courier Forschungsinstitut Senckenberg* 110, 195–230.

Sandberg, C.A., Ziegler, W., Dreesen, R. & Butler, J.L., 1988. Late Frasnian mass extinction: conodont event stratigraphy, global changes, and possible causes. *Courier Forschungsinstitut Senckenberg* 102, 267–307.

Sandberg, C.A. Ziegler, W. Leuteritz, K. & Brill, S.M., 1978. Phylogeny, speciation, and zonation of *Siphonodella* (Conodonta, Upper Devonian and Lower Carboniferous). *Newsletters on Stratigraphy* 7(2), 102–20.

Sandberg, C.A., Gutschick, R.C., Johnson, J.G., Poole, F.G. & Sando, W.J., 1983. Middle Devonian to Late Mississippian geologic history of the Overthrust belt region, western United States. *Rocky Mountain Association of Geologists, Geologic Studies of the Cordilleran Thrust Belt* 2, 691–719.

Sarjeant, W.A.S., 1979. Middle and Upper Jurassic dinoflagellate cysts: the world excluding North America. *American Association of Stratigraphic Palynologists, Contribution Series* 5B, 133–57.

Sastry, M.V.A., Rao, B.R.J. & Mamgain, V.D., 1968. Biostratigraphic zonation of the Upper Cretaceous formations of Trichinopoly District, South India. *Geological Society of India Memoir* 2, 10–17.

Sastry, V.V., Raju, A.T.R., Sinha, R.N. & Venkatachala, B.S., 1974. Evolution of the Mesozoic sedimentary basins on the east coast of India. *APEA Journal* 14, 29–41.

Sato, T. & Westermann, G.E.G., 1991. Japan and South-East Asia. *Newsletters on Stratigraphy* 24, 81–108.

Saunders, W.B. & Ramsbottom, W.H.C., 1993. Re-evaluation of two Early Pennsylvanian (Middle Namurian) ammonoids and their bearing on mid-Carboniferous correlations. *Journal of Paleontology* 67(6), 993–99.

Scheibnerová, V., 1986. Cretaceous biostratigraphy of the Great Australian Basin based on benthonic Foraminifera. *Geological Survey of New South Wales Record* 22, 98 pp.

Schimper, W.P., 1874. *Traité de Paléontologie Végétale*, J.B. Baillière et fils, Paris, 806 pp.

Schindler, E., 1993. Event-stratigraphic markers within the Kellwasser Crisis near the Frasnian–Famennian boundary (Upper Devonian) in Germany. *Palaeogeography, Palaeoclimatology, Palaeoecology* 104, 115–25.

Schönlaub, H.P. (ed.), 1980. Second European conodont symposium (ECOS II). Guidebook, abstracts. *Abhandlungen der Geologischen Bundesanstalt* 35.

Schönlaub, H.P., 1986. Significant geological events in the Paleozoic record of the Southern Alps (Austrian part). pp. 163–67 *in* Walliser, O.H. (ed.), *Global Bio-events*, Springer, Berlin, 442 pp.

Scott, A.C. & Playford, G., 1985. Early Triassic megaspores from the Rewan Group, Bowen Basin, Queensland. *Alcheringa* 9, 297–323.

Sedgwick, A., 1852. On the classification and nomenclature of the Lower Palaeozoic rocks of England and Wales. *Quarterly Journal of the Geological Society* 8, 136–68.

Sedgwick, A. & Murchison, R.I., 1835. On the Silurian and Cambrian systems exhibiting the order in which the older sedimentary strata succeeded each other in England and Wales. *London and Edinburgh Philosophical Magazine* 7, 483–85.

Sedgwick, A. & Murchison, R.I., 1839. Classification of the older stratified rocks of Devonshire and Cornwall. *London and Edinburgh Philosophical Magazine, Series 3*, 14, 241–60.

Seeley, H.G. 1891 On *Agrosaurus macgillivrayi* (Seeley), a saurischian reptile from the NE coast of Australia. *Quarterly Journal of the Geological Society of London* 47, 164–65.

Seguin, M.K. & Petryk, A.A., 1986. Palaeomagnetic study of the Late Ordovician–Early Silurian platform sequence of Anticosti Island, Quebec. *Canadian Journal of Earth Sciences* 23, 1880–90.

Selwyn, A.R.C., 1861. *Geology of the Colony of Victoria*, Catalogue for the Victorian Exhibition, Government Printer, Melbourne.

Sessarego, H.L. & Césari, S.N., 1989. An Early Carboniferous flora from Argentina. Biostratigraphic implications. *Review of Palaeobotany and Palynology* 57, 247–64.

Shackleton, N.J. & Opdyke, N.D., 1976. Oxygen isotope and paleomagnetic stratigraphy of Pacific core V28–239: Late Pliocene to latest Pleistocene. *Geological Society of America Memoir* 145, 449–64.

Shafik, S., 1973. Eocene–Oligocene nannoplankton biostratigraphy in the western and southern margins of Australia. *Abstracts, 45th Congress, Australian and New Zealand Association for the Advancement of Science, Section* 3, 101–103.

Shafik, S., 1978*a*. A new nannofossil zone based on the Santonian Gingin Chalk, Perth Basin, Western Australia. *BMR Journal of Australian Geology & Geophysics* 3, 211–26.

Shafik, S., 1978*b*. Paleocene and Eocene nannofossils from the Kings Park Formation, Perth Basin, Western Australia. *Bureau of Mineral Resources, Geology & Geophysics, Bulletin* 192, 165–72.

Shafik, S., 1981. Nannofossil biostratigraphy of the *Hantkenina* (Foraminifera) interval in the Upper Eocene of southern Australia. *BMR Journal of Australian Geology & Geophysics* 6, 108–16.

Shafik, S., 1983. Calcareous nannofossil biostratigraphy: an assessment of foraminiferal events in the Eocene of the Otway Basin, south-eastern Australia. *BMR Journal of Australian Geology & Geophysics* 8, 1–17.

Shafik, S., 1985*a*. Cretaceous coccoliths in the Middle Eocene of the western and southern margins of

Australia: evidence of a significant reworking episode. *BMR Journal of Australian Geology & Geophysics* 9, 353–59.

Shafik, S., 1985*b*. Calcareous nannofossils from the Toolebuc Formation, Eromanga Basin, Australia. *BMR Journal of Australian Geology & Geophysics* 9, 171–81.

Shafik, S., 1987. Tertiary nannofossils from offshore Otway Basin and off western Tasmania. *Bureau of Mineral Resources, Geology & Geophysics, Record* 1987/11, 67–96.

Shafik, S., 1990*a*. Late Cretaceous nannofossil biostratigraphy and biogeography of the Australian western margin. *Bureau of Mineral Resources, Geology & Geophysics, Report* 295, 164 pp.

Shafik, S., 1990*b*. Maastrichtian and Early Tertiary record of the Great Australian Bight Basin and its onshore equivalents on the Australian southern margin: A nannofossil study. *BMR Journal of Australian Geology & Geophysics* 11, 473–97.

Shafik, S., 1990*c*. BMR Cruise 95: Nannofossil biostratigraphy. *Bureau of Mineral Resources, Geology & Geophysics, Record* 1990/57, 30–39.

Shafik, S., 1990*d*. Calcareous nannofossil age determination of dredge samples, BMR Cruise 96, I. Jurassic to Late Oligocene. *Bureau of Mineral Resources, Geology & Geophysics, Record* 1990/85, 56–82.

Shafik, S., 1990*e*. Calcareous nannofossil age determination of dredge samples, BMR Cruise 96, II. Neogene. *Bureau of Mineral Resources, Geology & Geophysics, Record* 1990/85, 83–89.

Shafik, S., 1991. Upper Cretaceous and Tertiary stratigraphy of the Fremantle Canyon, South Perth Basin: a nannofossil assessment. *BMR Journal of Australian Geology & Geophysics* 12, 65–91.

Shafik, S., 1992*a*. Late Cretaceous–Paleogene nannofossil biostratigraphy of Challenger No. 1 (Challenger Formation type section), offshore Perth Basin, Western Australia. *BMR Journal of Australian Geology & Geophysics* 13, 19–29.

Shafik, S., 1992*b*. Eocene and Oligocene calcareous nannofossils from the Great Australian Bight: evidence of significant reworking episodes and surface-water temperature changes. *BMR Journal of Australian Geology & Geophysics* 13, 131–42.

Shafik, S., 1992*c*. Tertiary calcareous nannofossil biostratigraphy, offshore Otway Basin. *Bureau of Mineral Resources, Geology & Geophysics, Report* 306, 93–119.

Shafik, S., 1993. Albian and Maastrichtian nannofloral biogeographic provinces in Western Australia: implications for palaeolatitudes and pole positions for Australia. *AGSO Research Newsletter* 19, 15.

Shafik, S., 1994. Significance of calcareous nannofossil-bearing Jurassic and Cretaceous sediments on the Rowley Terrace, offshore North West Australia. *AGSO Journal of Australian Geology & Geophysics* 15, 71–88.

Shafik, S. & Chaproniere, G.C.H., 1978. Nannofossil and planktic foraminiferal biostratigraphy around the Oligocene–Miocene boundary in parts of the Indo–Pacific region. *BMR Journal of Australian Geology & Geophysics* 3, 135–51.

Shaver, R.H., 1984. Atokan Series concepts with special reference to the Illinois Basin and Iowa. pp.101–12 *in* P.K. Sutherland & W.L. Manger (eds), 1984*a*, q.v.

Shaw, A.B., 1964. *Time in Stratigraphy*, McGraw-Hill, New York, 365 pp.

Shergold, J.H., 1972. Late Upper Cambrian trilobites from the Gola Beds, western Queensland. *Bureau of Mineral Resources, Geology & Geophysics, Bulletin* 112, 126 pp.

Shergold, J.H., 1973. *Meneviella viatrix* sp. nov., a new conocoryphid trilobite from the Middle Cambrian of western Queensland. *Bureau of Mineral Resources, Geology & Geophysics, Bulletin* 126, 19–26.

Shergold, J.H., 1975. Late Cambrian and Early Ordovician trilobites from the Burke River Structural Belt, western Queensland, Australia. *Bureau of Mineral Resources, Geology & Geophysics, Bulletin* 153: vol. 1, 251 pp.; vol. 2, 58 pls.

Shergold, J.H., 1977. Classification of the trilobite *Pseudagnostus*. *Palaeontology* 20(1), 69–100.

Shergold, J.H., 1980. Late Cambrian trilobites from the Chatsworth Limestone, western Queensland. *Bureau of Mineral Resources, Geology & Geophysics, Bulletin* 186, 111 pp.

Shergold, J.H., 1981. Towards a global Late Cambrian agnostid biochronology. pp. 208–14 *in* M.E. Taylor (ed.), Short papers for the Second International Symposium on the Cambrian System. *US Geological Survey Open File Report* 81–743.

Shergold, J.H., 1982. Idamean (Late Cambrian) trilobites, Burke River Structural Belt, western

Queensland. *Bureau of Mineral Resources, Geology & Geophysics, Bulletin* 187, 69 pp.

Shergold, J.H., 1986. Review of the Cambrian and Ordovician palaeontology of the Amadeus Basin, central Australia. *Bureau of Mineral Resources, Geology & Geophysics, Report* 276, 21 pp.

Shergold, J.H., 1989. Australian Phanerozoic Timescales: 1. Cambrian biostratigraphic chart and explanatory notes. *Bureau of Mineral Resources, Geology & Geophysics, Record* 1989/31, 25 pp.

Shergold, J.H. (ed.), 1991*a*. Late Proterozoic and Early Palaeozoic palaeontology and biostratigraphy of the Amadeus Basin. *Bureau of Mineral Resources, Geology & Geophysics, Bulletin* 236, 97–111.

Shergold, J.H., 1991*b*. Late Cambrian (Payntonian) and Early Ordovician (Late Warendian) trilobite faunas of the Amadeus Basin, central Australia. *Bureau of Mineral Resources, Geology & Geophysics, Bulletin* 237, 15–75.

Shergold, J.H., 1993. The Iverian Stage (Late Cambrian) and its subdivision in the Burke River Structural Belt, western Queensland. *BMR Journal of Australian Geology & Geophysics* 13 (4), 345–58.

Shergold, J.H. & Nicoll, R.S., 1992. Revised Cambrian–Ordovician boundary biostratigraphy, Black Mountain, western Queensland. pp. 81–92 *in* B.D. Webby & J.R. Laurie (eds), q.v.

Shergold, J.H., Laurie, J.R. & Sun Xiaowen, 1990. Classification and review of the trilobite order Agnostida Salter, 1864: an Australian perspective. *Bureau of Mineral Resources, Geology & Geophysics, Report* 296, 92 pp.

Shergold, J.H., Southgate, P.N. & Cook, P.J., 1989. New facts on old phosphates: Middle Cambrian, Georgina Basin. *BMR Research Newsletter* 10, 14–15.

Shergold, J.H., Cooper, R.A., Jago, J.B. & Laurie, J.R., 1985. The Cambrian System in Australia, Antarctica and New Zealand. Correlation charts and explanatory notes. *International Union of Geological Sciences, Publication* 19, 85 pp.

Shergold, J.H., Gorter, J.D., Nicoll, R.S. & Haines, P.W., 1991*a*. Stratigraphy of the Pacoota Sandstone (Cambrian–Ordovician), Amadeus Basin, NT. *Bureau of Mineral Resources, Geology & Geophysics, Bulletin* 237, 1–14.

Shergold, J.H., Nicoll, R.S., Laurie, J.R. & Radke, B.M., 1991*b*. The Cambrian–Ordovician boundary at Black Mountain, western Queensland. Guidebook for Field Excursion 1, 6th International Symposium on the Ordovician System, Sydney. *Bureau of Mineral Resources, Geology & Geophysics, Record* 1991/48, 50 pp.

Sherrard, K., 1967. Tentaculitids from New South Wales, Australia. *Proceedings of the Royal Society of Victoria* 80, 229–46.

Sherwin, L., 1968. *Denckmannites* (Trilobita) from the Silurian of New South Wales. *Palaeontology* 11, 691–96.

Sherwin, L., 1971. Stratigraphy of the Cheeseman's Creek district, New South Wales. *Records of the Geological Survey of New South Wales* 13, 199–237.

Sherwin, L., 1980. Faunal correlations of the Siluro–Devonian units, Mineral Hill–Trundle, Peak Hill area. *Geological Survey of New South Wales, Quarterly Notes* 39, 1–14.

Sherwin, L., 1992. Siluro–Devonian biostratigraphy of central New South Wales. *Geological Survey of New South Wales, Quarterly Notes* 86, 1–12.

Shipboard Scientific Party, 1990. Section 2, Site reports. *Proceedings of the Ocean Drilling Program, Initial Reports* 122, 81–384.

Shive, P.N., Steiner, M.B. & Huycke, D.T., 1984. Magnetostratigraphy, paleomagnetism and remanence acquisition in the Triassic Chugwater Group of Wyoming. *Journal of Geophysical Research* 89, 1801–15.

Siesser, W.G. & Bralower, T.J., 1992. Cenozoic calcareous nannofossil biostratigraphy on the Exmouth Plateau, eastern Indian Ocean. *Proceedings of the Ocean Drilling Program, Scientific Results* 122, 601–31.

Silberling, N.J. & Tozer, E.T., 1968. Biostratigraphic classification of the marine Triassic in North America. *Geological Society of America, Special Paper* 110, 1–63.

Singh, G., Opdyke, N.D. & Bowler, J.M., 1981. Late Cainozoic stratigraphy, palaeomagnetic chronology and vegetational history from Lake George, NSW. *Journal of the Geological Society of Australia* 28, 435–52.

Singleton, F.A., 1941. The Tertiary geology of Australia. *Proceedings of the Royal Society of Victoria (New Series)* 53, 1–125.

Singleton, O.P., 1954. The Tertiary stratigraphy of Western Australia—a review. *Proceedings of the Pan Indian Ocean Science Congress*, 59–65.

Singleton, O.P., 1967. Otway region. Excursion handbook. *39th Congress of Australia and New Zealand Association for the Advancement of Science, Section C*, 171–81.

Singleton, O.P., McDougall, I. & Mallett, C.W., 1976. The Pliocene–Pleistocene boundary in south-eastern Australia. *Journal of the Geological Society of Australia* 23, 299–311.

Sissingh, W., 1977. Biostratigraphy of Cretaceous calcareous nannoplankton. *Geologie en Mijnbouw* 56, 37–56.

Skwarko, S.K., 1966. Cretaceous stratigraphy and palaeontology of the Northern Territory. *Bureau of Mineral Resources, Geology & Geophysics, Bulletin* 73, 135 pp.

Skwarko, S.K., 1983. Cenomanian (Late Cretaceous) mollusca from Mountnorris Bay, Arnhem Land, northern Australia. *Bureau of Mineral Resources, Geology & Geophysics, Bulletin* 217, 73–83.

Skwarko, S.K. (ed.), 1993. Permian palaeontology of Western Australia. *Bulletin of the Geological Survey of Western Australia* 136, 1–417.

Skwarko, S.K. & Kummel, B., 1974. Marine Triassic molluscs of Australia and New Guinea. *Bureau of Mineral Resources, Geology & Geophysics, Bulletin* 150, 111–28.

Sloan, R.E., Rigby, J.K., Van Valen, L.M. & Gabriel, D., 1986. Gradual dinosaur extinction and simultaneous ungulate radiation in the Hell Creek Formation. *Science* 232, 629–32.

Sluiter, I.R.K., 1991. Early Tertiary vegetation and climates, Lake Eyre region, north-eastern South Australia. pp. 99–118 *in* M.A.J. Williams, P. de Deckker & A.P. Kershaw (eds), The Cainozoic in Australia: a re-appraisal of the evidence. *Geological Society of Australia, Special Publication* 18.

Smethurst, M.A. & Briden, J.C., 1988. Palaeomagnetism of Silurian sediments in W Ireland: evidence for block rotation in the Caledonides. *Geophysical Journal* 95, 327–46.

Smethurst, M.A. & Khramov, A.N., 1992. A new Devonian palaeomagnetic pole for the Russian Platform and Baltica, and related apparent polar wander. *Geophysical Journal International* 108, 179–92.

Smith, J.P., 1904. The comparative stratigraphy of the marine Trias of western America. *Proceedings of the California Academy of Sciences, Series* 3, 1(10), 323–437.

Snelling, N.J. (ed.) 1985. The chronology of the geological record. *Geological Society, Memoir* 10, 343 pp.

Solodukho, M.G., Gusev, A.K., Ignatyev, V.I., Lukin, V.A., Boronin, V.P., Esaulova, N.K., Zharkov, I.Ya., Silantyev, V.V. & Bogov, A.V., 1993. Part 5—Volga Region. pp. 269–303 *in* B.I. Chuvashov & A.E.M. Nairn (eds), Permian System: guides to geological excursions in the Uralian type localities. *Occasional Publications, Earth Sciences and Resources Institute, University of South Carolina, New Series* 10.

Solovieva, M.N., 1985. Correction of the USSR general Carboniferous scale in connection with restudying in Moskovian Stage stratotype and a new model for the correlation of the lower Moskovian substage. *Compte Rendu 10ème Congrès International de Stratigraphie et de Géologie du Carbonifère, Madrid, 1983*, 1, 21–26.

Solovieva, M.N., Chizhova, V.A., Einor, O.L., Grigorieva, A.D., Reitlinger, E.A. & Vdovenko, M.V., 1985*a*. Review of recent data on the Carboniferous stratigraphy of the USSR. *Compte Rendu 10ème Congrès International de Stratigraphie et de Géologie du Carbonifère, Madrid, 1983*, 1, 3–10.

Solovieva, M.N., Fisunenko, O.P., Goreva, N.V., Barskov, I.S., Gubareva, V.S., Dzhenchuraeva, A.V., Dalmatskaya, I.I., Ivanova, E.A., Poletaev, V.I., Popov, A.V., Rumyantseva, Z.S., Teteryuk, V.K. & Shik, E.M., 1985*b*. New data on stratigraphy of the Moskovian Stage. *Compte Rendu 10ème Congrès International de Stratigraphie et de Géologie du Carbonifère, Madrid, 1983*, 1, 11–20.

Song Zhichen, Li Wenben & He Chengquan, 1983. Cretaceous and Palaeogene palynofloras and distribution of organic rocks in China. *Scientia Sinica* B 26, 538–49.

Southgate, P.N. & Shergold, J.H., 1991. Application of sequence stratigraphic concepts to Middle Cambrian phosphogenesis, Georgina Basin, Australia. *BMR Journal of Australian Geology & Geophysics* 12, 119–44.

Spell, T.I. & McDougall, I., 1992. Revisions to the age of the Brunhes–Matuyama boundary and the Pleistocene geomagnetic polarity timescale. *Geophysical Research Letters* 19, 1181–84.

Springer, M. & Lilje, A., 1988. Biostratigraphy and gap analysis: the expected sequence of biostratigraphic events. *Journal of Geology* 96, 228–36.

Stainforth, R.M., Lamb, J.L., Luterbacher, H., Beard, J.H. & Jeffords, R.M., 1975. Cenozoic planktonic foraminiferal zonation and characteristics of index forms. *University of Kansas Paleontological Contributions*, 62, 1–425.

Stait, B., 1980. *Gouldoceras* n. gen. (Cephalopoda, Nautiloidea) and a revision of *Hecatoceras* Teichert & Glenister from the Ordovician of Tasmania, Australia. *Journal of Paleontology* 54, 1113–18.

Stait, B., 1982. Ordovician Oncoceratida (Nautiloidea) from Tasmania, Australia. *Neues Jahrbuch für Geologie und Paläontologie Monatshefte* 1982, 607–61

Stait, B., 1983. Ordovician nautiloids of Tasmania, Australia—Ellesmerocerida and Tarphycerida. *Alcheringa* 7, 253–61.

Stait, B., 1984*a*. Re-examination and redescription of the Tasmanian species of *Wutinoceras* and *Adamsoceras* (Nautiloidea, Ordovician). *Geologica et Palaeontologica* 18, 53–57.

Stait, B. 1984*b*. Ordovician nautiloids of Tasmania—Gouldoceratidae fam. nov. (Discosorida). *Proceedings of the Royal Society of Victoria* 96, 187–207.

Stait, B. & Laurie, J., 1980. Lithostratigraphy and biostratigraphy of the Florentine Valley Formation in the Tim Shea area, south-west Tasmania. *Papers and Proceedings of the Royal Society of Tasmania* 114, 201–207.

Standard, J.C., 1962. A new study of the Hawkesbury Sandstone: preliminary findings. *Journal and Proceedings of the Royal Society of New South Wales* 95, 145–46.

Steiner, M.B. & Helsley C.E., 1974. Magnetic polarity sequence of the Upper Triassic Kayenta Formation. *Geology* 2, 191–98.

Steiner, M.B., Ogg, J.G. & Sandoval, J., 1987. Jurassic magnetostratigraphy, 3. Bathonian–Bajocian of Carcabuey, Sierra Harana and Campillo de Arenas (Subbetic Cordillera, southern Spain). *Earth and Planetary Science Letters* 82, 357–72.

Steiner, M.B., Ogg, J.G., Melendez, G. & Sequeiros, L., 1985. Jurassic magnetostratigraphy, 2. Middle–Late Oxfordian of Aguilon, Iberian Cordillera, northern Spain. *Earth and Planetary Science Letters* 76, 151–66.

Steiner, M.B., Ogg, J.G., Zhang, Z. & Sun, S., 1989. The Late Permian/Early Triassic magnetic polarity time scale and plate motions of South China. *Journal of Geophysical Research* 94, 7343–63.

Stephens, W.J. 1886. On some additional labyrinthodont fossils from the Hawkesbury Sandstone of New South Wales. *Proceedings of the Linnean Society of New South Wales NS* 1, 1175–95.

Stevens, G.R. & Speden, I.G., 1978. New Zealand. pp. 251–328 *in* M. Moullade & A.E.M. Nairn (eds), *The Phanerozoic Geology of the World II. The Mesozoic*, Elsevier, Amsterdam.

Stewart, I.R., 1988. Conodonts. pp. 79–81 in R.A.F. Cas & A.H.M. VandenBerg, q.v.

Stott, D.F., 1975. The Cretaceous System in northeastern British Columbia. *Geological Association of Canada Special Paper* 13, 441–67.

Stover, L.E. & Evans, P.R., 1973. Upper Cretaceous–Eocene spore–pollen zonation, offshore Gippsland Basin. *Geological Society of Australia, Special Publication* 4, 55–72.

Stover, L.E. & Helby, R.J., 1987. Some Australian Mesozoic microplankton index species. *Association of Australasian Palaeontologists, Memoir* 4, 101–34.

Stover, L.E. & Partridge, A.D., 1973. Tertiary and Late Cretaceous spores and pollen from the Gippsland Basin, south-eastern Australia. *Proceedings of the Royal Society of Victoria* 85, 237–86.

Strauss, D. & Sadler, P.M., 1989. Classical confidence intervals and Bayesian probability estimates for ends of local taxon ranges. *Mathematical Geology* 21, 411–27

Streel, M., & Loboziak, S., 1993. Confidence levels of correlations between miospore biohorizons and standard conodont zones during Middle and Late Devonian time. *The Gross Symposium, Gottingen, August 1993, Abstracts* (unpaginated).

Streel, M., Higgs, K., Loboziak, S., Riegel, W. & Steemans, P., 1987. Spore stratigraphy and correlation with faunas and floras in the type marine Devonian of the Ardenne–Rhenish regions. *Review of Palaeobotany and Palynology* 50, 211–29.

Struckmeyer, H.I.M. & Brown, P.J., 1990. Australian sealevel curves. Part 1: Australian inundation curves. *Bureau of Mineral Resources Geology & Geophysics, Record* 1990/11, 67 pp.

Strusz, D.L., 1961. Lower Palaeozoic corals from New South Wales. *Palaeontology* 4(3), 334–61.

Strusz, D.L., 1972. Correlation of the Lower Devonian rocks of Australasia. *Journal of the Geological Society of Australia* 18, 427–55.

Strusz, D.L., 1980. The Encrinuridae and related trilobite families, with a description of Silurian species from south-eastern Australia. *Palaeontographica, Abt* A 168(1–4), 1–68.

Strusz, D.L., 1982. Wenlock brachiopods from Canberra, Australia. *Alcheringa* 6(2), 105–42.

Strusz, D.L., 1983. Silurian *Maoristrophia* (Brachiopoda) from Canberra, Australia. *Alcheringa* 7, 163–68.

Strusz, D.L., 1984. Brachiopods of the Yarralumla Formation (Ludlovian), Canberra, Australia. *Alcheringa* 8, 123–50.

Strusz, D.L., 1985. Brachiopods from the Silurian of Fyshwick, Canberra, Australia. *BMR Journal of Australian Geology* & *Geophysics* 9(2), 107–19.

Strusz, D.L., 1989. Australian Phanerozoic Timescales: 3. Silurian—biostratigraphic chart and explanatory notes. *Bureau of Mineral Resources, Geology* & *Geophysics, Record* 1989/33, 28 pp.

Strusz, D.L., Webby, B.D., Wright, A.J. & Pickett J.W., 1988. Ordovician Silurian and Devonian corals and spongiomorphs of central New South Wales. *V International Symposium on Fossil Cnidaria including Spongiomorphs Excursion A3, Guide Book*, 63 pp.

Strzelecki, P.E. de, 1845. *Physical Description of New South Wales and Van Dieman's Land: Accompanied by a Geological Map, Sections and Diagrams and Figures of Organic Remains*, Longman, Brown, Green and Longman London, 462 pp.

Sun Xingjun, 1980. Late Cretaceous and Paleocene palynoflora of China. *Fifth International Palynological Conference (Cambridge 1980), Abstracts* 380.

Sutherland, P.K. & Manger, W.L. (eds), 1984*a*. The Atokan Series (Pennsylvanian) and its boundaries—a symposium. *Oklahoma Geological Survey Bulletin* 136, 1–198.

Sutherland, P.K. & Manger, W.L. (eds), 1984*b*. Biostratigraphy. *Compte Rendu 9ème Congrès International de Stratigraphie et de Géologie du Carbonifère, Washington and Champaign-Urbana, 1979.*

Sutherland, P.K. & Manger, W.L. (eds), 1992. Recent advances in Middle Carboniferous biostratigraphy—a symposium. *Oklahoma Geological Survey Circular* 94, 1–181.

Sweet, W.C., 1979. Graphic correlation of Permo–Triassic rocks in Kashmir, Pakistan and Iran. *Geologica et Palaeontologica* 13, 239–48.

Sweet, W.C., 1984. Graphic correlation of upper Middle and Upper Ordovician rocks, North American Mid-continent Province, USA. pp. 23–35 *in* D.L. Bruton (ed.), Aspects of the Ordovician System. *Palaeontological Contributions from the University of Oslo* 295.

Sweet, W.C. & Bergström, S.M., 1974. Provincialism exhibited by Ordovician conodont faunas. pp. 189–202 *in* C.A. Ross (ed.), Paleogeographic provinces and provinciality. *Society of Economic Paleontologists and Mineralogists Special Publication* 21.

Sweet, W.C. & Bergström, S.M., 1984. Conodont provinces and biofacies of the Late Ordovician. *Geological Society of America Special Paper* 196, 69–87.

Sweet, W.C., Mosher, L.C., Clark, D.L., Collinson, J.W. & Hasenmueller, W.A., 1971. Conodont biostratigraphy of the Triassic. *Geological Society of America, Memoir* 127, 441–65.

Swisher III, C.C., Grajales-Nishimura, J.M., Montanari, A., Cedillo-Pardo, E., Margolis, S.V., Claeys, P., Alvarez, W., Smit, J., Renne, P., Maurrasse, F.J. M.R. & Curtis, G.H., 1992. Chicxulub crater melt-rock and the K–T boudary tektites from Mexico and Haiti yield coeval $^{40}Ar/^{39}Ar$ ages of 654 Ma. *Science* 257, 954–58.

Taboada, A.C., 1989. La fauna de la Formation El Paso, Carbonifera Inferior de la Precordillera Sanjuanina. *Acta Geologica Lilloana* 17, 113–29.

Talent, J.A., 1989. Transgression–regression pattern for the Silurian and Devonian of Australia. pp. 201–19 *in* R.W. Le Maitre (ed.), *Pathways in Geology—Essays in Honour of Edwin Sherbon Hills*, Blackwell, Carlton.

Talent, J.A. & Yolkin, E.A., 1987. Transgression–regression patterns for the Devonian of Australia and southern West Siberia. *Courier Forschungsinstitut Senckenberg* 92, 235–49.

Talent, J.A., Berry, W.B.N. & Boucot, A.J., Packham, G.H. & Bischoff, G.C.O., 1975. Correlation of the Silurian rocks of Australia, New Zealand, and New Guinea. *Special Papers of the Geological Society of America* 150, 1–108.

Talent, J.A., Mawson, R., Andrew, A.S., Hamilton, P.J. & Whitford, D.J., 1993. Middle Palaeozoic extinction events: faunal and isotopic data.

Palaeogeography, Palaeoclimatology, Palaeoecology 104, 139–52.

Tarduno, J.A., 1990. Brief reversed polarity interval during the Cretaceous Normal Polarity Superchron. *Geology* 18, 683–86.

Tarduno, J.A., Lowrie, W., Sliter, W.V., Bralower, T.J. & Heller, F., 1992. Revised polarity characteristic magnetizations in the Albian Contessa section, Umbrian Apennines, Italy: implications for the existence of a mid-Cretaceous mixed polarity interval. *Journal of Geophysical Research* 97, 241–71.

Tarling, D.H. & Mitchell, J.G., 1976. Revised Cenozoic polarity time scale. *Geology* 4, 133–36.

Tasch, P., 1975. Non-marine Arthropoda of the Tasmanian Triassic. *Papers and Proceedings of the Royal Society of Tasmania* 109, 97–106.

Tasch, P., 1979. Permian and Triassic Conchostraca from the Bowen Basin (with a note on a Carboniferous leaiid from the Drummond Basin), Queensland. *Bureau of Mineral Resources, Geology & Geophysics, Bulletin* 185, 33–47.

Tasch, P. & Jones, P.J., 1979*a*. Carboniferous and Triassic conchostracans of Australia from the Canning Basin, Western Australia. *Bureau of Mineral Resources, Geology & Geophysics, Bulletin* 185, 3–20.

Tasch, P. & Jones, P.J., 1979*b*. Lower Triassic Conchostraca from the Bonaparte Gulf Basin, north-western Australia, with a note on *Cyzicus* (*Euestheria*) *minuta*[?] from the Carnarvon Basin. *Bureau of Mineral Resources, Geology & Geophysics, Bulletin* 185, 23–30.

Tate, R., 1896. Palaeontology. pp. 97–116 *in* B. Spencer (ed.), *Report on the Work of the Horn Scientific Expedition to Central Australia. Part III, Geology and Botany*, Dulau & Co., London, & Melville, Mullen & Slade, Melbourne.

Taylor, D.J., 1964. Foraminifera and the stratigraphy of the western Victorian Cretaceous sediments. *Royal Society of Victoria Proceedings* 77, 535–603.

Taylor, D.J., 1966. Esso Gippsland Shelf No. 1: the mid Tertiary foraminiferal sequence. *Bureau of Mineral Resources, Geology & Geophysics, Petroleum Search Subsidy Act Publication* 76, 31–46.

Taylor, D.W. & Hickey, L.J., 1990. An Aptian plant with attached leaves and flowers: implications for angiosperm origin. *Science* 247, 702–704.

Taylor, G., Truswell, E.M., McQueen, K.G. & Brown, C.M., 1990. Early Tertiary palaeogeography, land form evolution, and palaeoclimates of the southern Monaro, NSW, Australia. *Palaeogeography, Palaeoclimatology, Palaeoecology* 78, 109–35.

Teichert, C., 1939. The nautiloid *Bathmoceras* Barrande. *Transactions of the Royal Society of South Australia* 63, 384–91.

Teichert, C., 1940*a*. Marine Jurassic in the North-West Basin, Western Australia. *Journal of the Royal Society of Western Australia* 26, 17–27.

Teichert, C., 1940*b*. Marine Jurassic of East Indian affinities at Broome, north-western Australia. *Journal of the Royal Society of Western Australia* 26, 103–19

Teichert, C., 1941. Upper Paleozoic of Western Australia: correlation and paleogeography. *Bulletin of the American Association of Petroleum Geologists* 25 (3), 371–415.

Teichert, C., 1947. Early Ordovician cephalopods from Adamsfield, Tasmania. *Journal of Paleontology* 21, 420–28.

Teichert, C., 1950. Discovery of Devonian and Carboniferous rocks in the North-West Basin, Western Australia. *Australian Journal of Science* 12, 62–65.

Teichert, C. & Glenister, B.F., 1952. Fossil nautiloid faunas from Australia. *Journal of Paleontology* 26, 730–52.

Teichert, C. & Glenister, B.F., 1953. Ordovician and Silurian cephalopods from Tasmania, Australia. *Bulletin of American Paleontology* 34, 187–248.

Teichert, C. & Glenister, B.F., 1954. Early Ordovician cephalopod fauna from north-western Australia. *Bulletin of American Paleontology* 35, 1–112.

Teichert, C., Glenister, B.F. & Crick, R.E., 1979. Biostratigraphy of Devonian nautiloid cephalopods. *Special Papers in Palaeontology* 23, 259–62.

Teller, L., 1969. The Silurian biostratigraphy of Poland based on graptolites. *Acta Geologica Polonica* 19 (3), 393–501.

Théveniaut, H., Klootwijk, C., Foster, C. & Giddings, J., 1994. Magnetostratigraphy of the Late Permian coal measures of the Sydney and Gunnedah basins: A regional and global correlation tool. *Proceedings of the 28th Newcastle Symposium, Department of Geology, University of Newcastle*, pp. 11–23.

Thierstein, H.R., 1974. Calcareous nannoplankton—Leg 26, Deep Sea Drilling Project. *Initial Reports of the Deep Sea Drilling Project* 26, 619–67.

Thomas, A.T., Owens, R.M. & Rushton, A.W.A., 1984. Trilobites in British stratigraphy. *Special Report of the Geological Society of London* 16, 1–78.

Thomas, C., 1976. Palaeomagnetic and rock magnetic studies of the Lower Palaeozoic rocks of North Wales. Unpublished PhD thesis, Leeds University.

Thomas, D.E. 1960. The zonal distribution of Australian graptolites. *Journal and Proceedings of the Royal Society of New South Wales* 94, 1–58.

Thomas, D.E. & Henderson, Q.J., 1945. Some fossils from the Dundas Series, Dundas. *Papers and Proceedings of the Royal Society of Tasmania (for 1944)*, 1–8.

Thomas, D.E. & Singleton, O.P., 1956. The Cambrian stratigraphy of Victoria. pp. 149–63 *in* J. Rodgers (ed.), El Sistema Cambrico su paleogeografia y el problema de su base, 2. *20th Session International Geological Congress, Mexico.*

Thomas, G.A., 1957. Lower Carboniferous deposits in the Fitzroy Basin, Western Australia. *Australian Journal of Science* 19, 160–61.

Thomas, G.A., 1959. The Lower Carboniferous Laurel Formation of the Fitzroy Basin, Western Australia. *Bureau of Mineral Resources, Geology & Geophysics, Report* 38, 21–36.

Thomas, G.A., 1971. Carboniferous and Early Permian brachiopods from Western and northern Australia. *Bureau of Mineral Resources, Geology & Geophysics, Bulletin* 56, 276 pp.

Thomas, G.A. & Dickins, J.M., 1954. Correlation and age of the marine Permian formations of Western Australia. *Australian Journal of Science* 16 (6), 219–23.

Thompson, T.L., 1967. Conodont zonation of lower Osagean rocks (Lower Mississippian) of southwestern Missouri. *Missouri Division* of *Geological Survey and Water Resources, Report of Investigations* 39.

Thompson, T.L. & Fellows, L.D., 1970. Stratigraphy and conodont biostratigraphy of Kinderhookian and Osagean (Lower Mississippian) rocks of south-western Missouri and the adjacent area. *Missouri Division of Geological Survey and Water Resources, Report of Investigations* 45, 263 pp.

Thulborn, R.A., 1979. A proterosuchian thecodont from the Rewan Formation of Queensland. *Memoirs of the Queensland Museum* 19, 331–55.

Thulborn, R.A., 1983. A mammal-like reptile from Australia. *Nature* 306, 209.

Thulborn, R.A., 1986. Early Triassic tetrapod faunas of southeastern Gondwana. *Alcheringa* 10, 297–313.

Thulborn, R.A., 1990. Mammal-like reptiles of Australia. *Memoirs of the Queensland Museum* 28(1), 169.

Tidwell, W.D., Nishida, H. & Webster, N., 1989. *Oguracaulis banksii* gen. et sp. nov., a mid-Mesozoic tree-fern stem from Tasmania, Australia. *Papers and Proceedings of the Royal Society of Tasmania* 123, 15–25.

Tillyard, R.J., 1916. Mesozoic and Tertiary insects of Queensland and New South Wales. *Geological Survey of Queensland, Publication* 253, 11–48.

Tillyard, R.J., 1919. Mesozoic insects of Queensland. No. 5. Mecoptera, the new Order Paratrichoptera, and additions to Planipennia. *Proceedings of the Linnean Society of New South Wales* 44(1), 194–212.

Tillyard, R.J., 1922. Mesozoic insects of Queensland. No. 9. Orthoptera, and additions to the Protorthoptera, Odonata, Hemiptera and Planipennia. *Proceedings of the Linnean Society of New South Wales* 47(4), 447–70.

Tillyard, R.J., 1923. Mesozoic insects of Queensland. No. 10. Summary of the Upper Triassic insect fauna of Ipswich, Queensland. (With an appendix describing new Protorthoptera, Odonata, Hemiptera and Planipennia). *Proceedings of the Linnean Society of New South Wales* 48, 481–98.

Tillyard, R.J., 1925. A new fossil insect wing from Triassic beds near Deewhy, New South Wales. *Proceedings of the Linnean Society of New South Wales* 50, 374–77.

Tillyard, R.J., 1937. A small collection of cockroach remains from the Triassic beds of Mount Crosby, Queensland. *Proceedings of the Royal Society of Queensland* 48, 35–40.

Tillyard, R.J. & Dunstan, B., 1924. Mesozoic insects of Queensland. *Geological Survey of Queensland, Publication* 273, 1–506.

Tims, J.D. & Chambers, T.C., 1984. Rhyniophytina and Trimerophytina from the early land flora of Victoria, Australia. *Palaeontology* 27, 265–79.

Torsvik, T.H. & Trench, A., 1991. Ordovician magnetostratigraphy: Llanvirn–Caradoc limestones of the Baltic Platform. *Geophysical Journal International* 107, 171–84.

Torsvik, T.H., Trench, A., Svensson, I. & Walderhaug, H., 1993. Palaeogeographic significance of mid

Silurian palaeomagnetic results from southern Britain—major revision of the apparent polar wander path for Eastern Avalonia. *Geophysical Journal International* 113, 651–68.

Toumarkine, M. & Luterbacher, H., 1985. Paleocene and Eocene planktic Foraminifera. pp. 87–154 *in* H.M. Bolli, J.B. Saunders & K. Perch-Nielsen (eds), *Plankton Stratigraphy*, Cambridge University Press, Cambridge.

Towner, R.R. & Gibson, D.L., 1983. Geology of the onshore Canning Basin, Western Australia. *Bureau of Mineral Resources, Geology & Geophysics, Bulletin* 215, 51 pp.

Townrow, J.A., 1965. A new member of the Corystospermaceae Thomas. *Annals of Botany, NS* 29 (115), 495–511.

Townrow, J.A., 1966. On *Dicroidium odontopteroides* and *D. obtusilobium* in Tasmania. pp. 128–36 *in Symposium on Floristics of the Gondwanaland*, Birbal Sahni Institute of Palaeobotany, Lucknow.

Townrow, J.A., 1967*a*. On *Rissikia* and *Mataia*, podocarpaceous conifers from the Lower Mesozoic of southern lands. *Papers and Proceedings of the Royal Society of Tasmania* 101, 103–36.

Townrow, J.A., 1967*b*. On *Voltziopsis*, a southern conifer of Lower Triassic age. *Papers and Proceedings of the Royal Society of Tasmania* 101, 173–88.

Tozer, E.T., 1965. Lower Triassic stages and ammonoid zones of Arctic Canada. *Geological Survey of Canada, Paper* 65–12.

Tozer, E.T., 1967. A standard for Triassic time. *Geological Survey of Canada, Bulletin* 156, 1–103.

Tozer, E.T., 1984. The Trias and its ammonoids: the evolution of a time scale. *Geological Survey of Canada, Miscellaneous Report* 35, 1–171.

Tozer, E.T., 1986. Definition of the Permian–Triassic (P–T) boundary; the question of the age of the *Otoceras* beds. *Memorie della Societa Geologica Italiana* 34, 291–302.

Tozer, E.T., 1988. Toward a definition of the Permian–Triassic boundary. *Episodes* 11(4), 251–55.

Trench, A. & Torsvik, T.H., 1991. The Lower Palaeozoic apparent polar wander path for Baltica: palaeomagnetic data from Silurian limestones of Gotland, Sweden. *Geophysical Journal International* 107, 171–84.

Trench, A., Bluck, B.J. & Watts, D.R., 1988. Palaeomagnetic studies within the Ballantrae Ophiolite, southwest Scotland: magnetotectonic and regional tectonic implications. *Earth and Planetary Science Letters* 90, 431–48.

Trench, A., McKerrow, W.S. & Torsvik, T.H., 1991. Ordovician magnetostratigraphy: a correlation of global data. *Journal of the Geological Society* 148, 949–57.

Trench, A., Dentith, M.C., McKerrow, W.S. & Torsvik, T.H., 1992. The Ordovician magnetostratigraphic time scale: Reliability and correlation potential. pp. 69–77 *in* B.D. Webby & J.R. Laurie (eds), q.v.

Trench, A., McKerrow, W.S., Torsvik, T.H., Li, Z.X. & McCracken, S.R., 1993. The polarity of the Silurian magnetic field: indications from a global data compilation. *Journal of the Geological Society* 150, 823–31.

Truswell, E.M., 1980. Permo–Carboniferous palynology of Gondwanaland: progress and problems in the decade to 1980. *BMR Journal of Australian Geology & Geophysics* 2, 95–111.

Truswell, E.M. & Harris, W.K., 1982. The Cainozoic palaeobotanical record in arid Australia: fossil evidence for the origins of an arid-adapted flora. pp. 67–76 *in* W.R. Barker & P.J.M. Greenslade (eds), *Evolution of the Flora and Fauna of Arid Australia*, Peacock Publications, Adelaide.

Truswell, E.M. & Marchant, N.G., 1986. Early Tertiary pollen of probable Droseracean affinity from central Australia. *Special Papers in Palaeontology* 35, 163–78.

Truswell, E. M. & Owen, J.A., 1988. Eocene pollen from Bungonia, New South Wales. *Association of Australasian Palaeontologists, Memoir* 5, 259–84.

Truswell, E.M., Chaproniere, G.C.H. & Shafik, S., 1991. Australian Phanerozoic Timescales: 10. Cainozoic biostratigraphic chart and explanatory notes. *Bureau of Mineral Resources, Geology & Geophysics, Record* 1989/40, 1–16.

Truswell, E.M., Sluiter, I.R. & Harris, W.K., 1985. Palynology of the Oligo–Miocene sequence in the Oakvale-1 corehole, western Murray Basin, South Australia. *BMR Journal of Australian Geology & Geophysics* 9, 267–95.

Tschudy, R.H. & Tschudy, B.D., 1986. Extinction and survival of plant life following the Cretaceous/Tertiary boundary event, western interior, North America. *Geology* 4, 667–70.

Tsuchi, R., Takayanagi, Y. & Shibata, K., 1981. Neogene bio-events in the Japanese islands. pp. 15–32 *in* R. Tsuchi (ed.), *Neogene in Japan—Its Biostrati-*

graphy and Chronology, Kurofune Printing Co., Shizuoka.

Tucker, R.D., Ross, R.J. Jr, Williams, S.H. & Pharaoh, T.C., 1990. Time-scale calibration by high-precision U–Pb zircon dating of the Ercall Granophyre (p) and volcanic ashes in the Ordovician and Lower Silurian stratotypes of Britain. *Geological Society of Australia, Abstracts* 27; *Seventh International Conference on Geochronology, Cosmochronology and Isotope Geology, Canberra 1990*, 102.

Turner, N.J., Black, L.P. & Higgins, N.C., 1986. The St Marys Porphyrite—a Devonian ash-flow tuff and its feeder. *Australian Journal of Earth Sciences* 33, 201–18.

Turner, P., Turner, A., Ramos, A. & Sopeña, A., 1989. Palaeomagnetism of Permo–Triassic rocks in the Iberian Cordillera, Spain: acquisition of secondary and characteristic remanence. *Journal of the Geological Society, London* 146, 61–76.

Turner, S., 1982*a*. Thelodonts and correlation. pp. 128–32 *in* P.V. Rich & E. Thompson (eds), q.v.

Turner, S., 1982*b*. Middle Palaeozoic elasmobranch remains from Australia. *Journal of Vertebrate Paleontology* 2, 117–31.

Turner, S., 1982*c*. A catalogue of fossil fish in Queensland. *Memoirs of the Queensland Museum* 20, 599–611.

Turner, S., 1982*d*. *Saurichthys* (Pisces, Actinopterygii) from the Early Triassic of Queensland. *Memoirs of the Queensland Museum* 20, 545–51.

Turner, S., 1984. Studies on Palaeozoic Thelodonti (Craniata: Agnatha). Unpublished PhD thesis, University of Newcastle-upon-Tyne.

Turner, S., 1986. Vertebrate fauna of the Silverband Formation, Grampians, western Victoria. *Proceedings of the Royal Society of Victoria* 98, 53–62.

Turner, S., 1990. Early Carboniferous shark remains from the Rockhampton district, Queensland. *Memoirs of the Queensland Museum* 28, 65–73.

Turner, S., 1991. Palaeozoic vertebrate microfossils in Australasia. pp. 429–64 *in* P.V. Rich et al. (eds), q.v.

Turner, S., 1993. Early Carboniferous microvertebrates from the Narrien Range, central Queensland. *Association of Australasian Palaeontologists, Memoir* 15, 289–304.

Turner, S. & Dring, R.S., 1981. Late Devonian thelodonts (Agnatha) from the Gneudna Formation, Carnarvon Basin, Western Australia. *Alcheringa* 5, 39–48.

Turner, S. & Long, J.A., 1987. Lower Carboniferous palaeoniscoids (Pisces: Actinopterygii) from Queensland. *Memoirs of the Queensland Museum* 25, 193–200.

Turner, S. & Young, G.C., 1987. Shark teeth from the Early–Middle Devonian Cravens Peak Beds, Georgina Basin, Queensland. *Alcheringa* 11, 233–44.

Turner, S. & Young, G.C., 1992. Thelodont scales from the Middle–Late Devonian Aztec Siltstone, southern Victoria Land, Antarctica. *Antarctic Science* 4(1), 1–17.

Turner, S., Jones, P.J. & Draper, J.J., 1981. Early Devonian thelodonts (Agnatha) from the Toko Syncline, western Queensland, and a review of other Australian discoveries. *BMR Journal of Australian Geology & Geophysics* 6, 51–69.

Tuzhikova, V.I., 1985. *Miospores and Stratigraphy of Key Sections of the Triassic of the Urals*, Akademiya Nauk SSSR, Urals Science Centre, Sverdlovsk, 232 pp.

Vail, P.R., Mitchum, R.M. & Thomson, S., 1977. Seismic stratigraphy and global changes of sea level, Part 4: Global cycles of relative changes of sea level. *American Association of Petroleum Geologists Memoir* 26, 83–97.

Valencio, D.A., 1981. Reversals and excursions of the geomagnetic field as defined by palaeomagnetic data from Upper Palaeozoic–Lower Mesozoic sediments and igneous rocks from Argentina. *Advances in Earth and Planetary Sciences* 10, 137–42.

Valencio, D.A. & Mitchell, J., 1972. Palaeomagnetism and K–Ar ages of Permo–Triassic igneous rocks from Argentina and the international correlation of Upper Palaeozoic–Lower Mesozoic formations. *24th International Geological Congress, Section* 3, 189–95.

Valencio, D.A. & Vilas, J.F., 1972. Palaeomagnetism of Late Palaeozoic and Early Mesozoic rocks of South America. *Earth and Planetary Science Letters* 15, 75–85.

Valencio, D.A., Vilas, J.F.A. & Mendia, J.E., 1977. Palaeomagnetism of a sequence of red beds of the middle and upper sections of Paganzo Group (Argentina) and the correlation of Upper Palaeozoic–Lower Mesozoic rocks. *Geophysical Journal of the Royal Astronomical Society* 51, 59–74.

VandenBerg, A.H.M., 1976. Eastern Victoria. pp. 62–70 *in* J.G. Douglas & J.A. Ferguson (eds), q.v.

VandenBerg, A.H.M. & Cooper, R.A., 1992. The Ordovician graptolite sequence of Australasia. *Alcheringa* 16, 33–85.

VandenBerg, A.H.M., Garratt, M.J. & Spencer-Jones, D., 1976. Silurian–Middle Devonian. pp. 45–76 *in* J.G. Douglas & J.A. Ferguson (eds), q.v.

VandenBerg, A.H.M., Rickards, R.B. & Holloway, D.J., 1984. The Ordovician–Silurian boundary at Darraweit Guim, central Victoria. *Alcheringa* 8, 1–22.

Van den Berg, J., Klootwijk, C.T. & Wonders, A.A.H., 1978. The Late Mesozoic and Cenozoic movements of the Umbrian Peninsula: Further paleomagnetic data from the Umbrian sequence. *Geological Society of America Bulletin* 89, 133–55.

Van der Voo, R. & Johnson, R.J.E., 1985. Palaeomagnetism of the Dunn Point Formation (Nova Scotia): High palaeolatitudes for the Avalon terrane in the Late Ordovician. *Geophysical Research Letters* 12, 337–40.

Vandyke, A. & Byrnes, J.G., 1976. Palaeozoic succession beneath the Narragal Limestone, Oakdale Anticline near Mumbil. *Geological Survey of New South Wales, Record* 17(2), 123–34.

Van Hinte, J.E., 1976. A Jurassic time scale. *American Association of Petroleum Geologists Bulletin* 60, 489–97.

Van Hinte, J.E., 1978. A Cretaceous time scale. *American Association of Petroleum Geologists, Studies in Geology* 6, 269–87.

Varker, W.J. & Sevastopulo, G.D., 1985. The Carboniferous System: Part 1, Conodonts of the Dinantian Subsystem from Great Britian and Ireland. pp. 167–209 *in* A.C. Higgins & R.L. Austin (eds), *A Stratigraphical Index of Conodonts*, British Micropalaeontological Society.

Vdovenko, M.V., Aisenverg, D.Ye., Brazhnikova, N.Ye. & Poletaev, V.I., 1987. The problems of the Lower Carboniferous stratigraphy of the Russian Platform. *11th International Congress of Carboniferous Stratigraphy and Geology, Beijing*, 1987, *Abstracts of Papers* 1, 38–39.

Vearncombe, S. & Young, G.C., in prep. The Devonian palaeogeography of Australia. *AGSO Record.*

Veevers, J.J., 1959. Devonian brachiopods from the Fitzroy Basin, Western Australia. *Bureau of Mineral Resources, Geology & Geophysics, Bulletin* 45, 220 pp.

Veevers, J.J., 1970. Upper Devonian and Lower Carboniferous calcareous algae from the Bonaparte Gulf Basin, north-western Australia. *Bureau of Mineral Resources, Geology & Geophysics, Bulletin* 116, 173–88.

Veevers, J.J., 1988. Morphotectonics of Australia's north-western margin—a review. pp. 19–27 *in* P.G. Purcell & R.R. Purcell (eds), q.v.

Veevers, J.J. & Powell, C.McA., 1987. Late Paleozoic glacial episodes in Gondwanaland reflected in transgressive–regressive depositional sequences in Euramerica. *Geological Society of America Bulletin* 98, 475–87.

Veevers, J.J. & Wells, A.T., 1961. The geology of the Canning Basin, Western Australia. *Bureau of Mineral Resources, Geology & Geophysics, Bulletin* 60, 1–323.

Veevers, J.J., Roberts, J., White, M.E. & Gamuts, I., 1967. Sandstone of probable Lower Carboniferous age in the northeastern Canning Basin. *Australian Journal of Science* 29, 330–31.

Vissarionova, A.Ya. (ed.), 1975. *Field Excursion Guidebook for the Carboniferous Sections of the South Urals* (*Bashkiria*), Izdatelstvo `Nauka', Moscow, 183 pp.

Visscher, H. & Brugman, W.A., 1981. Ranges of selected palynomorphs in the Alpine Triassic of Europe. *Review of Palaeobotany and Palynology* 34, 115–28.

Visscher, H. & Van der Zwan, C.J., 1981. Palynology of the circum-Mediterranean Triassic: phytogeographical and palaeoclimatological implications. *Geologische Rundschau* 70(2), 625–35.

Voges, A., 1960. Die Bedeutung der Conodonten für die Stratigraphie des Unterkarbons I und II (*Gattendorfia*—und *Pericyclus*—Stufe) des Sauerlandes. *Fortschritte in der Geologie von Rheinland und Westfalen* 3 (1), 197–228.

Vogt, P.R. & Einwich, A.M., 1979. Magnetic anomalies and sea-floor spreading in the western North Atlantic, and a revised calibration of the Keithley (M) geomagnetic reversal chronology. *Initial Reports of the Deep Sea Drilling Project* 43, 857–76.

Von Alberti, F. 1834. *Beitrag zu einer Bunten Sandsteins, Muschelkalks und Keupers, und die Verbindung dieser Gebilde an einer Formation*, J.G. Cotta, Stuttgart und Tübingen.

Von Hillebrandt, A., 1984. The faunal relations of the Lower Jurassic ammonites of South America. pp. 716–29 *in* O. Michelsen & A. Zeiss (eds), *IUGS International Symposium on Jurassic Stratigraphy,*

Volume III (Erlangen 1984), Geological Survey of Denmark, Copenhagen.

Von Rad, U., Exon, N.F., Boyd, R. & Haq, B.U., 1992. Mesozoic palaeoenvironment of the rifted margin off NW Australia (ODP Legs 122/123). *Geophysical Monograph* 70, 157–84.

Wade, M., 1964. Application of the lineages concept to biostratigraphic zoning based on planktonic Foraminifera. *Micropaleontology* 10, 273–90.

Wade, R.T., 1935. *The Triassic Fishes of Brookvale*, British Museum (Natural History), London, 110 pp.

Wade, R.T., 1942. The Jurassic fish of New South Wales. *Journal of the Royal Society of New South Wales* 75, 71–84.

Wade, R.T., 1953. Triassic fossil fish from Leigh Creek, South Australia. *Transactions of the Royal Society of South Australia* 76, 80–81.

Wagner, R.H., 1984. Megafloral zones in the Carboniferous. pp. 109–34 *in* P.K. Sutherland & W.L. Manger (eds), 1984*b*, q.v.

Wagner, R.H. & Winkler Prins, C.F., 1985. Stratotypes of the two lower Stephanian stages, Cantabrian and Barruelian. *Compte Rendu 10ème Congrès International de Stratigraphie et de Géologie du Carbonifère, Madrid, 1983*, 4, 473–83.

Wagner, R.H. & Winkler Prins, C.F., 1991. Major subdivisions of the Carboniferous System. *Compte Rendu 11ème Congrés International de Stratigraphie et de Géologie du Carbonifère, Beijing, 1987*, 1, 213–45.

Wagner, R.H., Saunders, W.B. & Manger, W.L., 1985. Report of the IUGS Subcommission on Carboniferous Stratigraphy (General Assembly in Madrid, September 1983). *Compte Rendu 10ème Congrès International de Stratigraphie et de Géologie du Carbonifère, Madrid, 1983*, 1, 57–61.

Walkom, A.B., 1915. Mesozoic floras of Queensland, Part 1. The flora of the Ipswich and Walloon Series. (a) Introduction, (b) Equisetales. *Geological Survey of Queensland, Publication* 252, 1–54.

Walkom, A.B., 1917*a*. Mesozoic floras of Queensland, Part 1 continued. The flora of the Ipswich and Walloon Series. (c) Filicales &c. *Geological Survey of Queensland, Publication* 257, 1–66.

Walkom, A.B., 1917*b*. Mesozoic floras of Queensland, Part 1 concluded. The flora of the Ipswich and Walloon Series. (d) Ginkgoales, (e) Cycadophyta, (f) Coniferales. *Geological Survey of Queensland, Publication* 259, 1–48.

Walkom, A.B., 1919. Mesozoic floras of Queensland. Parts III and IV. The floras of the Burrum and Styx River Series. *Geological Survey of Queensland, Publication* 263, 7–77.

Walkom, A.B., 1921. On the occurrence of *Otozamites* in Australia, with descriptions of specimens from Western Australia. *Proceedings of the Linnean Society of New South Wales* 46, 147–53.

Walkom, A.B., 1924. On fossil plants from Bellevue, near Esk. *Memoirs of the Queensland Museum* 8 (1), 77–92.

Walkom, A.B., 1925. Fossil plants from the Narrabeen Stage of the Hawkesbury Series. *Proceedings of the Linnean Society of New South Wales* 50, 214–24.

Walkom, A.B., 1928*a*. Fossil plants from the Esk district, Queensland. *Proceedings of the Linnean Society of New South Wales* 53 (4), 145–50.

Walkom, A.B., 1928*b*. Fossil plants from the Upper Palaeozoic rocks of New South Wales. *Proceedings of the Linnean Society of New South Wales* 53, 255–69.

Walley, A.M., Strusz, D.L. & Yeates, A.N., 1990. *Palaeogeographic Atlas of Australia. Vol. 3—Silurian.* Bureau of Mineral Resources, Geology & Geophysics, Canberra, 27 pp.

Walliser, O.H., 1964. Conodonten des Silurs. *Abhandlungen des Hessischen Landesamtes für Bodenforschung* 41, 106 pp.

Walliser, O.H., 1984. Geologic processes and global events. *Terra Cognita* 4, 17–20.

Walter, M.R., Elphinstone, R. & Heys, G.R., 1989. Proterozoic and Early Cambrian trace fossils from the Amadeus and Georgina basins, central Australia. *Alcheringa* 13, 209–56.

Wang C.Y. (ed.), 1987. *Carboniferous Boundaries in China*, Science Press, Beijing, China, 180 pp.

Wang Qizheng, Mills, K.J., Webby, B.D. & Shergold, J.H., 1989. Late Cambrian (Mindyallan) trilobites from the Kayrunnera Group, western New South Wales. *BMR Journal of Australian Geology & Geophysics* 11, 107–18.

Wang Zhihao, 1991. Conodonts from Carboniferous–Permian boundary strata in China with comments on the boundary. *Acta Palaeontologica Sinica* 30 (1), 8–39.

Wang, S.T. & Turner, S. 1985. Vertebrate microfossils of the Devonian–Carboniferous boundary, Mumua section, Guizhou Province. *Vertebrata PalAsiatica* 23, 223–34.

Warner, R.A. & Harrison, J., 1961. Discovery of Middle Cambrian fossils in New South Wales. *Australian Journal of Science* 23, 268.

Warren, A.A., 1980. *Parotosuchus* from the Early Triassic of Queensland and Western Australia. *Alcheringa* 5, 273–88.

Warren, A.A., 1982. Australian fossil amphibians. pp. 145–57 *in* P.V. Rich & E.M. Thompson (eds), q.v.

Warren, A.A., 1991. Australian fossil amphibians. pp. 569–90 *in* P.V. Rich et al. (eds), q.v.

Warren, A.A. & Hutchinson, M.N., 1988. A new capitosaurid from the Early Triassic of Queensland, and the ontogeny of the capitosaurid skull. *Palaeontology* 31, 857–76.

Waterhouse, J.B., 1978. Chronostratigraphy for the world Permian. pp. 299–322 *in* G.V. Cohee et al., q.v.

Waterhouse, J.B., 1983. Systematic description of Permian brachiopods, bivalves and gastropods below Wall Sandstone Member, northern Bowen Basin. *University of Queensland Department of Geology Papers* 10 (3), 155–79.

Waterhouse, J.B., 1987. Late Palaeozoic Mollusca and correlations from the south-east Bowen Basin, East Australia. *Palaeontographica* A 198, 129–233.

Waterhouse, J.B. & Jell, J.S., 1983. The sequence of Permian rocks and faunas near Exmoor Homestead south of Collinsville, north Bowen Basin. pp. 231–67 *in Permian Geology of Queensland*, Geological Society of Australia, Queensland Division.

Watson, D.M.S., 1958. A new labyrinthodont (*Paracyclotosaurus*) from the Upper Trias of New South Wales. *British Museum (Natural History) Bulletin, Geology* 3, 233–63.

Webb, G.E., 1989. Late Viséan coral–algal bioherms from the Lion Creek Formation of Queensland, Australia. *Compte rendu 11ème Congrès International de Stratigraphe et de Géologie du Carbonifère, Beijing, 1987*, 3, 282–95.

Webb, G.E., 1990. Lower Carboniferous coral fauna of the Rockhampton Group, east-central Queensland. *Association of Australasian Palaeontologists, Memoir* 10, 1–167.

Webb, J.A., 1977. Stratigraphy and palaeontology of the Bukali area, Monto district, Queensland. *Department of Geology, University of Queensland Paper* 8 (1), 37–70.

Webb, J.A., 1992. Early Devonian sequence stratigraphy from Buchan and Bindi, eastern Victoria. *Geological Society of Australia, Abstracts* 32, 112–13.

Webby, B.D., 1969. Ordovician stromatoporoids from New South Wales. *Palaeontology* 12, 637–62.

Webby, B.D., 1971. The trilobite *Pliomerina* Chugaeva from the Ordovician of New South Wales. *Palaeontology* 14(4), 612–22.

Webby, B.D., 1972. The rugose coral *Palaeophyllum* Billings from the Ordovician of central New South Wales. *Proceedings of the Linnean Society of New South Wales* 97, 150–57.

Webby, B.D., 1973. *Remopleurides* and other Upper Ordovician trilobites from New South Wales. *Palaeontology* 16, 445–75.

Webby, B.D., 1974. Upper Ordovician trilobites from central New South Wales. *Palaeontology* 17, 203–52.

Webby, B.D., 1975. Succession of Ordovician coral and stromatoporoid faunas from central-western New South Wales, Australia. pp. 57–68 *in* B.S. Sokolov (ed.), *Drevniye Cnidaria*, vol. II, Nauka, Novosibirsk.

Webby, B.D., 1978. History of the Ordovician continental platform and shelf margin of Australia. *Journal of the Geological Society of Australia* 25, 41–63.

Webby, B.D., 1979. The oldest Ordovician stromatoporoids from Australia. *Alcheringa* 3, 237–51.

Webby, B.D., 1985. Influence of a Tasmanide island-arc on the evolutionary development of Ordovician faunas. *Records of the Geological Survey of New Zealand* 9, 99–101.

Webby, B.D., 1987. Biogeographic significance of some Ordovician faunas in relation to East Australian Tasmanide suspect terranes. pp. 103–17 *in* E.C. Leitch & E. Scheibner (eds), Terrane accretion and orogenic belts. *Geodynamics series* 19, American Geophysical Union, Washington, DC.

Webby, B.D. & Banks, M.R., 1976. *Clathrodictyon* and *Ecclimadictyon* (Stromatoporoidea) from the Ordovician of Tasmania. *Papers and Proceedings of the Royal Society of Tasmania* 110, 129–37.

Webby, B.D. & Laurie, J.R. (eds), 1992. Global perspectives on Ordovician geology. *Proceedings of the Sixth International Symposium on the Ordovician System, University of Sydney, 15–19 July 1991*, Balkema, Rotterdam, 513 pp.

Webby, B.D. & Rigby, J.K., 1985. Ordovician sphinctozoan sponges from central New South Wales. *Alcheringa* 9, 209–20.

Webby, B.D., Moors, H.T. & McLean, R.A., 1970. *Malongullia* and *Encrinuraspis*, new Ordovician trilobites from New South Wales, Australia. *Journal of Paleontology* 44 (5), 881–87, pls 125–26.

Webby, B.D., Wang Qizheng & Mills, K.J., 1988. Upper Cambrian–basal Ordovician trilobites from western New South Wales, Australia. *Palaeontology* 31, 905–38.

Webby, B.D., VandenBerg, A.H.M., Cooper, R.A., Stewart, I., Shergold, J.H., Nicoll, R.S., Burrett, C.F., Stait, B., Laurie, J. & Sherwin, L., 1991. The Ordovician System in Australasia: Subsystemic, series and stage subdivisions. pp. 47–57 *in* C.R. Barnes & S.H. Williams (eds), Proceedings of the Fifth International Symposium on the Ordovician System. *Geological Survey of Canada Paper* 90–99.

Webby, B.D., VandenBerg, A.H.M., Cooper, R.A., Banks, M.R., Burrett, C.F., Henderson, R.A., Clarkson, P.D., Hughes, C.P., Laurie, J., Stait, B., Thomson, M.R.A. & Webers, G.F., 1981. The Ordovician System in Australia, New Zealand and Antarctica: Correlation chart and explanatory notes. *International Union of Geological Sciences, Publication* 6, 1–64.

Webster, G.D., Chen, X. & Derewetzky, A., 1993. A preliminary evaluation of the Kinderhookian–Osagean boundary interval in Wyoming and Idaho, western USA. *Newsletter on Carboniferous Stratigraphy* 11, 27–29.

Wei J.Y. & Ji Q., 1989. Chapter 7, Geochemistry: stable isotope and trace element. pp. 48–65 *in* Ji Qiang et al., *The Dapoushang Section: An Excellent Section for the Devonian–Carboniferous Boundary*, Science Press, Beijing, 165 pp., 43 pls.

Westergärd, A.H., Agnostidea of the Middle Cambrian of Sweden. *Sveriges Geologiska Undersökning*, Series C, 477, 1–141.

Westermann, G.E.G., 1974. Ammonite succession of the Middle Jurassic in the southern Andes. *Mémoire du Bureau de Recherches Géologiques et Minières* 75, 423–30.

Westermann, G.E.G., 1981. Ammonite biochronology and biogeography of the circum-Pacific Middle Jurassic. pp. 459–98 *in* M.R. House & J.R. Senior (eds), The Ammonoidea. *Systematics Association Special Volume* 18, Academic Press, London.

Westermann, G.E.G., 1984. Gauging the duration of stages: a new approach for the Jurassic. *Episodes* 7/2, 26–28.

White, M.E., 1986. *The Greening of Gondwana*. Reed Books, Frenchs Forest, NSW, 256 pp.

Whitehouse, F.W., 1927. Notes accompanying an exhibit of fossils. *Proceedings of the Royal Society of Queensland* 39, vii–viii.

Whitehouse, F.W., 1936. The Cambrian faunas of north-eastern Australia. Part 1—Stratigraphic outline; Part 2—Trilobita (Miomera). *Memoirs of the Queensland Museum* 11 (1), 59–112.

Whitehouse, F.W., 1939. The Cambrian faunas of north-eastern Australia. Part 3—The polymerid trilobites (with supplement No. 1). *Memoirs of the Queensland Museum* 11(3), 179–282.

Whitehouse, F.W., 1945. The Cambrian faunas of northeastern Australia. Part 5—The trilobite genus *Dorypyge*. *Memoirs of the Queensland Museum* 12(3), 117–23.

Whitehouse, F.W., 1955. Appendix G: The geology of the Queensland portion of the Great Artesian Basin. pp. 1–20 *in* Dep. Co-ordinator General Public Works, Australian water supplies in Queensland. *Queensland Parliament Paper* A56–1955.

Whitelaw, M.J., 1991*a*. Magnetic polarity stratigraphy of the Fisherman's Cliff and Bone Gulch vertebrate fossil faunas from the Murray Basin, New South Wales, Australia. *Earth and Planetary Science Letters* 104, 417–23.

Whitelaw, M.J., 1991*b*. Magnetic polarity stratigraphy of Plio and Pleistocene fossil vertebrate localities in south-eastern Australia. *Geological Society of America Bulletin* 103, 1493–503.

Whitelaw, M.J., 1992. Magnetic polarity stratigraphy of three Pliocene sections and inferences for the ages of vertebrate fossil sites near Bacchus Marsh, Victoria, Australia. *Australian Journal of Earth Sciences* 39, 521–28.

Whittington, H.B., Dean, W.T., Fortey, R.A., Rickards, R.B., Rushton, A.W.A & Wright, A.D., 1984. Definition of the Tremadoc Series and the series of the Ordovician System in Britain. *Geological Magazine* 121, 17–33.

Wiedmann, J., 1980. Palaogeographie und stratigraphie im Grenzbereich Jura/Kreide Sudamerikas. *Munstersche Forschung zur Geologie und Paläontologie* 51, 27–61.

Wiedmann, J., Butt, A. & Einsele, G., 1982. Cretaceous stratigraphy, environment, and subsidence history at the Moroccan continental margin. pp. 366–95 *in* U. Von Rad, K. Hinz, M. Sarnthein & E. Seibold (eds), *Geology of the North-west*

African Continental Margin, Springer-Verlag, Berlin.

Wilde, G.L., 1975. Fusulinid evidence for the Pennsylvanian–Permian boundary. pp. 123–38 *in* J.A. Barlow (ed.), *The Age of the Dunkard, Proceedings of the First I.C. White Memorial Symposium*, West Virginia Geol. and Economic Survey, Morgantown.

Wilde, G.L., 1984. Systematics and the Carboniferous–Permian boundary. pp. 543–58 *in* P.K. Sutherland & W.L. Manger (eds), 1984*b*, q.v.

Wilde, G.L., 1990. Practical fusulinid zonation: the species concept; with Permian Basin emphasis. *West Texas Geological Society Bulletin* 29(7), 5–34.

Wilkins, R.W.J., 1963. Relationships between the Mitchellian, Cheltenhamian and Kalimnan stages in the Australian Tertiary. *Proceedings of the Royal Society of Victoria (New Series)* 76, 39–59.

Williams, G.D. & Stelck, C.R., 1975. Speculations on the Cretaceous palaeogeography of North America. *Geological Association of Canada Special Paper* 13, 1–20.

Williams, G.L., 1975. Dinoflagellate and spore stratigraphy of the Mesozoic–Cenozoic offshore eastern Canada. *Geological Survey of Canada Paper* 74–30, 107–61.

Williams, G.L., 1977. Dinocysts; their palaeontology, biostratigraphy and palaeoecology. pp. 1231–325 *in* A.T.S. Ramsay (ed.), *Oceanic Micropalaeontology*, Academic Press, London.

Williams, G.L. & Bujak, J.P., 1985. Mesozoic and Cenozoic dinoflagellates. pp. 847–964 *in* H.M. Bolli, J.B. Saunders & K. Perch-Nielsen (eds), *Plankton Stratigraphy*, Cambridge University Press, Cambridge.

Williams, I.S., Tetley, N.W., Compston, W. & McDougall, I., 1982. A comparison of K/Ar and Rb/Sr ages of rapidly cooled igneous rocks: two points in the Palaeozoic time scale reevaluated. *Journal of the Geological Society* 139, 557–68.

Williams, S.H. & Stevens, R.K., 1988. Early Ordovician (Arenig) graptolites of the Cow Head Group, western Newfoundland, Canada. *Palaeontographica Canadiana* 5, 1–167.

Williamson, P.E., Exon, N.F., Haq, B.U., Von Rad, U. & Leg 122 Scientific Shipboard Party, 1989. A North West Shelf Triassic reef play: results from ODP Leg 122. *APEA Journal* 29, 328–44.

Wilson, G.A., 1989. Documentation of conodont assemblages across the Lochkovian–Pragian (Early Devonian) boundary at Wellington, central New South Wales, Australia. *Courier Forschungsinstitut Senckenberg* 117, 117–71.

Wilson, G.J., 1984. New Zealand Late Jurassic to Eocene dinoflagellate biostratigraphy. *Newsletters on Stratigraphy* 13, 104–17.

Winchester-Seeto, T., 1993*a*. Chitinozoa from the Early Devonian (Lochkovian–Pragian) Garra Limestone, central New South Wales, Australia. *Journal of Paleontology* 67, 738–58.

Winchester-Seeto, T., 1993*b*. Chitinozoan assemblages from the Pragian of eastern Australia. *Proceedings of the Royal Society of Victoria* 105, 85–112.

Winchester-Seeto, T., 1993*c*. Expanded biostratigraphic perspectives for Devonian Chitinozoa. *Association of Australasian Palaeontologists, Memoir* 15, 249–54.

Winchester-Seeto, T., 1994. A preliminary survey of the biogeography of Early Devonian Chitinozoa. *Australasian Palaeontological Convention 1994, Abstracts & Programme*, p. 66.

Winchester-Seeto, T. & Paris, F., 1989. Preliminary investigation of some Devonian (Lochkovian–Eifelian) Chitinozoa from eastern Australia. *Alcheringa* 13, 167–73.

Winkler Prins, C.F., 1991 [imprint 1990]. SCCS Working Group on the subdivision of the Upper Carboniferous S.L. ('Pennsylvanian'): a summary report. *Courier Forschungsinstitut Senckenberg* 130, 297–306.

Wise, S.W. & Wind, F.H., 1977. Mesozoic and Cenozoic calcareous nannofossils recovered by DSDP Leg 36 drilling on the Falkland Plateau, southwest Atlantic sector of the Southern Ocean. *Initial Reports of the Deep Sea Drilling Project* 36, 269–492.

Wiseman, J.F., 1979. Neocomian eustatic changes—biostratigraphic evidence from the Carnarvon Basin. *APEA Journal* 19, 66–73.

Wiseman, J.F., 1980. Palynostratigraphy near the Jurassic–Cretaceous boundary in the Carnarvon Basin, Western Australia. *Fourth International Palynological Conference (Lucknow 1976–1977), Proceedings II*, 330–49.

Witte, W.K., Kent, D.V. & Olsen, P.E., 1991. Magnetostratigraphy and paleomagnetic poles from Late Triassic–earliest Jurassic strata of the

Newark Basin. *Geological Society of America Bulletin* 103, 1648–62.

Woodburne, M.O., Tedford, R.H., Archer, M., Turnbull, W.D., Plane, M.D. & Lundelius, E.L., 1985. Biochronology of the continental mammal record of Australia and New Guinea. pp. 347–63 *in* J.M. Lindsay (ed.), Stratigraphy, Palaeontology, Malacology. *Department of Mines and Energy, South Australia, Special Publication* 5.

Woodburne, M.O., MacFadden, B.J., Case, J.A., Springer, M.S., Pledge, N.S., Power, J.D., Woodburne, J.M. & Springer, K.B., 1993. Land mammal biostratigraphy and magnetostratigraphy of the Etadunna Formation (Late Oligocene) of South Australia. *Journal of Vertebrate Paleontology* 13, 483–515.

Woodward, A.S., 1890. The fossil fishes of the Hawkesbury Series at Gosford, New South Wales. *Geological Survey of New South Wales, Memoir (Palaeontology)* 4, 1–55.

Woodward, A.S., 1906. On a Carboniferous fish fauna from the Mansfield district, Victoria. *Memoirs of the National Museum* 1, 1–32.

Woodward, A.S., 1908. The fossil fishes of the Hawkesbury series at St Peters. *Geological Survey of New South Wales, Memoir (Palaeontology)* 10, 1–32.

Woollam, R. & Riding, J.B., 1983. Dinoflagellate cyst zonation of the English Jurassic. *Institute of Geological Sciences Report* 83/2.

Wright, A.J., Pickett, J.W., Sewell, D., Roberts, J. & Jenkins, T.B.H., 1990. Corals and conodonts from the Late Devonian Mostyn Vale Formation, Keepit, New South Wales. *Association of Australasian Palaeontologists, Memoir* 10, 211–54.

Wright, C.A., 1973. Planktonic foraminiferal biostratigraphy of the Palaeocene to Eocene interval, North West Shelf, Western Australia. Unpublished Report to BOC of Australia Ltd.

Wright, C.A., 1977. Distribution of Cainozoic Foraminiferida in the Scott Reef No. 1 well, Western Australia. *Journal of the Geological Society of Australia* 24, 269–77.

Wright, C.A. & Apthorpe, M., 1976. Planktonic foraminiferids from the Maastrichtian of the North West Shelf, Western Australia. *Journal of Foraminiferal Research* 6, 228–40.

Wright, C.W., 1963. Cretaceous ammonites from Bathurst Island, northern Australia. *Palaeontology* 6, 597–614.

Wu, W.S., Zhang, L.X., Zhao, X.H., Jin, Y.G. & Liao, Z.T., 1987. *Carboniferous Boundaries in China*, Science Press, Beijing, China, 180 pp.

Wyborn, D., Turner, B.S. & Chappell, B.W., 1987. The Boggy Plain Supersuite—a distinctive belt of I-type igneous rocks of potential economic significance in the Lachlan Fold Belt. *Australian Journal of Earth Sciences* 34, 21–43.

Wyborn, D., Owen, N., Compston, W. & McDougall, I., 1982. The Laidlaw Volcanics: a Late Silurian point on the geological time scale. *Earth and Planetary Science Letters* 59, 90–100.

Yabolkov, V.S. (ed.), 1975. *Field Excursion Guidebook for the Carboniferous Sections of the Moscow Basin*, Nauka, Moscow, 176 pp.

Yang, S.P., Hou, H.F., Gao, L.D., Wang, Z.J. & Wu, X.H., 1980. The Carboniferous System of China. *Acta Geologica Sinica* 54(3), 167–75.

Yang, S.P., Wu, W.S., Zhang, L.X., Liao, Z.T. & Ruan, Y.P., 1979. New classification of the Carboniferous System in China. *Acta Stratigraphia Sinica* 3(3), 180–92.

Yang, Zunyi, 1986. The Cretaceous System. pp. 153–67 in Yang, Z., Cheng, Y. & Wang, H. (eds), *The Geology of China*, Clarendon Press, Oxford.

Yaroshenko, O.P., 1978. Miospore assemblages and Triassic stratigraphy of the west Caucasus. *Academy of Sciences of the USSR, Transactions* 324, 3–128.

Yegoyan, V.L., 1975. Tithonian and Berriasian boundary is the boundary between the Jurassic and Cretaceous systems. *Mémoire du Bureau de Recherches Géologiques et Minières* 86, 363–69.

Yoo, E.K., 1988. Early Carboniferous Mollusca from Gundy, Upper Hunter, New South Wales. *Records of the Australian Museum* 40, 233–64.

Young, G.C., 1979. New information on the structure and relationships of *Buchanosteus* (Placodermi; Euarthrodira) from the Early Devonian of New South Wales. *Zoological Journal of the Linnean Society of London* 66, 309–52.

Young, G.C., 1982. Devonian sharks from south-eastern Australia and Antarctica. *Palaeontology* 25, 817–43.

Young, G.C., 1983. A new antiarchan fish (Placodermi) from the Late Devonian of south-eastern Australia. *BMR Journal of Australian Geology & Geophysics* 8, 71–81.

Young, G.C., 1984. An asterolepidoid antiarch (placoderm fish) from the Early Devonian of the

Georgina Basin, central Australia. *Alcheringa* 8, 65–80.

Young, G.C., 1985. New discoveries of Devonian vertebrates from the Amadeus Basin, central Australia. *BMR Journal of Australian Geology & Geophysics* 9, 239–54.

Young, G.C., 1988. New occurrences of phyllolepid placoderms from the Devonian of central Australia. *BMR Journal of Australian Geology & Geophysics* 10, 363–76.

Young, G.C., 1989*a*. Australian Phanerozoic Timescales 4. Devonian. Biostratigraphic chart and explanatory notes. *Bureau of Mineral Resources, Geology & Geophysics, Record* 1989/34, 17 pp.

Young, G.C., 1989*b*. The Aztec fish fauna of southern Victoria Land—evolutionary and biogeographic significance. pp. 43–62 *in* J.A. Crame (ed.), Origins and evolution of the Antarctic biota. *Geological Society of London Special Publication* 47.

Young, G.C., 1989*c*. New occurrences of culmacanthid acanthodians (Pisces: Devonian) from Antarctica and south-eastern Australia. *Proceedings of the Linnean Society of NSW*, 111, 12–25.

Young, G.C., 1990. New antiarchs (Devonian placoderm fishes) from Queensland, with comments on placoderm phylogeny and biogeography. *Memoirs of the Queensland Museum* 28, 35–50.

Young, G.C., 1993*a*. Middle Palaeozoic macrovertebrate biostratigraphy of eastern Gondwana. pp. 208–51 *in* J.A. Long (ed.), *Palaeozoic Vertebrate Biostratigraphy and Biogeography*, Belhaven Press, London

Young, G.C., 1993*b*. Marine/non-marine correlation using vertebrates: summary of recent studies in Australia and Antarctica. *IUGS Subcommission on Devonian Stratigraphy, Newsletter* 10, 58–61.

Young, G.C. 1995. Australian Phanerozoic Timescales 4. Devonian. Biostratigraphic chart and explanatory notes, second series. *Australian Geological Survey Organisation, Record* 1995/33, 1–47.

Young G.C. & Gorter, J.D., 1981. A new fish fauna of Middle Devonian age from the Taemas/Wee Jasper region of New South Wales. *Bureau of Mineral Resources Geology & Geophysics, Bulletin* 209, 83–147.

Young, G.C., Turner, S., Owen, M., Nicoll, R.S., Laurie, J. & Gorter, J., 1987. A new Devonian fish fauna, and revision of post-Ordovician stratigraphy in the Ross River Syncline, Amadeus Basin, central Australia. *BMR Journal of Australian Geology & Geophysics* 10, 233–42.

Yu, C.M. & Jell, J.S., 1990. Early Devonian rugose coral fauna from the Shield Creek Formation, Broken River Embayment, north Queensland. *Association of Australasian Palaeontologists, Memoir* 10, 169–209.

Yu Changmin (ed.), 1988. *Devonian–Carboniferous Boundary in Nanbiancun, Guilin, China—Aspects and Records*, Science Press, Beijing, China, 379 pp.

Yu Jingxian, 1982. Late Jurassic and Early Cretaceous dinoflagellate assemblages of eastern Heilongjiang Province, China. *Shenyang Institute of Geology and Mineral Resources, Chinese Academy of Geological Sciences, Bulletin* 5, 261–336.

Yu Jingxian & Zhang Wangping, 1980. Upper Cretaceous dinoflagellate cysts and acritarchs of western Xinjiang. *Chinese Academy of Geological Sciences, Bulletin* 2, 93–119.

Zachariasse, W.J., 1991. Neogene planktonic foraminifers from sites 761 and 762 off north-west Australia. *Proceedings of the Ocean Drilling Program, Scientific Results* 122, 665–75.

Zang Wenlong & Walter, M.R., 1992. Late Proterozoic and Cambrian microfossils and biostratigraphy, Amadeus Basin, central Australia. *Association of Australasian Palaeontologists, Memoir* 12, 132 pp.

Zeiss, A., 1974. Berechtigung und Gliederung der Tithon-Stufe und ihre Stellung im oberen Jura. *Mémoire du Bureau de Recherches Géologiques et Minières* 75, 283–91.

Zeissl, W. & Mauritsch, H., 1991. The Permian–Triassic of the Gartnerkofel-1 core (Carnic Alps, Austria): Magnetostratigraphy. pp. 193–207 *in* W.T. Holser & H.P. Schönlaub (eds), *The Permian–Triassic boundary in the Carnic Alps of Austria (Gartnerkofel Region)*, Abhandlungen Geologischen Bundesanstalt, Wien, 45.

Zhang, L.X. (ed.), 1987. *Carboniferous Stratigraphy in China*, Science Press, Beijing, 160 pp.

Zhang Wentang, 1988. The Cambrian System in eastern Asia. *International Union of Geological Sciences Publication* 24, 81 pp.

Zhao Jinke, Sheng Jinzhang, Yao Zhaoqi, Liang Xiluo, Chen Chuzhen, Rui Lin & Liao Zhuoting, 1981. The Changhsingian and the Permian–Triassic boundary of South China. *Nanjing Institute of Geology and Palaeontology Bulletin* 2, 1–85.

Zhao, X., Coe, R.S., Liu, C. & Zhou, Y., 1992. New Cambrian and Ordovician palaeomagnetic poles for the North China Block and their palaeogeographic implications. *Journal of Geophysical Research* 97, 1767–88.

Zhou, T.M., Sheng, J.Z. & Wang, Y.J., 1987. Carboniferous–Permian boundary beds and fusulinid zones at Xiaodushan, Guangnan, eastern Yunnan. *Acta Micropalaeontologica Sinica* 4(2), 123–60.

Zhuravlev, A.Yu. & Gravestock, D.I., 1994. Archaeocyaths from Yorke Peninsula, South Australia and archaeocyathan Early Cambrian zonation. *Alcheringa* 18, 1–54.

Ziegler, A.M., Rickards, R.B. & McKerrow, W.S., 1974. Correlation of the Silurian rocks of the British Isles. *Geological Society of America, Special Paper* 154.

Ziegler, B., 1974. Grenzen der Biostratigraphie im Jura und Gedanken zur stratigraphischen Methodik. *Mémoire du Bureau de Recherches Géologiques et Minières* 75, 35–67.

Ziegler, B., 1981. Ammonoid biostratigraphy and provincialism: Jurassic—Old World. pp. 433–57 *in* M.R. House & J.R. Senior (eds), The Ammonoidẹa. *Systematics Association Special Volume* 18, Academic Press, London.

Ziegler, W., 1962. Taxionomie und Phylogenie Oberdevonischer Conodonten und ihre stratigraphische Bedeutung. *Beihefte Hessiches Landesamt für Bodenforschung* 38, 1–166.

Ziegler, W., 1969. Eine neue Conodontenfauna aus dem höchsten Oberdevon. *Fortschritte in der Geologie von Rheinland und Westfalen* 17, 343–60.

Ziegler, W., 1971. Conodont stratigraphy of the European Devonian. *Geological Society of America, Memoir* 127, 227–84.

Ziegler, W., 1978. Devonian. pp. 337–39 *in* G.V. Cohee et al., q.v.

Ziegler, W. & Klapper, G., 1982. The *Disparilis* conodont zone, the proposed level for the Middle–Upper Devonian boundary. *Courier Forschungsinstitut Senckenberg* 55, 463–92.

Ziegler, W. & Klapper, G., 1985. Stages of the Devonian system. *Episodes* 8, 104–109.

Ziegler, W. & Lane, H.R., 1987. Cycles in conodont evolution from Devonian to mid-Carboniferous. pp. 147–64 *in* R.J. Aldridge (ed.), *Palaeobiology of Conodonts*, British Micropalaeontological Society.

Ziegler, W. & Sandberg, C.A., 1984. *Palmatolepis*-based revision of upper part of standard Late Devonian conodont zonation. *Geological Society of America, Special Paper* 196, 179–94.

Ziegler, W. & Sandberg, C.A., 1990. The Late Devonian standard conodont zonation. *Courier Forschungsinstitut Senckenberg* 121, 1–115.

Zittel, K.A. (translated M. Ogilvie-Gordon), 1901. *History of Geology and Palaeontology to the End of the Nineteenth Century*, Walter Scott, London. 562 pp.

FURTHER REFERENCES

McMahon, B.E. & Strangway, D.W., 1968. Investigation of Kiàmàn magnetic division in Colorado red beds. *Geophysical Journal of the Royal Astronomical Society* 5, 265–85.

Molinà-Gàrza, R.S., Geissmàn, J., Van der Voo, R., Lucà, S. & Hàyden, S., 1991. Paleomagnetism of the Moenkopì and Chinle Formations in central New Mexico: implications for the North American àppàrent polar wanderpath and Triassic magnetostratigraphy. *Journal of Geophysical Research* 96, 14239–62.

Index